Hamilton

Environmental Contamination Remediation and Management

Series Editor

Erin Bennett
SALEM, Massachusetts, USA

Iraklis Panagiotakis
Athens, Greece

There are many global environmental issues that are directly related to varying levels of contamination from both inorganic (e.g. metals, nutrients) and organic (e.g. solvents, pesticides, flame retardants) contaminants. These affect the quality of drinking water, food, soil, aquatic ecosystems, urban systems, agricultural systems and natural habitats. This has led to the development of assessment methods and remediation strategies to identify, reduce, remove or contain contaminant loadings from these systems using various natural, engineered or synthetic technologies. In most cases, these strategies utilize interdisciplinary approaches that rely on chemistry, ecology, toxicology, hydrology, modeling and engineering.

This book series provides an outlet to summarize environmental contamination and remediation related topics that provide a path forward in understanding the current state and the mitigation of environmental contamination both regionally and globally. Topic areas may include, but are not limited to, Water Re-use, Waste Management, Food Safety, Environmental Restoration, Remediation of Contaminated Sites, Analytical Methodology, and Climate Change.

More information about this series at http://www.springer.com/series/15836

Nidhi Nagabhatla • Christopher D. Metcalfe
Editors

Multifunctional Wetlands

Pollution Abatement and Other Ecological Services from Natural and Constructed Wetlands

Editors
Nidhi Nagabhatla
United Nations University - Institute for
Water, Environment and Health
Hamilton, ON, Canada

School of Geography and Earth Science
McMaster University
Hamilton, ON, Canada

Christopher D. Metcalfe
The School of the Environment
Trent University
Peterborough, ON, Canada

United Nations University - Institute for
Water, Environment and Health
Hamilton, ON, Canada

ISSN 2522-5847   ISSN 2522-5855 (electronic)
Environmental Contamination Remediation and Management
ISBN 978-3-319-67415-5   ISBN 978-3-319-67416-2 (eBook)
DOI 10.1007/978-3-319-67416-2

Library of Congress Control Number: 2017956211

© Springer International Publishing AG 2018
This work is subject to copyright. All rights are reserved by the Publisher, whether the whole or part of the material is concerned, specifically the rights of translation, reprinting, reuse of illustrations, recitation, broadcasting, reproduction on microfilms or in any other physical way, and transmission or information storage and retrieval, electronic adaptation, computer software, or by similar or dissimilar methodology now known or hereafter developed.
The use of general descriptive names, registered names, trademarks, service marks, etc. in this publication does not imply, even in the absence of a specific statement, that such names are exempt from the relevant protective laws and regulations and therefore free for general use.
The publisher, the authors and the editors are safe to assume that the advice and information in this book are believed to be true and accurate at the date of publication. Neither the publisher nor the authors or the editors give a warranty, express or implied, with respect to the material contained herein or for any errors or omissions that may have been made. The publisher remains neutral with regard to jurisdictional claims in published maps and institutional affiliations.

Printed on acid-free paper

This Springer imprint is published by Springer Nature
The registered company is Springer International Publishing AG
The registered company address is: Gewerbestrasse 11, 6330 Cham, Switzerland

Foreword

The book that you hold is an attempt to unpack a big "known unknown," that is, the role that wetlands, both natural and constructed, play in water management, with the main focus on pollution abatement. It is focused and comprehensive at the same time. Characteristic features of the book include, but are not limited to:

- The wide thematic and geographic spread of the contributions in relation to natural and constructed wetlands that provides the readership with extensive empirical evidence and practical interventions on how to suitably utilize the various ecological services of wetland ecosystems.
- New data, information and knowledge that illustrate the multi-functionality of wetland ecosystems with regard to various aspects of water quality management and beyond; for example, in storm water management, habitat restoration, recreation and disaster risk reduction.
- Aligning the ecosystem services of wetland ecosystems with international processes and governance frameworks related to aquatic ecosystems, such as the Ramsar Convention, water security, ecosystems-based management, smart cities and the urban agenda, and the Sustainable Development Goals of Agenda 2030.

This book is a collection of specific case studies, and as such, it adds multiple new dimensions to a broader concept of Nature-based Solutions (NbS), which is high on the sustainable water development agenda at present. Suffice to mention that the World Water Development Report produced annually by UN-Water focuses in 2018 entirely on NbS, and hence the publication of this book is very timely.

And finally, the book is, essentially, a call for the generation of more specific knowledge and better sharing of information on the ecosystem services of wetlands that can, in turn, assist significantly in developing resilient wetlands and contribute to effective and sustainable management of water resources, both globally and locally.

The United Nations University, Institute for Water, Environment and Health (UNU-INWEH) is pleased to support the development of this publication. We believe that the information contained in this book will be a valuable resource for water management practitioners, researchers and decision makers who are looking for innovative and effective ways to manage our water resources.

Vladimir Smakhtin, PhD
Director: United Nations University
Institute for Water Environment and Health
Hamilton, ON, Canada

Acknowledgements

The Editors wish to thank all of our colleagues who took the time to review the chapters in the book, including (in alphabetical order) Dr. Bruce Anderson of Queen's University (Canada), Dr. Mark Hanson of the University of Manitoba (Canada), Dr. Nava Jimenez Luzma of the International Institute for Applied Systems Analysis (Austria), Dr. Victor Matamoros of the Spanish National Research Council (Spain), Dr. Hisae Nagashima of the University of Tohoku (Japan), Dr. Eric Odada of the University of Nairobi (Kenya), Dr. Nelson O'Driscoll of Acadia University (Canada), Dr. Marco Antonio Leon Romero of Tohoku University (Japan) and Dr. Kela Weber of the Royal Military College (Canada). We also thank the United Nations University, Institute for Water, Environment and Health (UNU-INWEH) in Hamilton, Ontario, Canada, for its institutional contribution to this project, and in particular, we thank the Director of UNU-INWEH, Dr. Vladimir Smakhtin, for contributing the Foreword and Ms. Kelsey Anderson for preparing several figures in the book. Finally, we wish to thank all of our colleagues who contributed the various chapters in the book in a timely and efficient manner.

Contents

1. **Multifunctional Wetlands: Pollution Abatement by Natural and Constructed Wetlands** 1
   Chris D. Metcalfe, Nidhi Nagabhatla, and Shona K. Fitzgerald

2. **Using Natural Wetlands for Municipal Effluent Assimilation: A Half-Century of Experience for the Mississippi River Delta and Surrounding Environs** 15
   Rachael G. Hunter, John W. Day, Robert R. Lane, Gary P. Shaffer, Jason N. Day, William H. Conner, John M. Rybczyk, Joseph A. Mistich, and Jae-Young Ko

3. **Recommendations for the Use of Tundra Wetlands for Treatment of Municipal Wastewater in Canada's Far North** 83
   Gordon Balch, Jennifer Hayward, Rob Jamieson, Brent Wootton, and Colin N. Yates

4. **The Long-Term Use of Treatment Wetlands for Total Phosphorus Removal: Can Performance Be Rejuvenated with Adaptive Management?** 121
   John R. White, Mark Sees, and Mike Jerauld

5. **An Investment Strategy for Reducing Disaster Risks and Coastal Pollution Using Nature Based Solutions** 141
   Ravishankar Thupalli and Tariq A. Deen

6. **The Role of Constructed Wetlands in Creating Water Sensitive Cities** 171
   Shona K. Fitzgerald

7. **Methylmercury in Managed Wetlands** 207
   Rachel J. Strickman and Carl P.J. Mitchell

8 Ornamental Flowers and Fish in Indigenous Communities in Mexico: An Incentive Model for Pollution Abatement Using Constructed Wetlands 241
Marco A. Belmont, Eliseo Cantellano, and Noe Ramirez-Mendoza

9 Phytoremediation Eco-models Using Indigenous Macrophytes and Phytomaterials 253
Kenneth Yongabi, Nidhi Nagabhatla, and Paula Cecilia Soto Rios

10 Accumulation of Metals by Mangrove Species and Potential for Bioremediation 275
Kakoli Banerjee, Shankhadeep Chakraborty, Rakesh Paul, and Abhijit Mitra

Index 301

Contributors

Gordon Balch Centre for Alternative Wastewater Treatment, Fleming College, Lindsay, ON, Canada

Kakoli Banerjee Department of Biodiversity and Conservation of Natural Resources, Central University of Orissa, Koraput, Landiguda, Koraput, Odisha, India

Marco A. Belmont Toronto Public Health, City of Toronto, Toronto, ON, Canada

Eliseo Cantellano FES Zaragoza, Universidad Nacional Autónoma de México, México D.F., Mexico

Shankhadeep Chakraborty Department of Oceanography, Techno India University, Kolkata, West Bengal, India

William H. Conner Baruch Institute of Coastal Ecology and Forest Science, Clemson University, Clemson, SC, USA

Jason N. Day Department of Oceanography and Coastal Sciences, Louisiana State University, Baton Rouge, LA, USA

Comite Resources, Inc., Baton Rouge, LA, USA

John W. Day Department of Oceanography and Coastal Sciences, Louisiana State University, Baton Rouge, LA, USA

Comite Resources, Inc., Baton Rouge, LA, USA

Tariq A. Deen United Nations University—Institute for Water, Environment and Health, Hamilton, ON, Canada

Shona K. Fitzgerald United Nations University—Institute for Water, Environment and Health, Hamilton, ON, Canada

Sydney Water, Sydney, NSW, Australia

Jennifer Hayward Centre for Water Resources Studies, Dalhousie University, Halifax, NS, Canada

Rachael G. Hunter Comite Resources, Inc., Baton Rouge, LA, USA

Rob Jamieson Centre for Water Resources Studies, Dalhousie University, Halifax, NS, Canada

Mike Jerauld DB Environmental, Loxahatchee, FL, USA

Jae-Young Ko Department of Marine Sciences, Texas A&M at Galveston, Galveston, TX, USA

Robert R. Lane Department of Oceanography and Coastal Sciences, Louisiana State University, Baton Rouge, LA, USA

Comite Resources, Inc., Baton Rouge, LA, USA

Chris D. Metcalfe The School of the Environment, Trent University, Peterborough, ON, Canada

United Nations University—Institute for Water, Environment and Health, Hamilton, ON, Canada

Joseph A. Mistich Comite Resources, Inc., Baton Rouge, LA, USA

Carl P.J. Mitchell University of Toronto, Scarborough, ON, Canada

Abhijit Mitra Department of Marine Science, University of Calcutta, Kolkata, West Bengal, India

Nidhi Nagabhatla United Nations University—Institute for Water, Environment and Health, Hamilton, ON, Canada

School of Geography and Earth Science, McMaster University, Hamilton, ON, Canada

Rakesh Paul Department of Biodiversity and Conservation of Natural Resources, Central University of Orissa, Koraput, Landiguda, Koraput, Odisha, India

Noe Ramirez-Mendoza Cooperativa La Coralilla, Ixmiquilpan, Hidalgo, Mexico

Paula Cecilia Soto Rios United Nations University—Institute for Water, Health and Environment, Hamilton, ON, Canada

Graduate School of Life Sciences, Tohoku University, Sendai, Miyagi, Japan

John M. Rybczyk Huxley College of the Environment, Western Washington University, Bellingham, WA, USA

Mark Sees City of Orlando Wastewater Division, Orlando, FL, USA

Gary P. Shaffer Comite Resources, Inc., Baton Rouge, LA, USA

Biological Sciences, Southeastern Louisiana University, Hammond, LA, USA

Rachel J. Strickman University of Toronto, Scarborough, ON, Canada

Ravishankar Thupalli International Mangrove Management Specialist, Kakinada, India

Center for South and South East Asian Studies, School of Political and International Studies, University of Madras, Chennai, India

John R. White Department of Oceanography and Coastal Sciences, Louisiana State University, Baton Rouge, LA, USA

Brent Wootton Centre for Alternative Wastewater Treatment, Fleming College, Lindsay, ON, Canada

Colin N. Yates Ecosim Consulting Inc., St. Catharines, ON, Canada

Kenneth Yongabi Phytobiotechnology Research Foundation Institute, Catholic University of Cameroon, Bamenda, Bamenda, Cameroon

Chapter 1
Multifunctional Wetlands: Pollution Abatement by Natural and Constructed Wetlands

Chris D. Metcalfe, Nidhi Nagabhatla, and Shona K. Fitzgerald

Introduction

Natural wetlands are complex ecological systems that incorporate physical, biological and chemical processes. These wetlands play an important role in protecting freshwater and marine ecosystems from excessive inputs of nutrients, pathogens, silt, oxygen demand, metals, organics and suspended solids, as well as providing a buffer against storms, soil stabilization and wildlife habitat (Sierszen et al. 2012; Zedler and Kercher 2005; Engelhardt and Ritchie 2002). Attempts have been made to quantify the economic benefits of these ecological systems (Woodward and Wui 2001; Barbier et al. 1997), but it is also recognized that natural wetlands have cultural value (Papayannis and Pritchard 2008). The Convention on Wetlands of International Importance, commonly known as the Ramsar Convention, held in 1971 established a global framework for conservation of natural wetlands. Estimates of the area of wetland ecosystems on a global scale vary from 917 million hectares (Lehner and Doll 2004) to more than 1270 million hectares (Finlayson and Spiers 1999). The Ramsar Convention in Article 1.1 defines wetlands broadly as, "areas of marsh, fen, peatland or water, whether natural or artificial, permanent or temporary,

---

C.D. Metcalfe (✉)
The School of the Environment, Trent University, Peterborough, ON, Canada

United Nations University—Institute for Water, Environment and Health, Hamilton, ON, Canada
e-mail: cmetcalfe@trentu.ca

N. Nagabhatla
United Nations University—Institute for Water, Environment and Health, Hamilton, ON, Canada

School of Geography and Earth Science, McMaster University, Hamilton, ON, Canada

S.K. Fitzgerald
Sydney Water Corporation, Sydney, NSW, Australia

© Springer International Publishing AG 2018
N. Nagabhatla, C.D. Metcalfe (eds.), *Multifunctional Wetlands*, Environmental Contamination Remediation and Management, DOI 10.1007/978-3-319-67416-2_1

Fig. 1.1 The wetland continuum illustrating the extent of ecological services and the energy requirements for operation and maintenance across a range of wetland types. Figure adapted from Young et al. (1998)

with water that is static or flowing, fresh, brackish or salt, including areas of marine water the depth of which at low tide exceed six metres". To date, 1052 sites in Europe, 211 sites in North America, 175 sites in South America, 359 sites in Africa, 289 sites in Asia and 79 sites in Oceania have been recognized as wetlands of international importance (Ramsar Secretariat 2013).

Despite these international efforts, monitoring of 1000 Ramsar wetlands over the period from 1970 to 1999 showed that the area of these sites declined by an average of 40% (Finlayson and Spiers 1999). The loss of wetlands is especially acute in the most populous regions of the world, including India (Bassi et al. 2014) and China (Jiang et al. 2015; An et al. 2007). However, urbanization, industrialization and expansion of agriculture have threatened natural wetlands in all areas of the world, and especially in urban areas (Hettiarachchi et al. 2015). One approach to reversing this trend has been to restore degraded wetland ecosystems to full or partial functionality (Zedler and Kercher 2005; Jenkins et al. 2010; Davenport et al. 2010; Yang et al. 2016).

Alternatively, wetlands can be created to fulfil specific ecological services, such as providing wildlife habitat, retaining pollutants and treating wastewater (Guittonny-Philippe et al. 2014; Tournebize et al. 2013; Babatunde et al. 2008; Rousseau et al. 2006; Kivaisi 2001; Worrall et al. 1997). These specific ecosystem services do not have to be mutually exclusive, as both natural and constructed wetlands can carry out a variety of ecosystem functions (Hsu et al. 2011; Hansson et al. 2005; Bolund and Hunhammar 1999; Santer 1989). For instance, in urban areas, constructed wetlands can reduce the urban heat island effect, which can positively influence human health (Bolund and Hunhammar 1999). Where constructed wetlands are used for urban stormwater treatment, they not only improve downstream water quality but can also restore the natural hydrology of the urban catchment and reduce downstream erosion from large stormwater flows (Wong et al. 2012).

The concept of a continuum of functions across different types of wetlands (Young et al. 1998) is helpful for considering the degree to which a constructed wetland replicates the functions of the natural environment. As illustrated in Fig. 1.1, the range of ecosystem functions provided by wetlands increases across the continuum from artificial (or engineered) wetlands to natural wetlands, while the amount of energy applied to operate and maintain these wetlands increases in the opposite direction. Of course, the degree of energy required to sustain the wetland is a challenge; especially in small communities in remote locations (Wu et al. 2015a) or in developing countries that lack sufficient financial and/or human resources to maintain the wetland (Kivaisi 2001).

Constructed Wetlands

Constructed wetlands are purpose-built systems that are engineered to achieve one or more of the functions of natural wetlands. Constructed wetlands include surface flow wetlands, which mimic natural inundated wetlands, or subsurface flow wetlands where the flow passes through a media bed in which plants are established. In surface flow wetlands, long detention times of typically 5–14 days and a large surface area promote removal of particulate and organic matter (Ghermandi et al. 2007). Microbial processes, including oxidation of organic matter and transformation of nutrients, occur through the plant biomass, the sediment and the decomposing plant matter on the bed surface. In subsurface flow wetlands, the detention times are typically shorter (i.e. 1–2 days) and as illustrated in Fig. 1.2, functional microorganisms are associated with the surfaces of the substrate and with the root systems (i.e. rhizosphere) of plants established in the substrate (Vymazal 2009; Stottmeisteer et al. 2003). The porous substrate also acts as a filter for reducing levels of suspended solids. In both the surface flow and the subsurface flow wetlands, plants

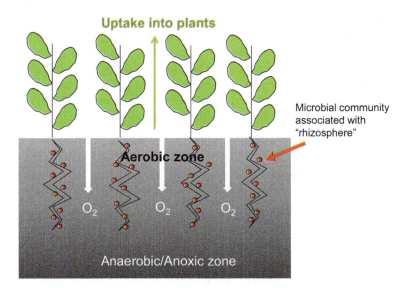

Fig. 1.2 Illustration of the functions of plant and microbial communities in natural and constructed wetlands with subsurface flow

function to oxygenate the surface layers of the sediments and thereby provide an aerobic environment for microbial activity (Valipour and Ahn 2016). In some cases, plants can accumulate and sequester nutrients (e.g. phosphorus) and pollutants (e.g. metals) from the surrounding substrate or from the water (Fig. 1.2). Recently, hybrid systems that include both vertical flow and subsurface flow systems have been shown to achieve superior pollutant removals (Vymazal 2013; Kabelo Gaboutloeloe et al. 2009). A variety of other constructed wetland designs have been proposed to enhance removals of pollutants (Wu et al. 2015b).

Constructed wetlands have been increasingly used for water reuse projects on a small scale, such as residential use for toilet flushing and gardening, or on a larger scale for irrigation of agricultural crops, golf courses and public parks, or to replenish natural wetlands and groundwater (Rousseau et al. 2006). The benefits of constructed wetlands are both social and environmental and need to be considered in the design of the wetland. The social benefits of wetlands include education and recreation, and so wetlands are often designed with interpretive signage, walking and bike paths and green space (Cunningham and Birtles 2013). These wetlands are often designed as part of an integrated urban design system with other features such as swales, grasslands and forest/shrub areas (Melbourne Water 2005).

While the benefits of constructed wetlands are many-fold, there are still challenges to their implementation, including availability of land, community support, maintenance and monitoring (Woods 1995) Given the relatively long hydraulic detention times and large surface areas required for constructed wetlands, there are often challenges in finding suitable land, especially in urban areas. Poor performance of constructed wetlands can occur where there is poor design or where the wetland is poorly maintained. Although maintenance and operation of constructed wetlands is less demanding than conventional wastewater treatment systems, they still require regular maintenance and monitoring. This includes ensuring even flow distribution, managing water levels, weed control, plant health, animal control (e.g. mosquitos, rodents, nutria) and removal of accumulated solids (Kadlec and Knight 1996). It is valuable to have community support for these projects as the community is often relied upon to help with the maintenance of the wetland. Challenges to wetland implementation can also be driven by capital costs, which will vary throughout the world, depending on material availability and labour costs.

Concepts and Context

The Water Security Agenda

The numerous challenges in managing aquatic ecosystems, and particularly wetlands are multifaceted, specific to certain regions and often, to certain places. Wetland solutions are often context specific, such as management options for pollution abatement, storm water control, disaster risk reduction or to reduce hydro-variability. The Economics of Ecosystem and Biodiversity (TEEB) report focusing on wetlands, launched on the occasion of World Wetlands Day by the Ramsar Convention, United Nations Environment Programme (UNEP), International Union for Conservation of Nature (IUCN), and Wetlands International, among others, called for the urgent need to focus on wetlands as natural solutions to the global water crisis (TEEB 2013).

Wetlands are key elements for increasing "Water Security", which is defined by UN-Water as, "the capacity of a population to safeguard sustainable access to adequate quantities of acceptable quality water for sustaining livelihoods, human wellbeing, and socio-economic development, for ensuring protection against water-borne pollution and water-related disasters, and for preserving ecosystems in a climate of peace and political stability" (UN Water 2013). The conceptual framework of UN Water for water security, illustrated in Fig. 1.3, provides for a "shared approach" that addresses current problems in the water sector. The key organizational elements in the framework include: transboundary management of shared water resources, good governance, financing of water management programs, and peace and political stability, and these elements contribute to access to safe drinking water, improvements to human health and wellbeing, protection of ecosystems and livelihoods, strengthening of water policies, institutions and knowledge systems, reduction of water-related hazards, and adaptation to climate change and resilient communities (Fig. 1.3).

This framework provides a common platform that incorporates both value and knowledge systems with technical, infrastructural, social and political interventions in order to manage aquatic ecosystems. This approach calls for a greater emphasis on understanding the multiple functions of aquatic ecosystems, in conjunction with the benefits related to health and wellbeing, livelihoods, food and energy security, and more recently, climate change mitigation and adaptation (Logan et al. 2013). This framework also emphasizes that interventions should aim to reduce the pollution that makes water unavailable or unsuitable for other uses and contributes to water insecurity. The lack of sufficient water supplies to meet the demands of water use is a situation common in countries with developing and emerging economies, leading to the current global situation where 1.2 billion people lack access to clean drinking water (WHO 2016). This framework on water security is directly applicable to the management of wetlands. An in-depth understanding of the ecosystem services of wetlands is needed to assure that business investment, conservation and restoration efforts are closely tied to policies that promote the long-term sustainability of natural wetlands, as well as the development of man-made wetlands.

The water security agenda benefits from the recent concept of "Nature-based Solutions" that has united researchers and practitioners in an innovative paradigm that addresses many water related challenges (Logan et al. 2013). This concept is described in more detail below. Past and current thinking related to sustainability planning has promoted the idea that "development investment" not only includes construction of new infrastructure, but also includes increasing the capacity of natural ecosystems (e.g. wetlands) to function as systems for risk reduction, climate adaptation, water and energy storage, enhanced aesthetic value, etc. (Hey 1994; Temmerman et al. 2013). The case studies presented in this book describe projects

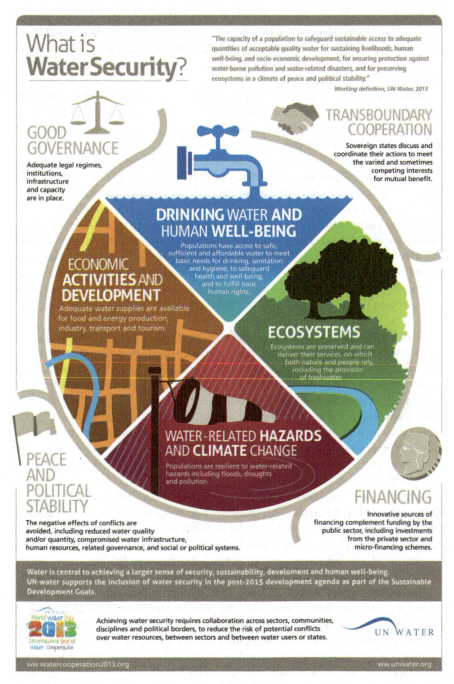

Fig. 1.3 The conceptual framework for water security outlined by UN Water, which highlights the multiple dimensions for managing aquatic systems to address future water needs; Source: UN Water (2013)

with both natural and constructed wetlands that provide evidence that wetlands are multifunctional and also provide many benefits to local communities.

Nature-Based Solutions

The term, "Nature-based Solutions" (NbS) refers to actions to alter or restore the local ecosystem and landscape to provide a solution to water management problems (Nesshöver et al. 2017). The importance of wetlands as natural infrastructure is widely discussed in the context of NbS, with a particular focus on nutrient and pollution retention, flow regulation and coastal protection (Thorslund et al. 2017). Wetland systems, whether they are "natural" or "constructed", or a combination of the two, are often more cost-effective systems for pollution control, storm water management and coastal zone protection than hard infrastructure solutions, and at the same time, provide multiple ecological functions and other benefits.

An interesting example from Yangtze River basin in China illustrates the potential of wetlands as an NbS for disaster risk reduction. Inhabited by more than 400 million people, this basin experienced a torrential storm in 1998, resulting in 4000 causalities and $25 billion USD in damage or loss of property and assets. As a disaster risk management strategy, the "32 Character Policy" in China resulted in the restoration of 2900 km^2 of floodplain wetlands with the capacity to retain 13 billion m^3 or 13 km^3 of water (Wang et al. 2007). A long-term wetland conservation network was established across the Yangtze River basin to manage water quality, preserve local biodiversity and expand wetland based nature reserves.

In the Americas, two case histories also illustrate the value of wetlands as an NbS for disaster risk reduction. In Chile, an earthquake and tsunami in 2010 resulted in $30 billion USD loss of assets and cruelly impacting lives, livelihoods and the assets of coastal communities. In the post-disaster planning, the Government of Chile made a decision to declare major portions of the coastal wetland (i.e. Yali National Reserve, Valparaiso) as a protected Ramsar wetland, while clearly recognizing the benefits of wetland ecosystems as a disaster risk reduction strategy (OECD 2016). Hurricane Katrina that flooded parts of the city of New Orleans and other areas of the state of Louisiana was the deadliest disaster in the modern history of the USA that left nearly a million people displaced. This disaster highlighted the inadequacy of hard engineered structures, floodwalls and levees for disaster risk reduction, and led to calls to investigate the services of natural wetlands as an NbS (Tibbetts 2006).

The Sustainable Development Goals

The UN has adopted the 2030 Agenda for Sustainable Development, which sets ambitious objectives for improving the lives of the global population and for protecting the environment. Sustainable Development Goal (SDG) 6 focuses on

Fig. 1.4 Individual objectives identified within SGD 6. The figure was modified from an illustration provided by UN Water

objectives to ensure access to clean water and sanitation for all (UN Water 2016). However, it is clear that water serves as a foundation for many of the other SDGs, as water security is essential for societal, economic and environmental development. Of the 17 SGD goals, there are key water-related targets embedded in the goals for reducing poverty (SDG1), improving health (SDG 3), sustainable cities (SDG 11), consumption-production (SDG 12), and protecting aquatic resources (SDG 14) and terrestrial ecosystems (SDG 15).

As illustrated in Fig. 1.4, there are several objectives that contribute to the goals of SDG 6 (UN Water 2016). SGD 6.1 aims to provide access in an equitable manner to adequate amounts of safe drinking water for 100% of the population in each country. Similarly, the objective of SGD 6.2 is to provide 100% access to facilities for sanitation and hygiene. All other objectives within SGD 6 are "aspirational" goals, meaning that individual participating countries can set their own targets and develop their own monitoring programs to assess their progress. Briefly, the objective of SDG 6.3 is to improve water quality by treating wastewater and minimizing the release of hazardous chemicals and materials. SGD 6.4 focuses on reducing water use and alleviating water scarcity, while SGD 6.5 addresses the need for integrated water resource management, as well as promoting transboundary cooperation to manage shared water resources. Finally, SGD 6.6 aims to protect terrestrial ecosystems that are key to water resource management. SGD 6.a and 6.b set targets for strengthening institutions to meet these SGD 6 objectives.

Many of the targets related to SGD 6.5 and SGD 6.6 have direct relevance to wetland management. The stated objective of SGD 6.6 is to, "protect and restore water-related ecosystems, including mountains, forests, wetlands, rivers, aquifers and lakes" (UN Water 2016), so countries that participate in this process will have to implement policies and practices to protect or restore natural wetlands. Integrated water resource management practices aimed at meeting the objectives of SGD 6.5 could include natural or constructed wetlands as part of an integrated approach to managing watersheds. Of course, wetlands can also be important systems for improving water quality (i.e. SGD 6.3) and reducing water use (i.e. SGD 6.4). Although SGD 6.3 appears to focus on technological solutions for treating urban wastewater, assimilation and constructed wetlands can be part of that solution, as well. Therefore, the SGD program may be an important incentive for countries to consider natural and constructed wetlands as a water management tool.

Pollution Abatement

The primary focus of the various chapters in this book is the ecosystem functions of natural and constructed wetlands related to the removal of pollutants from water, including nutrients, suspended solids, biological oxygen demand, and toxic metals. However, the various chapters also illustrate the other ecosystem services provided by wetlands, such as coastal protection from storms and tsunamis, flood control, habitat creation and recreational space.

Several books have reviewed the topic of pollution abatement by wetlands, including monographs by Crites et al. (2006), Kadlec and Wallace (2009), Stefanakis et al. (2014), and most recently, Scholtz (2015). These are all excellent references for water management practitioners interested in learning how both natural and constructed wetlands can be used to remove pollutants from water resources. The present book will contribute to the literature on the subject of pollution control using wetlands by describing a wide variety of case histories that span time periods of up to 30 years, as well as shorter periods of time. The contributions to this book systematically examine if and to what extent research and innovation so far has addressed the large-scale dynamics of wetland systems within the larger agenda of sustainable water management (i.e. water security), pollution abatement, potable water supply, disaster risk reduction and climate change adaptation. In addition to that, the case studies were selected to illustrate the multiple functions of wetlands over various scales and in different geographical contexts, thereby, contributing to the knowledge base for Nature-based Solutions.

The geographical scope of these case histories ranges from mangrove ecosystems in tropical marine environments to freshwater tundra wetlands, and the types of wetlands range from natural systems to constructed wetlands that have been extensively managed. The case histories describe both the successes and the challenges associated with using wetlands for water pollution control and sustainable water use. Another overarching theme of the book is to illustrate how wetlands can fulfil other ecosystem services, such as providing wildlife habitat, stormwater control and supporting recreational activities. Through the wide scope of the case histories described in the chapters in this book, the reader will gain an appreciation for the range of scenarios in which wetlands can be used to remove pollutants from water, and the challenges associated with using these natural and artificial systems.

The study from Louisiana, USA by Hunter and colleagues discusses how natural wetlands can be adapted as "assimilation" wetlands that receive treated municipal effluent. The chapter describes the benefits of these assimilation wetlands, including nutrient removal, environmental flow benefits, increased vegetation productivity, and decreased subsidence. Also, included is a description of some of the challenges that have been experienced over the long-term operation of these wetlands. In another case study of assimilation wetlands, but this time in Canada's far north, Balch and colleagues demonstrate that natural wetlands in cold environments can be used seasonally to treat municipal wastewater. Wastewater strength, hydrology and seasonal changes are key parameters influencing the performance of these systems. The authors identify knowledge gaps and discuss future research needs for maintaining these wetlands, as well as the monitoring challenges in the region.

White and colleagues describe a case study of the large constructed wetland used for "polishing" treated municipal wastewater generated by the City of Orlando and surrounding municipalities in central Florida, USA. The chapter describes the efficacy of phosphorus removal from wastewater and also discusses long-term performance challenges and adaptive strategies that have been used to manage this wetland since its construction in the 1980s. Strickman and Mitchell also examine the topic of urban wetlands, but the focus of their chapter is small wetlands that have been created in urban environments for stormwater control. These authors describe the role of these wetlands as sinks for mercury, but also describe how they can be net emitters of methylmercury produced *in situ* in the wetland. The chapter by Fitzgerald describes an urban wetland project in Sydney, Australia that demonstrates how constructed wetlands can be part of a strategy for creating "water sensitive cities". This author discusses the social and economic benefits of constructed wetlands in urban settings, and how urban wetlands can contribute to adopting and implementing the Sustainable Development Goals of the UN and the New Urban Agenda.

The chapter by Thupalli and Deen describes the benefits of mangrove ecosystems for disaster risk reduction in tropical and subtropical coastal areas, but also describes how these ecosystems contribute to pollution abatement and habitat protection, as well as protecting the livelihoods of local communities. A case is made for the need for regular consultations with communities to ensure long-term sustainability of mangrove ecosystems as green infrastructure that can accomplish comparable risk reduction role to, and hence possibly be a natural alternative to grey (built) infrastructure. The other chapter on mangroves in the Bay of Bengal region of India by Banerjee and colleagues describes the accumulation of metals by the roots and vegetation of mangroves and makes a case for the role of mangrove ecosystems for bioremediation of metal contaminated sites.

Constructed wetlands require monitoring and maintenance, and the time and labour required for these activities can be onerous. The study by Belmont and colleagues on a constructed wetland adapted for production of ornamental flowers and fish culture by an Indigenous community in central Mexico describes the critically relevant aspect of creating incentives for community participation in wetland projects. The treatment system yields fish and ornamental flowers that contribute to livelihoods in the community and are an economic incentive for maintaining the wetland. A study based in west Africa focuses on nature based solutions to achieve local-scale water security. This chapter by Yongabi and colleagues promotes the use of local aquatic plants and vegetation for pollution abatement as a low-cost alternative to conventional water treatment. Examples include using a species of water lily for removing heavy metals from water, and the use of phyto-materials from the seeds of a terrestrial plant in conjunction with sand filtration for treatment of potable water.

In aggregate, the chapters provide a synthesis of current knowledge related to natural and constructed wetlands, along with tested examples of interventions, methods and management strategies that include stakeholder participation.

Conclusions

The diverse benefits of natural and constructed wetlands make them suitable systems for improving water quality, while at the same time providing social and potentially economic benefits to local residents and long-term water security. These benefits must be balanced so that the design, operation and maintenance of the wetlands meet the requirements for pollution control, and also serve other ecological functions.

The chapters in this book are intended to inform water practitioners and researchers involved in watershed management of the range of options that are available for using natural and constructed wetlands for pollution abatement, and also some of the challenges associated with operating and maintaining these systems. The various case studies from around the world and in different settings (e.g. natural and constructed wetlands, urban to rural) will assist in understanding how technical and scientific knowledge can be integrated with community-based planning and policy development to reduce the impacts of pollution on the environment. In addition, the case histories will illustrate the importance of natural and manmade wetlands within the context of integrated water resource management, the water security agenda, the UN Sustainable Development Goals and the developing concept of Nature-based Solutions, as well as the socio-political actions required to support these management options.

References

An SQ, Li HB, Guan BH, Zhou CF, Wang ZS, Deng ZF, Zhi YB, Liu YH, Xu C, Fang SR (2007) China's natural wetlands: past problems, current status, and future challenges. Ambio 36:335–342

Babatunde AO, Zhao YQ, O'Neill M, O'Sullivan B (2008) Constructed wetlands for environmental pollution control: a review of developments, research and practice in Ireland. Environ Int 34:116–126

Barbier EB, Acreman M, Knowler M (1997) Economic valuation of wetlands—a guide for policy makers and planners. Ramsar Convention Bureau, Gland, Switzerland

Bassi N, Kumar MD, Sharma A, Pardha-Saradhi P (2014) Status of wetlands in India: a review of extent, ecosystem benefits, threats and management strategies. J Hydrol Reg Stud 2:1–19

Bolund P, Hunhammar S (1999) Ecosystem services in urban areas. Ecol Econ 29:293–230

Crites RW, Middlebrook EJ, Reed SC (2006) Natural wetland treatment systems. CRC-Taylor & Francis, Boca Raton, FL, 552 p

Cunningham D, Birtles P (2013) The benefits of constructing an undersized wetland. In: Water Sensitive Urban Design 2013: WSUD 2013. Engineers Australia, Barton, ACT, Australia

Davenport M, Bridges C, Mangun J (2010) Building local community commitment to wetlands restoration: a case study of the Cache River wetlands in southern Illinois, USA. Environ Manag 45:711–722

Engelhardt KAM, Ritchie ME (2002) The effect of aquatic plant species richness on wetland ecosystem processes. Ecology 83:2911–2924

Finlayson CM, Spiers AG (eds) (1999) Global review of wetland resources and priorities for wetland inventory. Supervising Scientist, Canberra, Australia

Ghermandi A, Bixo D, Thoeye C (2007) The role of free water surface constructed wetlands as polishing step in municipal wastewater reclamation and reuse. Sci Total Environ 380:247–258

Guittonny-Philippe A, Masotti V, Höhener P, Boudenne J-L, Viglione J, Laffont-Schwob I (2014) Constructed wetlands to reduce metal pollution from industrial catchments in aquatic Mediterranean ecosystems: a review to overcome obstacles and suggest potential solutions. Environ Int 64:1–16

Hansson L-A, Brönmark C, Nilsson PA, Abjörnsson K (2005) Conflicting demands on wetland ecosystem services: nutrient retention, biodiversity or both? Freshw Biol 50:705–714

Hettiarachchi M, Morrison TH, McAlpine C (2015) Forty-three years of Ramsar and urban wetlands. Glob Environ Chang 32:57–66

Hsu C-B, Hsieh H-L, Yang L, Wu S-H, Chang J-S, Hsiao S-C, Su H-C, Yeh C-H, Ho Y-S, Lin H-J (2011) Biodiversity of constructed wetlands for wastewater treatment. Ecol Eng 37:1533–1545

Hey RD (1994) Environmentally sensitive river engineering. In: Calow P, Petts GE (eds) The rivers handbook, vol 2. Blackwell Scientific, Oxford, UK, pp 337–362

Jenkins WA, Murray BC, Kramer RA, Faulkner SP (2010) Valuing ecosystem services from wetland restoration in the Mississippi alluvial valley. Ecol Econ 69:1051–1061

Jiang T-T, Pan J-F, Pu X-M, Wang B, Pan J-J (2015) Current status of coastal wetlands in China: degradation, restoration and future management. Estuar Coast Shelf Sci 164:265–275

Kabelo Gaboutloeloe G, Chen S, Barber ME, Stöckle CO (2009) Combinations of horizontal and vertical flow constructed wetlands to improve nitrogen removal. Water Air Soil Pollut 9:279–286

Kadlec RH, Wallace SD (2009) Treatment wetlands, 2nd edn. CRC Press, Boca Raton, FL, 106 p

Kadlec RH, Knight RL (1996) Treatment wetlands, 1st edn. CRC Press, Boca Raton, FL, 112 p

Kivaisi AK (2001) The potential for constructed wetlands for wastewater treatment and reuse in developing countries: a review. Ecol Eng 16:545–560

Lehner B, Doll P (2004) Development and validation of a global database of lakes, reservoirs and wetlands. J Hydrol 296:1–22

Logan LH, Karlsson EM, Gall HE, Park J, Emery N, Owens P, Niyogi D, Rao PSC (2013) Freshwater wetlands: balancing food and water security with resilience of ecological and social systems. In: Roger P (ed) Climate vulnerability. Academic Press, Oxford, UK, pp 105–116

Melbourne Water (2005) WSUD engineering procedures: stormwater. CSIRO Publishing, Melbourne, Australia

Nesshöver C, Assmuth T, Irvine KN, Rusch GM, Whalen KA, Delbaere B, Haase D, Jones-Walters L, Keune H, Kovacs E, Krauze K, Kulvik M, Rey F, van Dijk J, Vistad OI, Wilkinson ME, Wittmer H (2017) The science, policy and practice of nature-based solutions: an interdisciplinary perspective. Sci Total Environ 579:1215–1227

OECD (2016) Economic commission for Latin America and the Caribbean, environmental performance reviews. OECD Publishing, Paris, France

Papayannis T, Pritchard DE (2008) Culture and wetlands—a Ramsar guidance document. Ramsar Convention Bureau, Gland, Switzerland

Secretariat R (2013) The list of wetlands of international importance. The Ramsar Convention Bureau, Gland, Switzerland

Rousseau DPL, Lesage E, Story A, Vanrolleghem PA, De Pauw N (2006) Constructed wetlands for water reclamation. Desalination 218:181–189

Santer LK (1989) Ancillary benefit of wetlands constructed primarily for wastewater treatment. In: Hammer DA (ed) Constructed wetlands for wastewater treatment. Lewis Publishing, Chelsea, UK, pp 353–358

Scholtz M (2015) Wetlands for water pollution control, 2nd edn. Elsevier, Amsterdam, The Netherlands, 556 p

Sierszen ME, Morrice JA, Trebitz AS, Hoffman JC (2012) A review of selected ecosystem services provided by coastal wetlands of the Laurentian Great Lakes. Aquat Ecosyst Health 15:92–105

Stefanakis AL, Akratos CS, Tsihrintzis VA (2014) Vertical flow constructed wetlands: eco-engineering systems for wastewater and sludge treatment, 1st edn. Elsevier, Burlington, VT, 378 p

Stottmeisteer U, Weißner A, Kuschk P, Kappelmeyer U, Kästner M, Bederski O, Müller RA, Moormann H (2003) Effects of plants and microorganisms in constructed wetlands for wastewater treatment. Biotechnol Adv 22:93–117

Tibbetts J (2006) Louisiana's wetlands: a lesson in nature appreciation. Environ Health Perspect 114:A40–A43

Temmerman S, Meire P, Bouma TJ, Herman PMJ, Ysebaert T, de Vriend HJ (2013) Ecosystem-based coastal defence in the face of global change. Nature 504:79–83

TEEB (2013) The Economics of Ecosystems and Biodiversity (TEEB) in Business and Enterprise, Final Consultation Draft based on Russi D, ten Brink P, Farmer A, Badura T, Coates D, Förster J, Kumar R, Davidson N (2012), The Economics of Ecosystems and Biodiversity (TEEB) for Water and Wetlands, Earthscan, London, UK

Tournebize J, Passeport E, Chaumont C, Fesneau C, Guenne A, Vincent B (2013) Pesticide decontamination of surface waters as a wetland ecosystem service in agricultural landscapes. Ecol Eng 56:51–59

Thorslund J, Jarsjö J, Destouni G (2017) Wetlands as large-scale nature-based solutions: status and future challenges for research and management. Geophysical Research Abstracts Vol 19, EGU2017-13165 2017, European Geosciences Union General Assembly 2017, Copyright to authors, CC Attribution 3.0 License

UN Water (2013) Water Security and the Global Water Agenda: a UN Water Analytical Brief, United Nations University—Institute for Water, Environment and Health (UNU-INWEH) and United Nations Economic and Social Commission for Asia and the Pacific (United Nations ESCAP), 45 p

UN Water (2016) Integrated monitoring guide for SDG 6 targets and global indicators, Version July 19, 2016. http://www.unwater.org/app/uploads/2017/03/SDG-6-targets-and-global-indicators_2016-07-19.pdf

Valipour A, Ahn Y-H (2016) Constructed wetlands as sustainable ecotechnologies in decentralized practices: a review. Environ Sci Pollut Res 23:1180–1197

Vymazal J (2013) The use of hybrid constructed wetlands for wastewater treatment with special attention to nitrogen removal: a review of recent developments. Water Res 47:4795–4811

Vymazal J (2009) The use of constructed wetlands with horizontal sub-surface flow for various types of wastewater. Ecol Eng 35:1–17

Wang Y, Li L, Wang X, Yu X, Wang Y (2007) Taking stock of integrated river basin management in China. World Wildlife Fund China, Science Press, Beijing, Peoples Republic of China

WHO (2016) World health statistics 2016: monitoring health for the SDGs, sustainable development goals. World Health Organization, Geneva, Switzerland, 121 p

Wood A (1995) Constructed wetlands in water pollution control: fundamentals to their understanding. Water Sci Technol 32:21–29

Woodward RT, Wui YS (2001) The economic value of wetland services: a meta-analysis. Ecol Econ 37:257–270

Wong THF, Allen R, Beringer J, Brown RR, Deletic A, Fletcher TD, Gangadharan L, Gernjak W, Jakob C, O'Loan T, Reeder M, Tapper N, Walsh C (2012) Blueprint 2012—Stormwater Management in a Water Sensitive City. Centre for Water Sensitive Cities, Melbourne, Australia, ISBN 978-1-921912-01-6

Worrall P, Peberdy KJ, Millet MC (1997) Constructed wetlands and nature conservation. Water Sci Technol 35:2015–2213

Wu H, Zhang J, Ngo HH, Guo W, Hu Z, Liang S, Fan J, Liu H (2015b) A review on the sustainability of constructed wetlands for wastewater treatment: design and operation. Bioresour Technol 175:584–601

Wu H, Fan J, Zhang J, Ngo HH, Guo W, Liang S, Hu Z, Liu H (2015a) Strategies and techniques to enhance constructed wetland performance for sustainable wastewater treatment. Environ Sci Pollut Res 22:14637–14650

Yang W, Liu Y, Ou C (2016) Examining water quality effects of riparian wetland loss and restoration scenarios in a southern Ontario watershed. J Environ Manag 174:26–34

Young R, White W, Brown M, Burton J, Atkins B (eds) (1998) The constructed wetlands manual, vol 1. Department of Land and Water Conservation, Goulburn, NSW, Australia

Zedler JB, Kercher S (2005) Wetland resources: status, trends, ecosystem services, and restorability. Ann Rev of Environ Resour 30:39–74

Chapter 2
Using Natural Wetlands for Municipal Effluent Assimilation: A Half-Century of Experience for the Mississippi River Delta and Surrounding Environs

Rachael G. Hunter, John W. Day, Robert R. Lane, Gary P. Shaffer, Jason N. Day, William H. Conner, John M. Rybczyk, Joseph A. Mistich, and Jae-Young Ko

Introduction

The ability of wetlands to improve water quality is well established, with hundreds, if not thousands, of scientific studies published in peer-reviewed journals and books (e.g., Godfrey et al. 1985; Moshiri 1993; Lane et al. 1999, 2002, 2004, 2010; Hunter and Faulkner 2001; Mitsch and Jorgensen 2003; Kangas 2004; Kadlec and Wallace 2009; Hunter et al. 2009a, b; Seo et al. 2013; Shaffer et al. 2015). Use of natural ecosystems for assimilation of nutrients and suspended sediments in treated municipal

---

R.G. Hunter (✉) • J.A. Mistich
Comite Resources, Inc., PO Box 66596, Baton Rouge, LA 70896, USA
e-mail: rhuntercri@gmail.com

J.W. Day • R.R. Lane • J.N. Day
Department of Oceanography and Coastal Sciences, Louisiana State University, Baton Rouge, LA 70803, USA

Comite Resources, Inc., PO Box 66596, Baton Rouge, LA 70896, USA

G.P. Shaffer
Comite Resources, Inc., PO Box 66596, Baton Rouge, LA 70896, USA

Biological Sciences, Southeastern Louisiana University, Hammond, LA 70402, USA

W.H. Conner
Baruch Institute of Coastal Ecology and Forest Science, Clemson University, Clemson, SC 29634, USA

J.M. Rybczyk
Huxley College of the Environment, Western Washington University, Bellingham, WA 98225, USA

J.-Y. Ko
Department of Marine Sciences, Texas A&M at Galveston, Galveston, TX 77553, USA

© Springer International Publishing AG 2018
N. Nagabhatla, C.D. Metcalfe (eds.), *Multifunctional Wetlands*, Environmental Contamination Remediation and Management,
DOI 10.1007/978-3-319-67416-2_2

effluent is neither new nor strictly non-traditional (Day et al. 2004). There are thousands of wetland treatment systems worldwide with hundreds of years of operational experience (Kadlec and Wallace 2009). Because wetlands naturally occupy lower landscape positions within a watershed, they are ideally located to serve as biological filters, removing nutrients and sediment from water running off the surrounding landscape before it enters an open water body such as a river or lake.

Studies throughout the world have shown that wetlands chemically, physically, and biologically remove pollutants, sediments and nutrients from water flowing through them (Zhang 1995; Day et al. 2004; Alexander and Dunton 2006; Conkle et al. 2008; Meers et al. 2008; Kadlec and Wallace 2009; Vymazal 2010; Shaffer et al. 2015). Some questions remain as to the ability of wetlands to serve as long-term storage nutrient reservoirs, but examples of long-term sustainability are cypress systems in Florida that continue to remove major amounts of nutrients in treated effluent even after 20–45 years (Boyt et al. 1977; Ewel and Bayley 1978; Lemlich and Ewel 1984; Nessel and Bayley 1984), and the Breaux Bridge and Amelia assimilation wetlands that have received treated effluent for 70 and 47 years, respectively (Hesse et al. 1998; Blahnik and Day 2000; Ko et al. 2004; Day et al. 2006; Hunter et al. 2009b).

With regard to water quality, the primary constituents of interest in treated municipal effluent are nitrogen, phosphorus, and suspended solids, which includes both mineral sediments and particulate organic matter. The basic principle underlying wetland assimilation of these constituents is that the rate of effluent application must balance the rate of removal. The primary mechanisms by which this balance is achieved are physical settling and filtration, chemical precipitation and adsorption, and biological processes that result in burial, storage in vegetation, and denitrification (Reddy and DeLaune 2008). Treated effluent typically introduces nutrients as a combination of inorganic (e.g., nitrate + nitrite (NO_x), ammonia (NH_3), and phosphate (PO_4)) and organic forms, both dissolved and particulate. Nitrogen and/or phosphorus from treated effluent can be removed by short-term processes such as plant uptake, long-term processes such as peat and sediment accumulation, and permanently by denitrification (Reddy and DeLaune 2008).

In the Mississippi River Delta, there are ten assimilation wetlands currently receiving discharge of secondarily-treated, disinfected municipal effluent and four others awaiting permits or under review, as of July 2017 (Fig. 2.1). The assimilation systems in the Mississippi River Delta are not constructed wetlands, however, they are also not "natural" wetlands because they have been highly impacted by anthropogenic activity.

The Mississippi River Delta is a profoundly altered regional ecosystem covering over 10,000 km^2. Over 25% of coastal wetlands in the Mississippi River Delta were lost in the twentieth century. One of the primary causes is the almost complete isolation of the delta plain from the Mississippi River by levees that prevent regular riverine input that occurred under natural conditions before human alterations (Day et al. 2007, 2014). The river provided fresh water, mineral sediments, and nutrients during annual floods. This annual flooding maintained a salinity gradient and provided sediments to promote wetland formation and nutrients to enhance productiv-

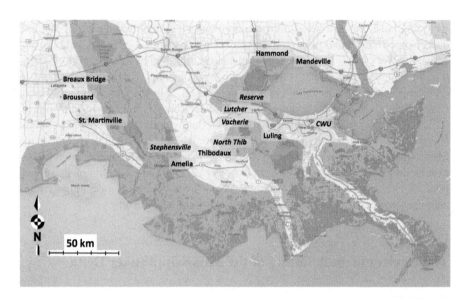

Fig. 2.1 Location of wetland assimilation projects in coastal Louisiana. Municipalities in italics indicate recently completed or ongoing ecological baseline studies. Note that Breaux Bridge, Broussard, and St. Martinville are not impacted by coastal water levels. All the other sites are at or near sea level and are impacted by sea level rise. This inhibits the ability of these sites to drain and have dry periods. Shading indicates areas with a high proportion of wetlands

ity. In addition, there has been a pervasive alteration of hydrology both in the horizontal plane due to spoil banks and canals, as well as vertically caused by enhanced subsidence due to fluid withdrawal (mainly oil and gas), compaction, and drainage. All of the wetland assimilation systems discussed here are in areas where the natural hydrology has been fundamentally altered by human activities.

Wetland assimilation in Louisiana can achieve sustainable low cost tertiary treatment of secondarily-treated municipal effluent while benefiting and restoring wetlands (Day et al. 2004; Hunter et al. 2009a, b). A properly designed wetland assimilation system can be a more economical and sustainable means of tertiary treatment compared to conventional engineering options. The cost of tertiary treatment is a concern as the U.S. Environmental Protection Agency (USEPA) is requiring increasingly stringent limits in discharge permits for wastewater treatment plants. Out of 105 major wastewater treatment facilities in Louisiana, only 12% (13 plants) monitor for nitrogen and phosphorus concentrations, compared with an average of 57% in the 12 states included in the Mississippi River/Gulf of Mexico Watershed Nutrient Task Force. Of the 13 treatment facilities monitoring nutrients in Louisiana, 10 discharge into assimilation wetlands (Hypoxia Task Force 2016).

Freshwater resources, including treated effluent, should be used in a manner that results in the greatest benefits to society. However, municipalities cannot be expected to bear all costs for wetland assimilation projects, so when possible they should be integrated into larger restoration efforts where a variety of funding sources are used.

By doing so, Louisiana benefits from improved water quality with lower cost and energy investments, and from restoration of degraded, yet valuable, wetland ecosystems. The work that is being done in Louisiana is informed by a rich history of scientific and applied experience, taking advice from leading scientists in the field in designing wetland assimilation systems such as Drs. John Day, Jr., William Mitsch, Curtis Richardson, Michael Odgen, Robert Kadlec, Robert Knight, and Scott Wallace.

In this chapter, we discuss the history of wetland assimilation in the Mississippi River Delta. We first provide a background on the environmental setting of the Mississippi River Delta and then discuss steps involved in establishing an assimilation wetland, approaches for ensuring project success, and benefits of these systems. Finally, we review monitoring data from several currently operating systems and discuss recent controversy concerning assimilation wetlands.

The Environmental Setting of the Mississippi River Delta

The functioning and status of assimilation wetlands in the Mississippi River Delta cannot be fully understood without considering the environmental setting of the delta itself—a system formed over the past 6000–7000 years by flooding from the Mississippi River that is on a rapid non-sustainable trajectory of deterioration. Flood control levees and the pervasive alteration of hydrology have isolated wetlands from annual flooding from the river with other deleterious affects including prolonged flooding, saltwater intrusion, and conversion to open water (Kesel 1988, 1989; Mossa 1996; Roberts 1997; Day et al. 2007). Wetland loss in the twentieth century was catastrophic, with approximately 25% of coastal wetlands lost since the middle of the twentieth century (Barras et al. 2008; Couvillion et al. 2011).

A central cause of wetland loss in the delta is subsidence. Subsidence is a natural geologic process due to the compaction, consolidation, and dewatering of sediments. Under natural conditions, sediment deposition from the river and *in situ* organic soil formation balanced subsidence in much of the delta. Now, relative sea level rise, the combination of subsidence plus eustatic sea-level rise, is greater than accretion in much of the delta leading to progressively increased flooding. This has an important impact on several of the assimilation wetlands we review in this chapter. Breaux Bridge, Broussard, and St. Martinville (Fig. 2.1) are located 4–5 m above sea level and are not affected by coastal water levels, which allows them to drain during dry periods. These wetlands also are far enough inland that they are not flooded by hurricane storm surges. All the other assimilation wetlands are affected by coastal water levels, which leads to prolonged and sometimes permanent flooding that prevents them from having dry periods and makes them susceptible to storm surge. Thus forested wetland sites in the coastal zone are generally permanently flooded, which prevents recruitment of baldcypress and water tupelo seedlings that need several months of dry ground to germinate (Allen et al. 1996). These freshwater coastal forested wetlands are also threatened by saltwater intrusion.

Other factors exacerbating wetland loss include altered hydrology due to the proliferation of dredged canals and deep-well fluid withdrawal associated with the oil and gas industry (Turner et al. 1994; Day et al. 2000; Morton et al. 2002; Chan and Zoback 2007), intentional and unintentional impoundments (Day et al. 1990; Boumans and Day 1994; Cahoon 1994), and herbivory by nutria and other herbivores (Shaffer et al. 1992, 2015; Evers et al. 1998). Almost a third of the delta has been isolated or semi-isolated through the purposeful or accidental construction of various types of impoundments (Day et al. 1990). Throughout this paper we will show how human impacts have negatively impacted areas where wetland assimilation projects are established.

Sustainability in the Mississippi River Delta

Sustainability in the Mississippi River Delta is difficult in the face of increasingly severe climate impacts. Climate change will impact the delta through accelerated eustatic sea-level rise (Meehl et al. 2007; FitzGerald et al. 2008; Pfeffer et al. 2008; Vermeer and Rahmstorf 2009; IPCC 2013; Koop and van Leeuwen 2016; Deconto and Pollard 2016), more severe hurricanes (Emanuel 2005; Webster et al. 2005; Hoyos et al. 2006; Goldenberg et al. 2001; Kaufmann et al. 2011; Mei et al. 2014), drought (IPCC 2007; Shaffer et al. 2015), more erratic and extreme weather (Min et al. 2011; Pall et al. 2011; Royal Society 2014) and increased Mississippi River discharge (Tao et al. 2014). Recent estimates are that sea level will rise between 1 and 2 m by 2100 (Horton et al. 2014; Koop and van Leeuwen 2016; Deconto and Pollard 2016). The combination of accelerated sea-level rise, more intense hurricanes, and drought will lead to increased wetland loss and enhanced saltwater intrusion. Coastal baldcypress—water tupelo swamps are especially susceptible to climate change impacts that increase salinity.

Decreasing energy availability and higher energy prices will limit options for restoration of deltas and complicate human response to climate change (Day et al. 2005, 2007, 2014, 2016a; Tessler et al. 2015). The implication of future energy scarcity is that the cost of energy will increase during the coming decades (Campbell and Laherrere 1998; Deffeyes 2001; Bentley 2002; Hall and Day 2009; Murphy and Hall 2011; Day et al. 2005, 2016a) and the cost of energy-intensive activities will also increase significantly. In a future characterized by scarce and expensive energy, maintaining traditional infrastructure will likely become increasingly unsustainable. Advanced conventional municipal wastewater treatment is very energy intensive, and the use of wetlands offers an energy efficient means to achieve tertiary treatment.

Establishing an Assimilation Wetland

Process Overview

In the State of Louisiana, LDEQ, with oversight from the USEPA, regulates wastewater treatment and the discharge of treated municipal effluent. Over the past 25 years, scientists, regulatory personnel, and dischargers have worked closely to

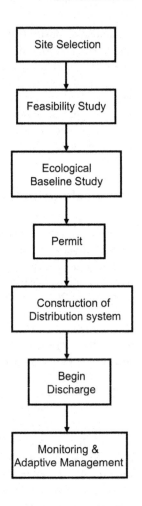

Fig. 2.2 Steps in the establishment of a wetland assimilation project

develop an approach where wetland assimilation systems meet water quality goals while protecting and restoring wetlands (Day et al. 2004; Louisiana Department of Environmental Quality 2010, 2015; Shaffer et al. 2015). Wetlands are carefully selected and monitored prior to discharge of treated effluent and these actions are part of a process to ensure project success (Fig. 2.2).

Site Selection

The process of establishing a wetland assimilation project begins with identification of a suitable candidate wetland (Fig. 2.2). All wetland ecosystems are not created equal and some are clearly unsuited for wetland assimilation. The LDEQ has recognized several wetland types that are not appropriate for assimilation, including seasonally flooded pine flatlands with carnivorous plants and areas heavily used for recreation and oyster production. There are a number of factors taken into consideration for site selection including location, size, hydrology, ecological condition, land ownership, and competing uses. Later in this chapter we address the issue of freshwater herbaceous wetlands.

Feasibility Study

After a candidate wetland is selected, a feasibility study is conducted to determine if the discharge of municipal effluent into the candidate wetland is possible (Fig. 2.2). The feasibility study usually lasts 2–4 months, depending upon the size and complexity of the wetland. During the feasibility study, wetland characteristics (hydrology, soils, vegetation, fauna) are described, along with assessment of surrounding landscape uses, expected nutrient loading rates, and presence of protected flora and fauna and archaeological or historical sites. A preliminary conceptual design of the treated effluent distribution system also is included.

Ecological Baseline Study

If the feasibility study finds the candidate wetland suitable for assimilation, a year-long ecological baseline study (EBS) is conducted (Fig. 2.2). The purpose of the EBS is to describe in detail the baseline ecological conditions of the candidate site, including hydrology, soil and water chemistry, accretion rate, and vegetative species composition and productivity. In addition, a preliminary engineering design and cost analysis are conducted. The EBS then forms part of the permit application, which is the fourth step in the process (Fig. 2.2). The EBS may be carried out at the same time that the permit applications are submitted and under review, however, the EBS data must be completed before final approval of the permits.

Regulatory and Permitting

Several permits may be required for a wetland assimilation project. An LPDES permit is required under the authority of the Federal Clean Water Act and the Louisiana Environmental Quality Act. These two acts require criteria (as set forth in the permit) to protect the beneficial uses (e.g., fish and wildlife propagation) and contain an anti-degradation policy that limits lowering of water quality. The LPDES permit designates biochemical oxygen demand (BOD), total suspended solids (TSS), and fecal coliform effluent limits for discharge to the wetland (Table 2.1) and also outlines monitoring requirements (discussed in next section) and nutrient

Table 2.1 Design characteristics of wastewater treatment plants and LPDES mean monthly and weekly permit limits for BOD, TSS and fecal coliform at Louisiana assimilation wetlands

Municipality	LPDES permit #	Year 1st Issued	Treatment system	Design capacity (MGD[a])	BOD (mg/L)	TSS (mg/L)	Fecal coliform (colonies/100 mL)	Population (2010 census)
Amelia	90999	2007	Oxidation pond with chlorination-dechlorination	0.9	30/45	90/135	200/400	2459
Breaux Bridge	30578	2003	3 oxidation ponds and chlorination-dechlorination system	1.27	30/45	90/135	200/400	8139
Broussard	33786	2007	3 oxidation ponds and chlorination-dechlorination system	1.0	30/45	90/135	200/400	8197
Guste Island	122552	2008	2-cell facultative lagoon with chlorination-dechlorination system	0.6	30/45	90/135	200/400	Private subdivision—Data not available
Hammond	19578	2010	3 cell oxidation lagoon and chlorination-dechlorination system	8.0	30/45	90/135	200/400	20,019
Luling	43356	2008	Facultative oxidation pond with UV disinfection	3.5	30/45	90/135	200/400	12,119
Mandeville	19420	2003	3 aerated lagoon cells, 3-celled rock reed filter, and UV disinfection	4.0	CBOD 10/15	15/23	200/400	11,560
Riverbend	19244	Pending	Oxidation pond with chlorination-dechlorination system	0.7	30/45	90/135	200/400	5000 homes, Population unknown
St. Martinville	19216	2006	63.7 ha facultative lagoon; UV disinfection; cascade aeration structure	1.5	30/45	90/135	1000/2000	6114
Thibodaux	19012	2004	Aerated lagoon and high-rate trickling filter; UV disinfection	4.0	30/45	30/45	200/400	14,566

[a]Million gallons per day

loading rates. Generally, effluent limits are somewhat less restrictive than for direct discharge to open water bodies because of the ability of wetlands to process and assimilate nutrients and organic matter without deleterious effects (Day et al. 2004). The permit requires disinfection so that pathogens are not discharged to wetlands and toxic materials must be below state and federal limits.

Water Quality Standards (WQS) are provisions of Louisiana State Law and these standards are applied to each assimilation wetland. Water Quality Standards consist of policy statements pertinent to water quality necessary to preserve or achieve the objectives of the standards, designated uses for which public waters of the state are to be protected, and criteria which specify general and numerical limitations for various water quality parameters that are required for designated water uses (Louisiana Department of Environmental Quality 2015). Water Quality Standards for assimilation wetlands in Louisiana serve to protect and preserve the biological and aquatic community integrity.

A CUP may be required for an assimilation wetland project, usually for the discharge pipeline installation. The CUP process is part of the Louisiana Coastal Resources Program (LCRP) that works to preserve, restore, and enhance Louisiana's valuable coastal resources. The purpose of the CUP application process is to make certain that any activity affecting Louisiana's coastal zone, such as a project that involves either dredging or filling, is performed in accordance with guidelines established in the LCRP. The guidelines are designed so that development in the coastal zone can be accomplished with the greatest benefit and the least amount of damage. Section 404(b)(1) of the Clean Water Act serves to regulate the alteration or discharge of dredged and/or fill material into U.S. waters, including wetlands. Many wetland assimilation projects require a 404 permit, particularly for the placement and installation of the effluent discharge pipeline.

Construction and Monitoring

After the necessary permits are issued, the discharger begins construction of the distribution pipeline and, upon completion, discharge and monitoring begin (Fig. 2.2). Monitoring of vegetation, soils, water, and hydrology is required in the LPDES permit for the life of the project and annual monitoring reports are submitted to LDEQ. Continual cooperation among those involved (e.g., municipality personnel and/or dischargers, ecological monitoring team, and regulatory agencies) is essential to ensure proper management over the life of the project.

Approaches to Ensure Success of an Assimilation Wetland

To ensure a healthy and sustainable assimilation wetland and a successful project, treated effluent must be discharged into the assimilation wetland at an appropriate loading rate, which is explained in detail below. The effluent must be disinfected and free from toxins to avoid endangering fauna or flora. Strict policy guidelines must be adhered to as well as long-term monitoring to detect potential problems and to achieve water quality goals, as well as to maintain a healthy wetland.

Appropriate Loading Rate

The basic principal underlying the use of wetlands for municipal effluent assimilation is that the rate of effluent discharge to the wetland must balance the rate of nutrient removal. Therefore, one of the most important factors in designing wetland assimilation systems is the loading rate. In general, loading rate refers to the rate per unit of area at which a material (e.g., a constituent in the effluent) is discharged into a system over a given time period. High nutrient loading rates to wetland systems may not allow for sufficient processing time, resulting in a wetland that is overloaded in nitrogen and/or phosphorus and that has a reduced capacity for assimilation of nutrients in the future. Conversely, at low loading rates, the wetland may have a higher capacity to remove nutrients than at high loading rates.

Specific to wetland systems receiving secondarily-treated municipal effluent, loading rates are normally calculated using nutrient concentrations (i.e., total nitrogen and phosphorus) of the municipal effluent, the volume of the discharge, and the area of the receiving wetland. For wetland assimilation systems in Louisiana, typical loading rates for total nitrogen (TN) range from 2 to 15 g/m^2/year and for total phosphorus (TP) from 0.4 to 4 g/m^2/year (Day et al. 2004). Removal efficiencies for TN and TP at these loading rates average between 65 and 90%, while NO_x removal is between 90 and 100%. Nutrient removal efficiency is the percentage of nutrients removed from the overlying water column and retained within the wetland ecosystem or released into the atmosphere. Richardson and Nichols (1985) reviewed a number of wetlands receiving municipal effluent and found a clear relationship between loading rate and nutrient removal efficiency (Fig. 2.3). The relationship between nutrient removal efficiency and loading rate is not linear, with very effi-

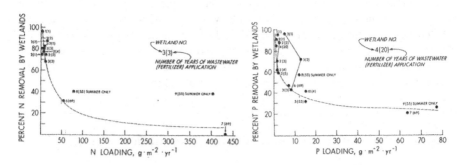

Fig. 2.3 Nitrogen and phosphorus removal efficiency as a function of loading rate in various municipal effluent assimilation wetlands (taken from Richardson and Nichols 1985)

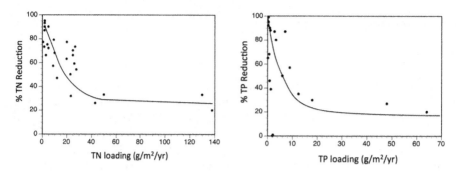

Fig. 2.4 Nitrogen and phosphorus removal efficiency as a function of loading rate in Louisiana wetlands receiving secondarily treated municipal effluent, stormwater, or diverted Mississippi River water

cient nutrient removal at low loading rates, and rapidly decreasing removal efficiency as loading rates rise. Mitsch et al. (2001) found a similar loading-uptake relationship for wetlands in the upper Mississippi basin.

The curves generated by Richardson and Nichols (1985) are derived from data of wetland assimilation systems located in many different parts of the United States. Data from assimilation wetlands, stormwater wetlands, and coastal wetlands receiving diverted Mississippi River water showed that this relationship was generally true for wetlands in Louisiana (Fig. 2.4). Nutrient uptake also has been reported in coastal wetlands receiving Atchafalaya River water (Lane et al. 2002) and Mississippi River water (Lane et al. 1999, 2004).

Effluent Disinfection

Contamination by human pathogens is an important issue that must be considered in wetland assimilation since these pathogens can be transferred to other animal species as well as to humans. For this reason municipal effluent is disinfected prior to release into the wetlands. Nevertheless, studies have shown that pathogens are rapidly degraded in wetlands, much more so than in open water bodies such as lakes, streams or bayous (Kadlec and Wallace 2009). Proper disinfection is a particular concern for all municipal effluent treatment plants, and dischargers are responsible for regularly monitoring the effectiveness of disinfection systems. Commonly used disinfection methods include chlorination followed by dechlorination, UV radiation, and ozone. Chlorination is the most commonly used method of disinfection but, although dechlorination reduces toxic chlorine residuals, it increases operator costs and may introduce hazardous chemicals into the aquatic environment depending on the method used. Ozone and UV radiation are the cleanest disinfection methods, but have operator costs as well (City of New York Department of Environmental Protection and HydroQual, Inc. 1997).

Contaminants

Metals

The LDEQ currently requires cadmium, chromium, copper, iron, lead, magnesium, nickel, selenium, silver, and zinc concentrations to be measured at specific time increments in surface water, soils, and vegetation of wetlands receiving treated municipal effluent. Metal concentrations of surface waters at assimilation wetlands in Louisiana have been very low, with most concentrations below the detectable limit. There also have been no detectable differences in metal concentrations in sediments or vegetation between the assimilation wetlands and reference wetlands (Table 2.2). Similar results have been obtained for all assimilation wetlands in Louisiana. In general, there is little evidence that metal contamination is a problem for assimilation wetlands.

Contaminants of Emerging Concern

The impact of pharmaceuticals and personal care products (PPCPs) and endocrine disrupting compounds (EDCs) in natural systems has become an important issue (Koplin et al. 2002; Boyd et al. 2004). These compounds are excreted into sewage systems and enter the aquatic environment with the discharge of treated municipal wastewater (Li et al. 2014). Impacts from PPCPs are most pronounced in smaller streams where effluent discharge makes up a large proportion of the flow. Conventional wastewater treatment plants have limited ability to remove PPCPs due to short retention times while natural and constructed wetlands can promote removal through a number of mechanisms, including photolysis, plant uptake, microbial

Table 2.2 Mean metal concentrations in soils and vegetation at effluent discharge sites and nearby reference sites

Analyte	Soils concentration (mg/kg)		Vegetation concentration (mg/kg)	
	Discharge	Reference	Discharge	Reference
Cadmium	1.31 ± 0.25	1.08 ± 0.15	0.62 ± 0.06	0.79 ± 0.20
Chromium	14.44 ± 1.64	16.64 ± 2.20	0.93 ± 0.16	0.63 ± 0.08
Copper	19.18 ± 1.88	21.35 ± 3.18	3.29 ± 0.72	5.09 ± 0.96
Iron	14,343 ± 1538	12,067 ± 1567	146 ± 25	125 ± 23
Lead	16.20 ± 1.72	20.20 ± 2.52	1.23 ± 0.29	0.72 ± 0.10
Magnesium	3081 ± 397	2725 ± 326	2195 ± 408	2719 ± 468
Nickel	13.69 ± 1.18	16.02 ± 2.43	1.90 ± 0.34	2.20 ± 0.38
Selenium	2.99 ± 0.23	3.11 ± 0.44	3.65 ± 0.47	3.02 ± 0.31
Silver	0.90 ± 0.16	0.44 ± 0.07	0.67 ± 0.02	0.68 ± 0.13
Zinc	78.36 ± 7.77	82.80 ± 11.66	27.45 ± 3.38	46.74 ± 8.32

Data shown are means (± standard error) for samples collected at Breaux Bridge, Broussard, Hammond, Luling, and Mandeville assimilation wetlands degradation, hydrolysis, and sorption to soil (White et al. 2006; Li et al. 2014; Verlicchi and Zambello 2014). Verlicchi and Zambello (2014) reviewed 47 studies of constructed wetlands used for reducing concentrations of contaminants of emerging concern and concluded that these systems have the potential to remove many contaminants, including naproxen, salicylic acid, ibuprofen and caffeine. Removal efficiency was related to a variety of factors, including type of constructed wetland (e.g., free surface, sub-surface), hydraulic retention time, type of pre-treatment, redox potential, and environmental conditions.

Recent research has shown that most PPCP parent compounds are removed rapidly from the water column through various degradation or removal pathways (Boyd et al. 2003; Batt et al. 2007). For example, a recent study of feedlot wastes found a decrease of 83–93% in estrogenic activity as the wastewater flowed through a constructed wetland system (Shappell et al. 2007). Some municipalities are currently using wetlands to remove PPCPs from effluent and percent reduction is related to residence time in the treatment system. Treatment systems using ponds with long residence times (10–30 day) are more effective in removing PPCPs than highly engineered systems with a short residence time (12–18 h). Low loading rates, long residence times, and diverse microbial habitats in wetland assimilation systems should further promote the breakdown of PPCPs.

One group of personal care products known as nonylphenol ethoxylates (NPEOs) enter the environment due to their use in paints, inks, detergents, pesticides and cleaners. NPEOs can be as much as 10% of the total dissolved organic carbon entering a wastewater treatment plant (Ahel et al. 1994). Vegetated treatment wetlands have demonstrated the ability to remove up to 75% of NPEOs from domestic wastewater (Belmont et al. 2006).

A study at the Mandeville wastewater treatment facility, which consists of both a series of aeration lagoons, a constructed wetland, and natural wetlands, showed that the treatment system decreased the concentrations of nine types of PPCPs by 90% or more, dependent on the compound (Conkle et al. 2008). For most compounds reduction in concentration occurred over a 30-day treatment period in the aerated lagoons. However, the adjacent forested wetland also showed significant (6–52%) removal for several common pharmaceuticals (Conkle et al. 2010).

Little research has focused on sorption in wetland soils though this may be an important removal mechanism since many compounds are likely to bind to soils that have charged binding sites such as clays and organic matter found in wetland soils. Three estrogenic compounds, Bisphenol-A, 17ß-estradiol and 17α-ethinylestradiol (Clara et al. 2004), and three antibiotic compounds, ciprofloxacin, norfloxacin and ofloxacin (Conkle et al. 2010), have been shown to have high sorption coefficients, indicating that sorption is a major pathway for compound removal from the water column. Estrogenic sorption was studied using sewage sludge, while the antibiotics were tested on a wetland soil containing 20% organic matter. Research to date has not addressed the fate of these compounds once bound in soil or their effect on microfauna.

Concentrations of these compounds of concern are highest closest to the point of effluent discharge. The experience with the Mandeville wastewater treatment system suggests that in order to minimize exposure of the biota to higher concentrations of these compounds, discharging into a settling pond is necessary prior to discharge into the receiving wetland. This allows for initial removal as well as providing time-averaged concentrations before discharge into the wetland system.

Policy Considerations

The use of wetlands for municipal effluent assimilation has important implications for total maximum daily loads (TMDLs) and nutrient limits. A TMDL is a calculation of the maximum amount of a pollutant that a water body can receive and still meet EPA and state environmental water quality standards. In the case of water quality problems related to over-enrichment and eutrophication, the pollutants of interest are nutrients and non-toxic organic compounds. One problem that may arise for small municipalities in a watershed dominated by other pollutant sources (such as agriculture or a large city) is that the TMDL allocation for the municipality will be very low, necessitating greater wastewater treatment prior to discharge to receiving water bodies. The use of wetland assimilation provides an economical means for such additional water quality improvement (Ko et al. 2004, 2012).

Land Ownership

Communities in Louisiana have employed a variety of strategies to work with landowners when utilizing wetlands for assimilation of nutrients and sediments in various wastewaters. These strategies range from outright land purchase to a memorandum of understanding (MOU) and flowage easements (Table 2.3).

Ecological Monitoring

Monitoring of vegetation, hydrology, water quality and soils is a vital component of any wetland assimilation project. Requirements for monitoring are outlined in the LPDES permit for discharge of treated effluent into a wetland (Table 2.4). Vegetation data provide information on the health and vigor of the plant community, and whether vegetative species composition or dominance is being altered due to effluent addition. Water gauge data provide information about hydrology and changes in the depth and duration of inundation. Metals and nutrient data of soils and vegetation determine if there is an accumulation of these materials that could become problematic. Surface water quality data provide information of the efficiency of the system in removing nutrients from the water column. Data are collected from the assimilation wetland and from an ecologically similar reference wetland that is not

Table 2.3 Assimilation wetland land use agreements

Community	Type of agreement	Collaborating entity
Mandeville	Purchase	City owns land
Hammond	Memorandum of understanding	LDWF and City owns 230 acres
St. Charles—Luling	Flowage easement	Private landowner
Thibodaux	Flowage easement	Private landowner
Amelia	Flowage easement	Private landowner
Broussard	Purchase	NA
St. Martinville	Purchase	NA
Breaux Bridge	Memorandum of understanding	Nature conservancy

Table 2.4 LPDES monitoring requirements for a typical wetland assimilation project in Louisiana

Parameter	Wetland component			
	Flora	Sediment	Surface water	Effluent
Species classification	P			
Percentage of whole cover (for each species)	P			
Growth studies	A			
Water stage			M	
Metals: Mg, Pb, Cd, Cr, Cu, Zn, Fe, Ni, Ag, Se	P	P	P	S
Metals analysis: Hg, As		P		
Nutrient analysis I: TKN, TP	P	P	S	
Nutrient analysis II: NH_3N, NO_2N, NO_3N, PO_4		P	S	
Others: BOD5, TSS, pH, Dissolved Oxygen			P	
Accretion Rate		P		

P: Periodically—Sampling must be made once during March through May and once during September through November in the fourth year of the permit period for three Assimilation areas and one Reference area
A: Annually—Sample once per year at three Assimilation areas and one Reference area
M: Monthly—Samples should be taken at three Assimilation areas and one Reference area each month.
S: Semi-annually—Sample twice per year. Once during September through February and once during March through August at three Assimilation areas and one Reference area

impacted by the treated effluent. By comparing data between the assimilation and reference wetlands, as well as pre- and post-discharge data at the assimilation wetland, it is possible to determine if the assimilation wetland is being positively or negatively impacted by effluent addition.

According to the LPDES discharge permit, if wetland monitoring indicates that there is: (A) more than a 20% decrease in naturally occurring litterfall or stem growth; or (B) significant decrease in the dominance index or stem density of baldcypress; then, the permittee shall conduct such studies and tests as to determine if the impact to the wetland was caused by the effluent. Thus, monitoring provides a mechanism for evaluating the impacts of treated effluent on an assimilation wetland. It is important to note that wetland monitoring requirements may be modified by LDEQ if data indicate that changes are necessary.

Nutria Management

Nutria (*Myocaster coypus*), an introduced rodent, can severely impede attempts to restore baldcypress swamps (Myers et al. 1995) and herbaceous wetlands (Shaffer et al. 1992; Evers et al. 1998; Shaffer et al. 2015) in coastal Louisiana. This has certainly been the case for the Manchac land bridge, Jones Island, Sawgrass Bayou, and Big Branch National Wildlife Refuge in the mid and upper Lake Pontchartrain Basin, where nutria have killed tens of thousands of baldcypress seedlings and hundreds of hectares of herbaceous marsh (Effler et al. 2007; McFalls et al. 2010). Nutria appear to be able to detect the higher protein content of wetland vegetation with higher nutrient content, whether from fertilizer (Shaffer et al. 2009; Ialeggio and Nyman 2014) or treated municipal effluent (Lundberg 2008; Shaffer et al. 2015). Very few nutria were observed during the pre-discharge data collection phase at the Hammond assimilation wetland. However, within 12 months of discharge initiation, nutria numbers increased dramatically (Shaffer et al. 2015). Nutria are very prolific and can breed any time of year, producing at least 2 litters per year with an average of 4.5 young per litter (http://www.nutria.com/site).

Nutria herbivory can be controlled by several methods, such as nutria exclusion devices for seedlings, which are very effective in preventing herbivory of individual seedlings (Myers et al. 1995). For large areas, however, the only effective protection is population reduction. In an attempt to manage nutria populations in coastal Louisiana, the Coastal Wetlands Planning, Protection, and Restoration Act (CWPPRA) established the Coastwide Nutria Control Program (CNCP). The goal of the CNCP, which is managed by the Louisiana Department of Wildlife and Fisheries, is to encourage the harvest of nutria by paying trappers $5 per tail. In the 2015–2016 trapping season, 349,235 nutria tails, worth $1,746,175 in incentive payments, were collected from 274 participants (http://www.nutria.com/control_program). As part of the adaptive management at the Hammond assimilation wetland, one trapper and several hunters shot over 2000 nutria in one season (Shaffer et al. 2015). This is discussed in more detail below.

Benefits of Wetland Assimilation

Wetland Restoration

The introduction of treated municipal effluent into degraded forested wetlands of Louisiana is a major step towards their ecological restoration. The nutrient component of municipal effluent increases wetland vegetation productivity (Rybczyk et al. 1996; Hesse et al. 1998; Lundberg 2008; Hunter et al. 2009b; Shaffer et al. 2015), which helps offset regional subsidence by increasing organic matter deposition on the wetland surface, thereby decreasing flooding duration and producing a positive feedback loop of increased ecosystem vigor and resiliency (Fig. 2.5). The

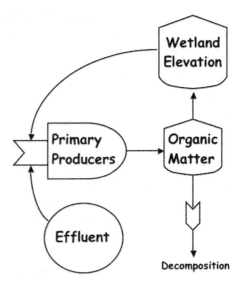

Fig. 2.5 Conceptual model of the effects of effluent application to wetlands

freshwater component of effluent also provides a buffer against saltwater intrusion events, especially during periods of drought, which are predicted to increase in frequency in the future due to global climate change (IPCC 2001). For example, the prolonged drought of 2000–2001 led to widespread death of baldcypress in the Lake Pontchartrain basin when saltwater intruded into these areas (Shaffer et al. 2009; Day et al. 2012).

Recent efforts to restore and enhance wetlands in the subsiding delta region have focused on attempts to decrease vertical accretion deficits by either physically adding sediments to wetlands or by installing sediment trapping mechanisms (e.g., sediment fences), thus increasing elevation and relieving the physio-chemical flooding stress (Day et al. 1992, 1999, 2004; Boesch et al. 1994). Breaux and Day (1994) proposed an alternate restoration strategy by hypothesizing that adding nutrient rich secondarily-treated municipal effluent to hydrologically isolated and subsiding wetlands could promote vertical accretion through increased organic matter production and deposition. Their work, and other studies, has shown that treated municipal effluent does stimulate productivity and accretion in wetlands (Rybczyk 1997; Hesse et al. 1998; Brantley et al. 2008; Hunter et al. 2009b; Shaffer et al. 2015). Rybczyk et al. (2002) reported that effluent discharge into the Thibodaux assimilation wetlands increased accretion rates by a factor of three (Fig. 2.6). DeLaune et al. (2013) reported that the Davis Pond river diversion in southern Louisiana led to accretion rates in receiving wetlands of more than 1 cm/year.

Over the past several decades, many attempts have been made to restore degraded baldcypress-water tupelo swamps in coastal Louisiana. In general, four primary interacting factors have been responsible for the very limited success of these restoration attempts: saltwater intrusion, persistent flooding, nutria, and lack of nutrients.

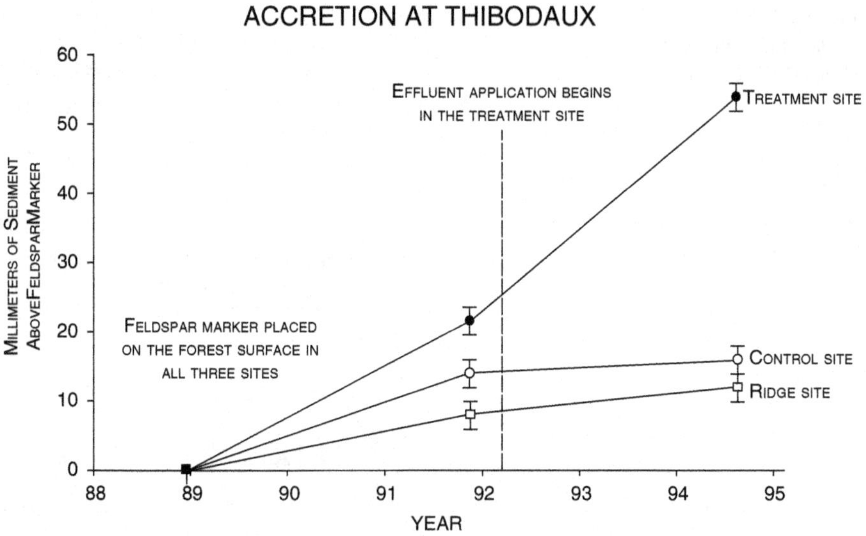

Fig. 2.6 Sediment accretion measured using feldspar horizon markers at the Thibodaux assimilation wetland (treatment site) and reference wetland (control site); From Rybczyk et al. 2002)

Carefully implemented wetland assimilation projects solve all four problems. Municipal effluent is a source of fresh water to otherwise hydrologically isolated wetlands, buffering saltwater intrusion events and providing fresh water during droughts. Hurricane storm surge can cause long-term changes in porewater salinity in coastal swamps and marshes but salt water can be flushed from soils by discharge of municipal effluent, river diversions, or other sources of fresh water (Steyer et al. 2007). Although input of treated effluent does not reduce persistent flooding, consistent input of municipal effluent decreases the residence time of water in a wetland, pushing out toxins (e.g., sulfides) that accumulate under stagnant conditions. Through stringent management nutria populations can be held in check. Finally, nutrient rich municipal effluent addition promotes increased rates of primary production and soil accretion, an important part of any restoration plan for wetlands in coastal Louisiana.

Enhanced Productivity

Secondarily-treated effluent delivers nutrient rich water to wetlands, stimulating vegetative productivity. While this could lead to eutrophication in some aquatic systems, many regions of the Gulf Coast are isolated from historic pulses of nutrients and sediments by dams, dikes, and levees, and, thus, are nutrient limited. Treated effluent can be used to enhance and restore productivity to these areas (Day et al. 2004).

Table 2.5 Mean litterfall, stem growth, and total net primary productivity (NPP) of forested wetlands receiving discharge of treated effluent and reference wetlands in Louisiana

Site	Discharge site			Reference site		
	Litterfall (g/m^2/year)	Stem growth (g/m^2/year)	Total NPP (g/m^2/year)	Litterfall (g/m^2/year)	Stem growth (g/m^2/year)	Total NPP (g/m^2/year)
Breaux Bridge (2002–2013)	56.9 ± 8.7	12.2 ± 2.1	69.2 ± 10.7	34.5 ± 4.1	23.3 ± 3.7	57.8 ± 7.8
Broussard (2007–2013)	39.7 ± 7.5	32.2 ± 6.0	72.0 ± 13.5	14.7 ± 3.2	17.4 ± 1.3	32.1 ± 4.6
Luling (2008–2013)	31.9 ± 2.7	21.7 ± 2.0	53.6 ± 4.7	17.6 ± 0.9	24.5 ± 2.3	42.1 ± 3.2

Data shown are means of post-discharge monitoring data collected by Comite Resources, Inc. (monitoring time period in parenthesis). Data are based on mean productivity per tree

Net primary productivity is generally higher at wetlands receiving discharge compared to corresponding reference wetlands (Table 2.5).

Hesse et al. (1998) conducted a tree ring analysis to document long-term effects of discharge of treated municipal effluent on the growth rate of baldcypress at the Cypriere Perdue assimilation wetland near Breaux Bridge, Louisiana. Treated effluent has been discharged into this wetland since 1954, but long-term monitoring has only been conducted since the city was issued an LPDES permit specific to assimilation wetlands in 2001. Growth chronologies from 1920 to 1992 were developed from cross-dated tree core samples taken from Discharge (Treated) and Reference (Control) sites with similar size and age classes. Significant differences in growth response between sites showed a consistent pattern of growth enhancement in the site receiving treated effluent (Fig. 2.7).

Shaffer et al. (2015) found that growth of baldcypress seedlings at the Hammond assimilation wetland was greatest where treated effluent was discharged and growth followed a linear decrease to 700 m from discharge. The diameter increase of mature baldcypress trees located along the effluent discharge pipe was five times greater than that of the Maurepas swamp and tenfold higher then trees at the Reference site. Baldcypress seedlings planted within 20 m of the effluent outfall system in 2008 averaged over 8 m tall in 2010 and were growing 2.01 cm/year (+0.08 cm/year S.E.) in diameter (Shaffer et al. 2015). There have been numerous studies showing either increased growth or no effect to baldcypress that are exposed to highly nitrified water. For example, Brantley et al. (2008) found significantly higher baldcypress growth downstream of effluent discharged from the Mandeville Bayou Chinchuba wastewater treatment plant. Shaffer et al. (2009) found increased growth rates in the Maurepas basin in areas receiving regular non-point source inputs, as did Effler et al. (2007) for trees given nutrient amendments. At the Amelia assimilation wetland, total NPP was higher at the Discharge site than at the corresponding Reference site (Day et al. 2006).

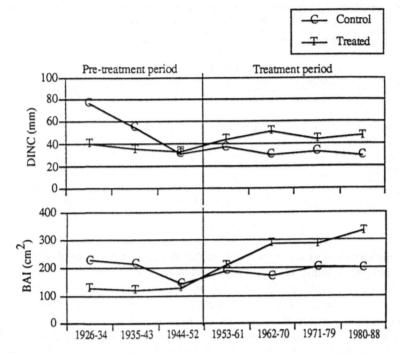

Fig. 2.7 Average periodic diameter increment (DINC) and basal area increment (BAI) growth/tree for each 9-year interval for baldcypress in the Cypriere Perdue Swamp; Taken from Hesse et al. (1998)

Nutrient Reduction

Water quality improvement of municipal effluent has been well documented in the assimilation wetlands at Amelia (Day et al. 2006), Breaux Bridge (Blahnik and Day 2000; Hunter et al. 2009b), Hammond (Shaffer et al. 2015), Luling (Hunter et al. 2009a), Mandeville (Brantley et al. 2008), St. Bernard (Day et al. 1997b), and Thibodaux (Zhang et al. 2000; Izdepski et al. 2009). Reduction of NO_x, NH_3, total Kjeldahl nitrogen (TKN), and total phosphorus typically range between 60 and 100% (Table 2.6). At most sites, concentrations of nitrogen and phosphorus were reduced to background levels by the time surface water left the wetland.

Burial in sediments integrates numerous processes that remove nutrients from treated effluent, including the settling of organic and inorganic sediments from the water column, microbial uptake, and the incorporation of organic matter (e.g., leaf litter or roots) into the sediment. Plant uptake cannot be considered a long-term loss unless the nitrogen and phosphorus are stored in persistent woody tissue and then ultimately harvested or buried in the wetland. Nutrients assimilated by herbaceous plants, however, can remain unavailable for long periods if they are asso-

Table 2.6 Percent nutrient reductions of effluent entering and leaving the assimilation wetlands in coastal Louisiana

Site	Parameter	Discharge concentration (mg/L)	Outlet concentration (mg/L)	% Reduction
Amelia[a]	TKN	2.98	1.00	66
	Total P	0.73	0.06	92
	NO_x	0.80	<0.01	100
Breaux Bridge[b]	PO_4	1.00	0.20	80
	Total P	2.90	0.30	87
Hammond[c]	NH_3	9.85	0.50	95
	NO_x	6.18	0.01	100
	PO_4	3.55	0.01	100
	Total P	4.04	0.04	99
Luling[c]	NO_x	0.52	0.19	63
	PO_4	0.62	0.20	67
Mandeville[c]	NH_3	1.90	0.60	68
	NO_x	5.86	1.09	81
	PO_4	3.31	1.64	57
	Total P	3.82	1.64	57
St. Bernard[d]	TKN	13.60	1.40	90
	Total P	3.29	0.23	95
Thibodaux[e]	NO_x	8.70	<0.10	100
	TKN	2.90	0.90	69
	PO_4	1.90	0.60	68
	Total P	2.46	0.85	66

[a]Day et al. (1997a)
[b]Day et al. (1994)
[c]Comite Resources, Inc. monitoring data
[d]Day et al. (1997b)
[e]Zhang et al. (2000)

ciated with refractory organic matter that becomes incorporated in the soils (Morris et al. 2013).

Increased nutrient inputs at many wetland treatment systems leads to a growth of algae. When light and nutrients are not limiting, algae can contribute significantly to the food web and nutrient cycling (Kadlec and Wallace 2009). However, because most of the assimilation wetlands in Louisiana are forested and have a closed canopy, particularly where effluent discharge occurs, algal blooms do not occur. Two exceptions include the Hammond and Thibodaux assimilation wetlands. At the Hammond assimilation wetland, treated effluent is discharged into an emergent freshwater wetland. Nutria herbivory of the wetland vegetation caused the system to largely degrade to open water which was subsequently colonized by algae. As the system recovered, emergent marsh species colonized the area and shaded out the growth of the algae and floating aquatic species. At the Thibodaux assimilation wetland, the system had very sparse growth of degraded baldcypress when discharge of treated effluent began. Over time, a floating marsh emerged and, again, shaded out the growth of algae. Both of these assimilation wetlands are discussed in more detail later in this chapter.

Carbon Sequestration

The term 'carbon sequestration' describes removal of atmospheric carbon dioxide (CO_2), usually by plants, and permanent storage of the fixed carbon in the ecosystem. Carbon sequestration is mostly viewed in relation to mitigating CO_2 released during the burning of fossil fuels (Williams 1999; Lal 2004; Euliss et al. 2006). Wetlands located in the Louisiana coastal zone have the potential to permanently store carbon due to high regional geological subsidence of 2–10 mm/year (Penland et al. 1988). Rybczyk et al. (2002) found that the Thibodaux assimilation wetlands had significantly higher accretion rates compared to an adjacent reference wetland. Because this accretion was due primarily to an increase in organic matter (OM) rather than mineral sediments, significant carbon burial occurred (OM generally consists of 50% carbon by weight). Estimates of carbon burial pre-(1375 kg C/ha/year) and post-effluent addition (3680 kg C/ha/year) indicate that 2305 kg C/ha/year of additional carbon was sequestered due to the discharge of municipal effluent (Day et al. 2004).

Global warming has become a major worldwide concern that has facilitated significant growth in emissions trading programs collectively referred to as carbon markets. Projects that reduce greenhouse gas emissions generate 'carbon offsets'. A carbon offset (mt CO_2e), also referred to as a carbon credit, is a metric ton reduction in emissions of CO_2 or greenhouse gases made to compensate for, or to offset, an emission made elsewhere (Murray et al. 2011). For a variety of financial, environmental, and political reasons, substantial interest exists for carbon offsets derived from terrestrial landscapes, including wetland ecosystems. The carbon sequestered in coastal and marine ecosystems has been termed 'blue carbon' (Mcleod et al. 2011; Sifleet et al. 2011). Allowing entities to privately invest in wetland restoration projects to offset greenhouse gas emissions elsewhere holds promise as a new carbon offset sector. In the future, the ability to sell carbon credits may provide an important source of revenue for municipalities in Louisiana using wetland assimilation of municipal effluent. Lane et al. (2017) recently documented net carbon sequestration at the Luling wetland assimilation system.

Mitigation of Impacts of Global Climate Change

There are two important global trends that should be considered as part of an analysis of wetland assimilation. These trends are global climate change and the cost and availability of energy, specifically oil. Three climate trends have important implications for wetlands located along coastal Louisiana: accelerated sea level rise; greater frequency of strong hurricanes; and more frequent and longer durations of drought, all of which lead to saltwater intrusion. The introduction of treated municipal effluent to wetlands directly counters the last of these trends (via freshwater addition), and indirectly counters the others through increased vegetative productivity (providing hurricane protection) and accretion (which increases wetland surface elevation).

Energy and Economic Savings

The availability and cost of energy will likely become an important factor affecting society in the near future. Over the past decade, increasing information has appeared in the scientific literature suggesting that world oil production is peaking or will peak within a decade or two, implying that demand will consistently be greater than supply, and that the cost of energy will increase significantly in the coming decades. Conventional sewage treatment is expensive and highly energy intensive compared with wetland assimilation. Economic cost benefit analyses of wastewater treatment operations at the Breaux Bridge and Thibodaux assimilation wetlands (Breaux 1992; Breaux and Day 1994; Breaux et al. 1995; Ko et al. 2004, 2012) conservatively estimated capitalized cost savings using wetland assimilation rather than conventional tertiary treatment (Table 2.7).

A study of the feasibility of using wetlands for assimilation of shrimp processing wastewater also demonstrated significant cost savings (Day et al. 1998; Cardoch 2000). The avoided cost estimate approach was used to compare costs of conventional on-site treatment of the shrimp processing effluent by the dissolved air flotation method with the cost of wetland assimilation. The annualized cost of the conventional treatment calculated to $214,000 per year, as compared to wetland assimilation costs of $63,000 per year, for a potential cost savings of $1,500,000 over 25 years (Day et al. 2004).

In conventional treatment, for every unit of carbon of organic matter oxidized in BOD reduction two to three units of carbon are released to the atmosphere as CO_2 from the burning of fossil fuels. Virtually no fossil fuels are needed for wetland assimilation. Thus, wetland assimilation has a much lower greenhouse gas impact

Table 2.7 Cost comparisons for three wetland assimilation projects (Day et al. 2004)

Site	Cost of conventional treatment	Cost of assimilation wetland	Cost savings
Breaux Bridge[a]	$3,300,000	$664,000	$2,636,000
Thibodaux[b]	$1,650,000	$1,150,000	$500,000
Dulac[c]	$2,200,000	$700,000	$1,500,000

[a]Costs reported in 2000 dollars. Capitalized costs are discounted at 7% for 20 years
[b]Costs reported in 1992 dollars. Capitalized costs are discounted at 9% for 30 years
[c]Costs reported in 1995 dollars. Capitalized costs are discounted at 8% for 25 years

than conventional tertiary treatment systems. Wetland assimilation also offsets CO_2 production through significant carbon sequestration from increased above- and belowground production. Sequestration of carbon, as soil organic matter, is especially significant in subsiding areas like the Mississippi River Delta.

An Overview of Wetland Assimilation Systems in Louisiana

Selected Case Studies: Thibodaux and Amelia, Louisiana

Thibodaux

Although the Thibodaux assimilation wetland is not the longest functioning system, it is one of the most intensively studied wetlands, with 3 years of baseline study (1989–1992) before discharge began. Additional studies and monitoring following the guidelines outlined in the LPDES permit have continued to the present.

The Pointe au Chene wetland, located 10 km southwest of Thibodaux, Louisiana, is a 231-ha subsiding baldcypress-water tupelo swamp on the back slope of Bayou Lafourche, a former distributary of the Mississippi River that was cut off in 1904 (Fig. 2.8). Historically, Bayou Lafourche carried an average of 12% of the

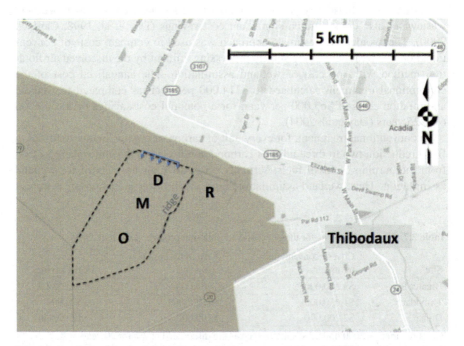

Fig. 2.8 Location of monitoring sites at the Thibodaux assimilation wetland. *R* Reference site, *D* Discharge site, *M* Mid site, *O* Out site. The *blue lines* indicate location of the discharge pipe and outlets, and the *black dashed line* the assimilation wetland

Mississippi River, or about 1100 m^3/s. Prior to the dredging of the Terrebonne-Lafourche drainage canal the wetland received upland runoff from the natural levee of Bayou Lafourche. Now, the spoil bank along the canal prevents any upland runoff from entering the site. The canal is directly connected to coastal waters so that water levels at the site are affected by coastal water levels (Conner and Day 1988). The area was further altered by the construction of a road for access to an oil drilling site on the western side of the assimilation wetland. The area is also bisected by a minor distributary ridge from Bayou Lafourche which separates the assimilation wetland from the reference wetland. The Reference area is about 10 cm higher and the ridge is about 40 cm higher than the assimilation wetland. Because of the hydrological changes, this area only received rainwater prior to effluent discharge. All flow from the area leaves via a 100-m wide shallow wetland channel between the access road and the ridge. There are three monitoring sites in the assimilation wetland that follow the flow of surface water from the discharge of treated effluent to where surface water leaves the wetland (termed Discharge, Mid, and Out sites), along with a reference wetland monitoring site.

The soils are classified as Fausse (very fine, montmorillonitic, nonacid, thermic typic fluvaquents) and effectively restrict groundwater exchange (Zhang et al. 2000). Over the past decades the study area experienced increased flooding due to subsidence and isolation from outside freshwater inputs and transitioned from bottomland hardwood forest to baldcypress-water tupelo swamp. The area immediately adjacent to the effluent input was a shallow, treeless, open water area cleared for the construction of a power line right of way. The dominant woody vegetation is baldcypress (*Taxodium distichum*) and water tupelo (*Nyssa aquatica*), with some black willow (*Salix nigra*) and swamp red maple (*Acer rubrum* var. *drummondii*). Because it is impounded and permanently flooded, there has been no forest regeneration in the Thibodaux assimilation wetland. Forested wetlands in the Verret basin, of which the Thibodaux wetland is part, began flooding in the early 1970s and by the late 1980s, most forested wetlands in the basin were permanently flooded (Conner and Day 1988). Relative sea level rise in the basin now exceeds 1.0 cm/year.

In a recent paper, Conner et al. (2014) analyzed water level changes in the Verret basin from 1986 to 2009 and concluded that the combination of rising water levels, hurricanes, and altered hydrology is fundamentally changing the structure of the forested wetland community. They concluded that the number of baldcypress trees is decreasing over time due to the loss of adult trees and lack of recruitment, and as Chinese tallow (*Triadica sebifera*) and red maple (not the major forest canopy trees one commonly associates with forested wetlands) continue to become common in the canopy through time, the system is losing its "swamp" character. Thus, it is important to understand that at the beginning of effluent discharge, the site was a fundamentally altered system with permanent flooding, stagnant conditions, and a dying bottomland hardwood community. In the 25 year history of the site, relative water level rise has been about 27 cm.

During the EBS study, surface water height was measured monthly at Discharge and Reference sites and mean water levels were about 10 cm deeper at the Discharge site, reflecting the difference in base elevations of the two sites. Measurements were not made for 2 years during the permit application and review process, but were restarted when effluent discharge began. Water levels at both the Discharge and Reference sites increased by 15–20 cm during this time. This was a reflection of hurricane Andrew in August 1992, ongoing relative water level rise, and most importantly, to higher rainfall in the several years after discharge began. A Before-After-Control-Impact (BACI) analysis showed that there was a significant increase in water levels at both sites but no difference between the Discharge and Reference areas, indicating that the increase was not due to the discharge of treated effluent (Rybczyk et al. 2002).

Mean net primary productivity (NPP) was higher at the Reference area compared to the Discharge area during the 2 years prior to effluent discharge. Productivity was likely affected by higher surface water levels and periods of inundation at the Discharge area than at the Reference area, but the most important factor was higher tree density at the Reference area. Although NPP decreased at both Discharge and Reference areas during the 2 years following discharge, the Discharge area still had lower productivity. Rybczyk et al. (1995) showed that decreased productivity at both sites was due to Hurricane Andrew, the eye of which passed within 80 km of the site. Tree mortality was higher and litterfall was lower at both sites due to the hurricane. NPP declined slightly but not significantly at the Reference site, while NPP increased slightly at the Discharge site. Because of the robust monitoring and additional measurements carried out after the hurricane, the impacts of the hurricane and the discharge of treated effluent were able to be separated.

Rybczyk et al. (2002) measured litter decomposition and accretion over feldspar marker horizons before and after discharge began. BACI statistical analysis revealed that neither leaf litter decomposition rates nor initial leaf litter nitrogen and phosphorus concentrations were affected by discharge of treated effluent. A similar analysis revealed that final nitrogen and phosphorus leaf litter concentrations did significantly increase in the Discharge site relative to the Reference site after effluent discharge began. Total pre-effluent accretion, measured 34 months after feldspar horizon markers were laid down, averaged 22.3 ± 3.2 mm and 14.9 ± 4.6 mm at the Discharge and Reference sites, respectively, and were not significantly different. However, total accretion measured 68 months after the markers were installed and 29 months after effluent discharge began averaged 54.6 ± 1.5 mm at the Discharge site and 19.0 ± 3.2 mm at the Reference site and were significantly different. Additionally, after discharge of treated effluent began, the estimated rate of accretion in the Discharge site (11.4 mm year^{-1}) approached the estimated rate of relative sea level rise (12.3 mm year^{-1}). Most of this increased accretion was attributed to organic matter inputs, as organic matter accumulation increased significantly at the Discharge site after effluent application began, while mineral accumulation rates remained constant. These findings indicate that there is a potential for using treated effluent to balance accretion deficits in subsiding wetland systems.

Ten years after the discharge of treated effluent began in 1992, switchgrass (*Panicum virgatum*) became established along the wetland boundary where effluent was discharged and by 2003 began to extend into the wetland. Within 4 years a highly productive emergent wetland developed with floating marsh characteristics.

Izdepski et al. (2009) studied the dynamics of this floating marsh community. They reported that NPP, total belowground biomass, NO_3, and plant-tissue $\delta^{15}N$ ratios varied significantly along a 75-m marsh transect, while mean plant-tissue $\delta^{13}C$ values differed between the dominant species. The area nearest the effluent discharge had the highest NPP (3876 g/m²/year), total belowground biomass (4079 ± 298.5 g/m²), and mean NO_3 (5.4 ± 2.9 mg/L). The mean $\delta^{15}N$ of pennywort (*Hydrocotyle umbellate*) floating marsh was less enriched at 0–75 m (9.7 ± 1.9%) compared to 100–200 m (21.0 ± 3.8%). The $\delta^{13}C$ of the belowground peat mat of the floating marsh was similar to switchgrass but not pennywort, indicating that switchgrass was forming the mat. Nutrient availability affected NPP and $\delta^{15}N$. NPP was greater than most reported values for floating marsh from 0 to 45 m then decreased along with NO_3 concentrations and $\delta^{15}N$ further from the effluent source. The herbaceous wetlands still persist. These results suggest that nutrient rich fresh water can promote restoration of some floating marshes.

Rybczyk et al. (1998) developed a wetland elevation/sediment accretion model for the site to determine if addition of treated municipal effluent could stimulate organic matter production and deposition to the point that sediment accretion would balance relative sea level rise. They simulated the effect of predicted increases in eustatic sea level rise (ESLR) on wetland stability and determined the amount of additional mineral sediment that would be required to compensate for relative sea level rise. The model also simulated primary production (roots, leaves, wood, and floating aquatic vegetation) and mineral matter deposition, both of which contribute to changes in elevation. Simulated wetland elevation was more sensitive to estimates of deep subsidence and future ESLR rates than to other processes that affect wetland elevation (e.g., rates of decomposition and primary productivity). The model projected that although the addition of treated effluent would increase long-term accretion rates from 0.35 to 0.46 cm/year, it would not be enough to offset the current rate of relative sea level rise. A series of mineral input simulations revealed that, given no increase in ESLR rates, an additional 3000 g/m²/year of mineral sediments would be required to maintain a stable elevation.

Keim et al. (2012) used tree-ring analysis to evaluate the combined effects of rising water levels and 13 years of municipal effluent addition on baldcypress growth at the Thibodaux assimilation wetland. Trees at the Discharge, downstream outflow, and adjacent Reference areas all experienced increased growth coinciding with a period of widespread rapid subsidence and water level increases in the late 1960s. Tree growth at the Discharge and outflow sites began to decrease before discharge of treated effluent began in 1992, and afterward was apparently unaffected by effluent discharge. In contrast, trees at the Reference site have not experienced growth declines. Hydrological changes caused by subsidence have apparently overwhelmed any effect of treated effluent on baldcypress growth. Increasing inundation may have increased growth initially by eliminating competition from species less tolerant of inundation; however, after a decade of sustained flooding, growth declined steadily. Release of baldcypress from competition continues at the topographically higher Reference site, but growth will likely subsequently decrease as ongoing subsidence and ESLR cause more prolonged inundation. These data suggest that short-term increases in water level and nutrients stimulated growth of baldcypress, but long-term increased inundation was a net stressor, and was more important than nutrient limitations in controlling growth at the Discharge site.

From 2010 to 2014, researchers at Nicholls State University, LA carried out a series of detailed studies on productivity, biogeochemistry, and benthic population dynamics (Minor 2014). They reported that water levels were higher at the Discharge site compared to the Reference, but were similar at the Out site. The initial studies in the early 1990s showed that the Discharge site had higher water levels because the soil surface was lower. The development of the flotant marsh at the site impedes flow resulting in a higher water level near the inlet (Izdepski et al. 2009). Monitoring indicates that the wetland is still reducing nutrients to background levels after 25 years of operation. Forest productivity measured on an aerial basis was lower at the Discharge site but was similar at Mid, Out, and Reference sites. The lower productivity at the Discharge site was due to low tree density. Minor (2014) noted that by 2014 almost all bottomland hardwood species had died, a process that had begun before effluent discharge. If the flotant marsh is taken into consideration, the productivity of the Discharge site is higher than the Reference and Out sites. Macrofaunal assemblages were different between the Reference and Discharge sites, likely as a result of differences in surface water levels and the resulting vegetation community structure.

Amelia

The Ramos Swamp assimilation wetland is a continuously flooded tidal freshwater forested wetland located south of Lake Palourde, approximately 2 km north of Amelia, Louisiana (Fig. 2.9). Although treated municipal effluent discharge began in 1973, the system was not permitted until 2007. The City uses a 13.4 ha oxidation pond with a chlorination/dechlorination system for sewage treatment. Since this wetland is over 40 km from the coast, daily water level fluctuations are less than a few cm. The swamp area directly affected by the effluent flow is 77 ha within a larger forested wetland area of over 1000 ha. The dominant vegetation is baldcypress, water tupelo, black willow, swamp red maple, and green ash (*Fraxinus pennsylvanica*). Because the forest does not have a completely closed canopy, there is floating aquatic vegetation dominated by mosquito fern (*Salvinia minima*), duckweed (*Lemna minor*), watermeal (*Wolffia* sp.), pennywort (*Hydrocotyle ranunculoides*), and water lettuce (*Pistia stratiotes*). Submerged aquatic vegetation is predominantly hornwort (*Ceratophyllum* sp.). The flooded soils consist of well consolidated riverine clay, overlain by high organic clays and topped by a poorly consolidated 30–60 cm layer of plant detritus (Lytle et al. 1959). There are three monitoring sites in the assimilation wetland along the flow of surface water from the discharge of treated effluent to where surface water leaves the wetland (termed Discharge, Mid, and Out sites) and a nearby Reference wetland that is not affected by treated effluent.

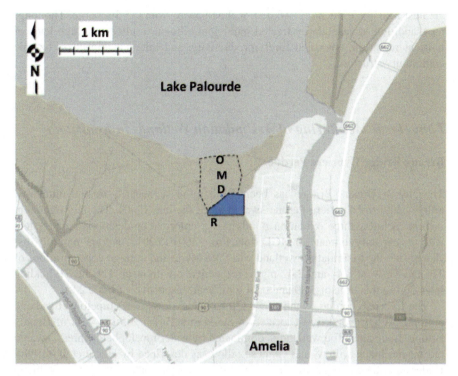

Fig. 2.9 Location of monitoring sites at the Amelia Ramos Swamp assimilation wetland. *R* Reference site, *D* Discharge site, *M* Mid site, *O* Out site. The water body north of the site is Lake Palourde. The *blue* polygon indicates the location of the oxidation pond, and the *black dashed line* the assimilation wetlands

Day et al. (2008) measured surface water nutrient concentrations at Discharge and Reference sites. TKN concentrations (2.0–4.0 mg/L) accounted for almost 75% of TN. NH_4-N was about 25% of TN and ranged from 0.4–1.0 mg/L. NO_x-N was generally less than 1% of TN. Within the wetland, PO_4 concentrations (0.1–0.9 mg/L) were about 50% of TP. TN and TP were reduced by about 79% and 88%, respectively, as water flowed through the wetland. These removal rates are consistent with low loading rates of 9.4 g TN and 1.2 g TP/m^2/year.

Day et al. (2008) also measured vegetation productivity and benthic community structure at assimilation and Reference wetlands. Litterfall was significantly greater at the Discharge site (717 g/m^2/year) compared to the Reference site (412 g/m^2/year). Stem growth ranged from 302 to 776 g/m^2/year and was not statistically different among the Reference, Discharge, and Out sites. Total NPP was highest at the Discharge and Out sites (1467 and 1442 g/m^2/year, respectively) and these values were significantly higher than NPP values at the Reference site (714 g/m^2/year). Total individuals, total species, and species richness of macroinvertebrates was greatest near the effluent outfall and declined away from the discharge. The long-term addition of secondarily treated municipal effluent resulted in a high level of nutrient retention, enhanced forest productivity, and minimal impact on benthic community structure.

Long-Term Monitoring of Assimilation Wetlands in Louisiana

Breaux Bridge Cypriere Perdue

The city of Breaux Bridge has been discharging secondarily-treated municipal effluent into the Cypriere Perdue swamp since the late 1940s. The city treatment system includes three oxidation ponds and a chlorination-dechlorination system with the capacity to treat 1.0 MGD flow. From 2001 to 2013, average monthly discharge into the assimilation wetland was 0.96 MGD and average concentrations of TN and TP were 8.44 and 2.42 mg/L, respectively, with mean TN and TP loading rates during this time of 1.89 and 0.24 g/m^2/year, respectively (Table 2.8).

The Cypriere Perdue swamp is a 1470 ha baldcypress-water tupelo wetland and bottomland hardwood forest located in St. Martin Parish, 3.5 km west of Breaux Bridge, Louisiana. The wetland is dominated by water tupelo, baldcypress, swamp red maple, black willow, and Chinese tallow, as described by Hesse et al. (1998). Under natural conditions, flow from the area was to the south with some flow likely going to the Vermillion River. During high water periods, backwater flooding from the Vermillion can raise water levels at the site by over 2 m. The Ruth Canal now connects the Vermillion River with Bayou Teche, a former distribuary of the Mississippi River, and almost no flow from the site goes south of the Canal. The original location of the Reference site was moved because high water levels caused effluent to flow into the area and the Out site was moved because short-circuiting caused effluent to by-pass the area (Fig. 2.10).

Blahnik and Day (2000) studied the hydrology and nutrient loading rates at the Breaux Bridge assimilation wetland. Pond discharge, surface water elevations, and fluorescent dye travel times were recorded to assess surface water hydrology, and water samples were collected for NO_x, NH_3, PO_4, and TSS analyses. Wetted surface area increased with pond discharge rate, and 58–66% of surface water flow was concentrated in shallow channels covering only 10–12% of the total study area. Hydraulic retention time was much longer (0.9–1.1 days) than minimum dye travel times (2–3 h) through the 4 ha study area. Higher pond discharge rates created more treatment surface area, and higher constituent loading rates produced higher removal rates. They concluded that higher nutrient loads could be assimilated without requiring significant increases in wetland area.

Monitoring between 2001 and 2013 showed that mean TN concentrations declined from the Discharge site to the Out site and concentrations at the Out site were actually lower than at the Reference site (Fig. 2.11). The type of nitrogen, or

Table 2.8 Mean effluent discharge, TN and TP concentrations in effluent, and loading rates for assimilation wetlands in Louisiana

Municipality	Years for data summary	Mean discharge (MGD)	Mean TN (mg/L)	Mean TP (mg/L)	Mean TN loading rate (g/m²/year)	Mean TP loading rate (g/m²/year)	% of TN in treated EFFLUENT		
							NH_3	NO_x	Org N[c]
Breaux Bridge	2001–2013	0.96	8.44	2.42	1.89	0.24	49.6	6.1	44.3
Broussard	2007–2013	0.59	24.64	3.45	14.75	2.62	69.3	9.2	21.5
Hammond	2007–2013	3.90	17.91	3.64	2.39	0.48	52.6	32.5	14.9
Luling	2006–2013	1.58	7.06	2.34	2.52	0.84	29.2	9.3	61.5
Mandeville BC[a]	2006–2013	1.19	14.36	3.31	56.50	13.90	43.2	40.2	16.6
Mandeville TM[b]	2009–2013	1.44	15.52	3.02	7.48	1.46			
St. Martinville	2011–2013	0.74	5.40	1.85	8.70	3.00	29.3	20.0	50.7

[a]Mandeville Bayou Chinchuba
[b]Mandeville Tchefuncte Marsh
[c]Organic Nitrogen

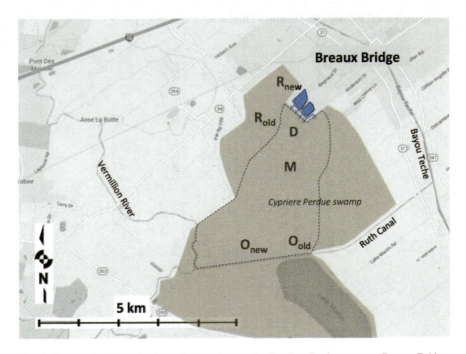

Fig. 2.10 Location of wetland monitoring sites at the Cypriere Perdue swamp, Breaux Bridge, Louisiana. R_{old} Original Reference site, R_{new} New Reference site, D Discharge site, M Mid site, O_{old} Original Out site, and O_{new} New Out site. The *blue* polygons indicate the location of the oxidation ponds, *blue lines* the discharge pipe and outlets, and the *black dashed line* the assimilation wetlands

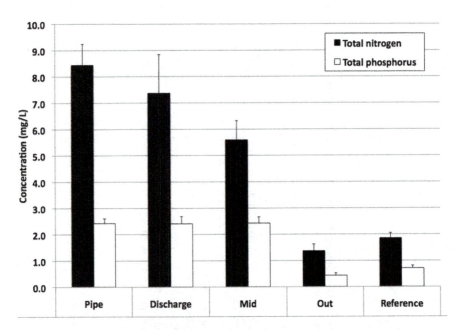

Fig. 2.11 Mean total nitrogen and total phosphorus concentrations at the effluent pipe and in surface water at the Breaux Bridge assimilation wetland between 2001 and 2013. Error bars represent standard error of the mean

nitrogen species, impacts nutrient removal because NO_x rapidly diffuses into anoxic soil layers where it is used as an electron acceptor and reduced to gaseous end products such as nitrous oxide and nitrogen gas (Reddy and DeLaune 2008). Thus, NO_x is removed more rapidly than NH_3, which must first be nitrified before being denitrified. Of the TN concentration, 49.6% is NH_3, 6.1% is NO_x, and 44.3% is organic nitrogen. Since NH_3 is a much larger percentage of TN than NO_x at this assimilation wetland, TN concentrations did not drop as rapidly as seen in other assimilation wetlands where the effluent is highly nitrified (i.e., Hammond, Mandeville Bayou Chinchuba and Tchefuncte Marsh assimilation wetlands). TP concentrations decreased to background conditions at the Out site (Fig. 2.11). Phosphorus is typically removed through vegetation uptake and abiotic retention in soils but it has no permanent removal mechanism such as denitrification for nitrogen (Reddy and DeLaune 2008).

Because of differences in tree density among sites, mean woody or litterfall productivity for each site is divided by the number of trees to determine mean productivity per tree. Mean litterfall and woody productivity are added together to determine mean NPP per tree. At the Breaux Bridge assimilation wetland, mean annual NPP in the Discharge and Reference wetlands ranged between about 30 and 135 g/m²/year per tree from 2002 to 2015 (Fig. 2.12). In general, litterfall productivity was a higher percentage of NPP than woody productivity.

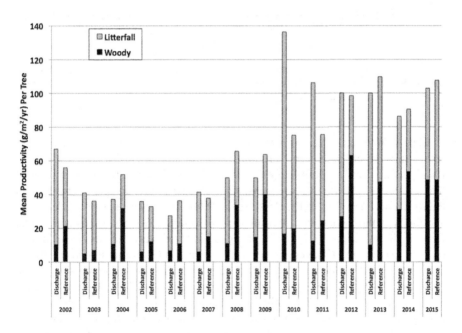

Fig. 2.12 Mean annual litterfall and woody productivity at the Breaux Bridge Cypriere Perdue assimilation wetland

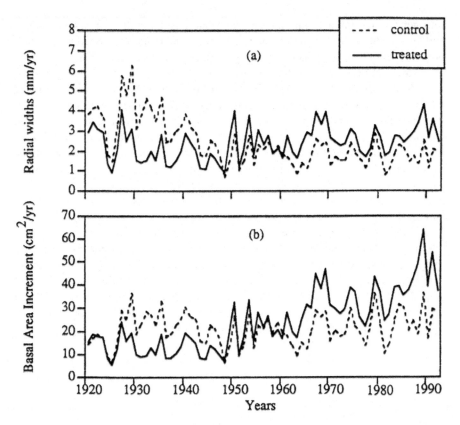

Fig. 2.13 Mean annual ring width chronologies of (**a**) diameter increment (DINC); and (**b**) basal area increment (BAI) for baldcypress in the Cypriere Perdue swamp (Hesse et al. 1998). Discharge of treated effluent began in the late 1940s

Long-term tree ring analysis of baldcypress growth at the site provides a context within which short-term variability can be interpreted (Hesse et al. 1998). Results of tree ring analysis show that after effluent discharge to the swamp began the late 1940's, growth rate was consistently higher in the Discharge area than in the Reference area (Fig. 2.13; Hesse et al. 1998). There was also a high degree of variability in these data, indicating that swamp productivity can vary dramatically. Growth of baldcypress is lower during periods of warm spring weather and drought (Stahle and Cleaveland 1992; Cleaveland 2006; Keim and Amos 2012). Day et al. (2012) reported that growth rates of baldcypress in two sites in the Pontchartrain Basin were most strongly correlated with May Palmer Drought Severity Index (PDSI). This high variability in year-to-year growth rates of baldcypress must be taken into consideration when interpretating forested wetland growth rates at assimilation wetlands because they are so strongly affected by climatic variability.

Broussard Cote Gelee

The City of Broussard discharges secondarily-treated effluent into 300 acres of the Cote Gelee forested wetlands via two outlets that can be operated independently (Fig. 2.14). The city treatment system includes three oxidation ponds and a chlorination-dechlorination system with the capacity to treat 1.0 MGD flow. From 2007 to 2013, average monthly discharge into the assimilation wetland was 0.59 MGD and average concentrations of TN and TP were 24.64 and 3.45 mg/L, respectively, with mean TN and TP loading rates during this time of 14.75 and 2.62 g/m^2/year, respectively (Table 2.8).

The Broussard Cote Gelee assimilation wetland is primarily a baldcypress-water tupelo swamp, but in the slightly more elevated parts of the area there is a mixed forest with bottomland hardwood species, such as pumpkin ash (*Fraxinus profunda*), water hickory, swamp red maple, and water elm (*Planera aquatica*). Under natural conditions, surface water from uplands adjacent to the site flowed

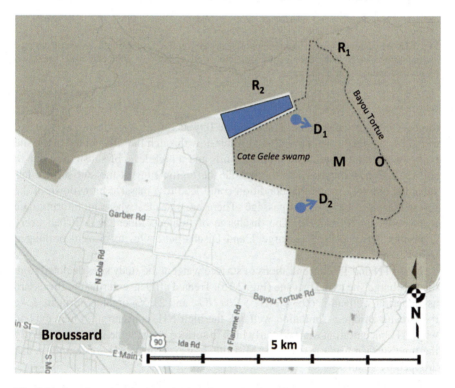

Fig. 2.14 Location of wetland monitoring sites at the Cote Gelee swamp, Broussard, Louisiana. R_1 First Reference site, R_2 Second Reference site, D_1 First Discharge site, D_2 Second Discharge site, *M* Mid site, *O* Out site. The *blue* polygon indicates the location of the oxidation ponds, the *blue arrows* the outlet pipes, and the *black dashed line* the assimilation wetlands

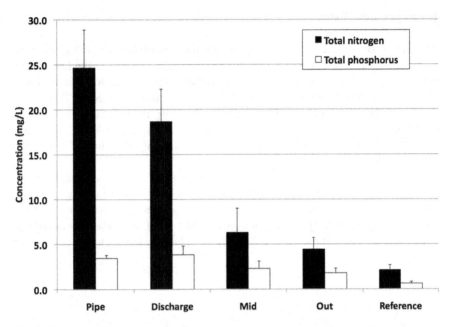

Fig. 2.15 Mean total nitrogen and total phosphorus concentrations measured at the effluent pipe and in surface water at the Broussard assimilation wetland between 2007 and 2013. Error bars represent standard error of the mean

through the forested wetlands and then into Bayou Tortue. A road embankment and a shallow dredged canal have altered water flow through the site and led to the site being over-drained with well-oxidized soils and a high level of subsidence due to soil oxidation. Exposed roots throughout the region suggest the soil surface has subsided one to 2 ft and this condition could lead to a massive blow-down of the forest during a major storm passage. There are six sites where monitoring data were collected (Fig. 2.14). The discharge of treated effluent is switched every 2 months between the Discharge 1 and Discharge 2 sites to prevent prolonged inundation.

Mean TN and TP concentrations of surface water at the study sites declined from the Discharge site to the Out site (Fig. 2.15). Treated effluent entering the Broussard assimilation wetland is 69.3% NH_3, 9.2% NO_x, and 21.5% organic nitrogen. Like the Breaux Bridge assimilation wetland, the high NH_3 concentration may explain why TN concentration was still fairly high in surface water at the Discharge site. By the time surface water reached the Mid site, however, mean TN concentration had dropped by about 75%.

Mean annual NPP was typically much higher at the Discharge site than at the Reference site at the Broussard assimilation wetland (Fig. 2.16). Data shown for the Reference site are the annual average for the two Reference sites that are monitored.

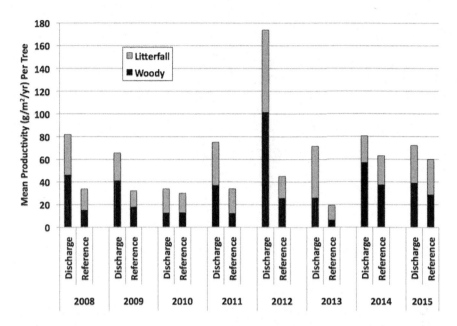

Fig. 2.16 Mean annual litterfall and woody productivity at the Broussard Cote Gelee assimilation wetland

St. Martinville

The City of St. Martinville began discharging secondarily-treated municipal wastewater from the city treatment facility to the Cypress Island Coulee wetlands in 2011. The St. Martinville wastewater treatment facility has a maximum design flow of 1.5 MGD, and consists of a 63.7-ha facultative lagoon with ultraviolet disinfection and a cascade aeration structure. From 2011 to 2013, the system had an annual average flow of 0.74 MGD and average concentrations of TN and TP were 5.40 and 1.85 mg/L, respectively, with mean TN and TP loading rates during this time of 8.70 and 3.00 g/m^2/year, respectively (Table 2.8).

The Cypress Island Coulee wetlands are located adjacent to the treatment facility and consist primarily of baldcypress-water tupelo swamp and red maple. These wetlands have been degraded by urbanization and conversion of surrounding areas to agriculture and are characterized by over-drained soils and subsidence. Because these wetlands were used for rice and crawfish production, they consist of a number of shallow "ponds" separated by low levees (Fig. 2.17). Secondarily-treated effluent is discharged at six different locations around the South basin of the wetlands and surface water drains into the Cypress Island Coulee after flowing through the wetlands (Fig. 2.17). To monitor the effects of this discharge on the vegetation of the

Fig. 2.17 Location of discharge pipes and wetland monitoring sites at the St. Martinville assimilation wetland. *P* pond number, *D* Discharge sites, *M* Mid sites, and *O* Out sites. The *blue* polygon indicates the location of the facultative lagoon, and the *black dashed line* the assimilation wetlands

receiving wetlands, two sets of three study sites (Treatment, Mid, and Out) were established, as well as three Reference sites located in nearby wetlands (One near Breaux Bridge (Fig. 2.10) and two near Broussard, LA (Fig. 2.14)).

Mean TN concentration discharged into the wetland declined from the Discharge site to the Out site (Fig. 2.18) and concentration at the Out site was similar to that measured at the Reference sites. Of the TN treated effluent discharged into the St. Martinville assimilation wetland, 29.3% is NH_3, 20% is NO_x, and 50.7% is organic nitrogen.

Mean annual NPP was typically higher at the Discharge site than at the Reference 2 site and lower than the Reference 1 site at the St. Martinville assimilation wetland (Fig. 2.19). Data shown for the Reference 2 site are the annual average for the two Reference sites monitored at the Broussard assimilation wetland. During most of the years when biomass was monitored, litterfall made up a higher percentage of NPP than woody biomass.

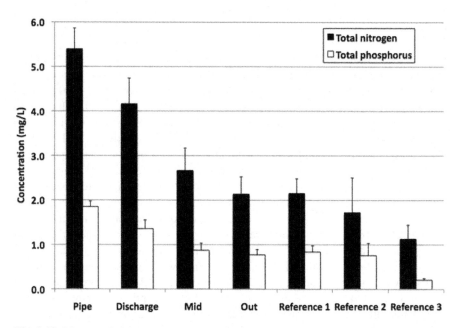

Fig. 2.18 Mean total nitrogen and total phosphorus concentrations at the effluent pipe and in surface water at the St. Martinville assimilation wetland and Reference sites between 2011 and 2015. Error bars represent standard error of the mean

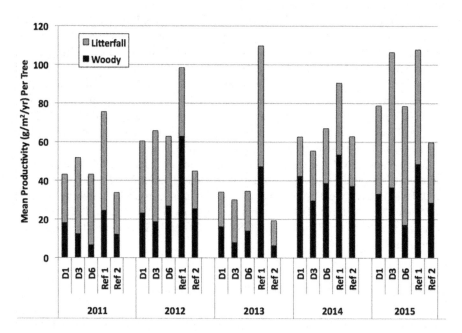

Fig. 2.19 Mean annual litterfall and woody productivity at the Discharge sites (D1, 3, 6) and Reference sites (Ref 1 and 2) at the St. Martinville assimilation wetland

Luling

The city of Luling discharges secondarily-treated effluent into forested wetlands adjacent to the wastewater oxidation pond. The city's treatment system consists of a facultative oxidation pond with ultraviolet disinfection. From 2007 to 2013, average monthly discharge into the assimilation wetland was 1.58 MGD. Effluent had mean concentrations of 7.06 and 2.34 mg/L for TN and TP, respectively, and mean TN and TP loading rates were 2.52 and 0.84 g/m^2/year, respectively (Table 2.8).

The assimilation wetland is located directly to the east of the oxidation pond. The 608-ha wetland is a continuously flooded freshwater forested wetland dominated by water tupelo and baldcypress. The site is located within the Davis Crevasse, a large crevasse splay that was formed in the nineteenth century when the river broke through the flood control levees and deposited a depositional splay of about 150 km^2 (Day et al. 2016b). Under natural conditions, water from the site where the oxidation pond is now located flowed in a southeasterly direction towards Lake Cataouatche through forested wetlands and freshwater marsh. The dredging of Cousins Canal short circuited water flow directly to the lake. The current discharge is a partial return to more normal water flow with the exception of continuous flooding. Three sites were established at the Luling assimilation wetland, including Discharge, Mid, and Out sites, and a Reference site was located nearby. The Discharge, Mid, and Reference sites are forested while the Out site is a freshwater emergent marsh (Fig. 2.20). The marsh Reference site is the same as the Hammond marsh Reference site (described below).

Mean TN and TP concentrations of surface water at the Luling assimilation wetland did not decline between the effluent pipe and the Discharge site but did decrease between the Discharge and Out sites (Fig. 2.21). Like the Breaux Bridge and Broussard assimilation wetlands, nitrogen entering the Luling assimilation wetland is higher in NH$_3$ than NO$_x$. The Luling effluent is 29.2% NH$_3$, 9.3% NO$_x$, and 61.5% organic nitrogen. Organic nitrogen is removed more slowly than NH$_3$ or NO$_x$ because it must be decomposed before these constituents are released. Both mean TN and TP concentrations of surface water at the Out site were very similar to those measured at the Forested and Marsh Reference sites.

Mean annual NPP was higher every year for the Discharge site compared to the Reference site at the Luling assimilation wetland (Fig. 2.22). End-of-season-live (EOSL) biomass collection began at the marsh Out site in 2008. Between 2008 and 2015, the Out site typically had higher productivity than the marsh Reference site (Fig. 2.23). During 2013, however, EOSL biomass at the Reference site was greater than at the Out site.

Mandeville Bayou Chinchuba and Tchefuncte Marsh

The City of Mandeville's wastewater treatment system includes three aerated lagoon cells, a three-celled rock reed filter, and an ultraviolet disinfection system. Effluent was discharged into the Bayou Chinchuba assimilation wetland starting in 1998, but

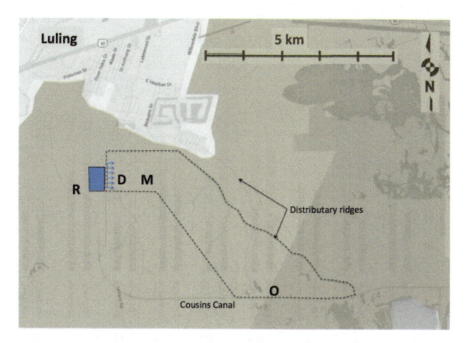

Fig. 2.20 Location of wetland monitoring sites at the Luling assimilation wetland. *R* Reference site, *D* Discharge site, *M* Mid site, *O* Out site. The marsh reference is located at the Hammond assimilation wetland. The distributary ridges were formed in the nineteenth century when the Davis Crevasse deposited river sediments in the area. The *blue* polygon indicates the location of the oxidation pond, *blue lines* the discharge pipe and outlets, and the *black dashed line* the assimilation wetlands

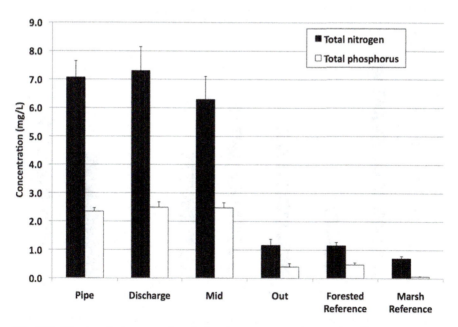

Fig. 2.21 Mean total nitrogen and total phosphorus concentrations at the effluent pipe and in surface water at the Luling assimilation wetland between 2006 and 2013. Error bars represent standard error of the mean

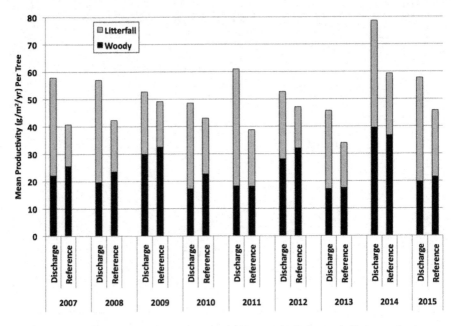

Fig. 2.22 Mean annual litterfall and woody productivity at the Luling assimilation wetland

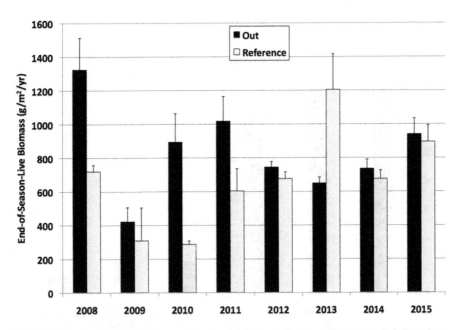

Fig. 2.23 Mean annual end-of-season-live biomass at the Luling assimilation wetland. Error bars represent standard error of the mean

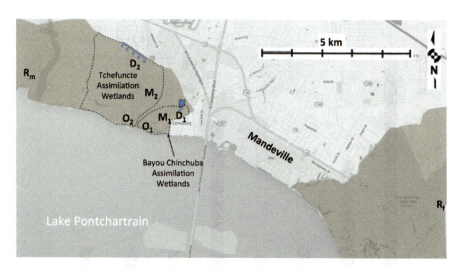

Fig. 2.24 Location of wetland monitoring sites at the Tchefuncte Marsh (TM) and Bayou Chinchuba (BC) assimilation wetlands. R_m Marsh Reference site, R_f Forested Reference site, D_1 BC Discharge site, M_1 BC Mid site, O_1 BC Out site, D_2 TM Discharge site, M_2 TM Mid site, O_2 TM Out site. The *blue* polygon indicates the location of the aerated lagoons, *blue lines* the discharge pipe and outlets, and the *black dashed line* the assimilation wetlands

the LPDES permit was not issued until 2003. Due to high loading rates a second assimilation wetland was established at the Tchefuncte Marsh in 2009 (Fig. 2.24) and effluent was then split between the two assimilation wetlands. Bayou Chinchuba and Bayou Castine (where the forested Reference site is located) are a combination of swamp and bottomland hardwood forest dominated by water tupelo, baldcypress, and swamp blackgum (*Nyssa biflora*). At the Tchefuncte Marsh, the Discharge site is forested, while the Mid, Out, and Reference sites are emergent marsh dominated by creeping panic (*Panicum repens*), dotted smartweed (*Polygonum punctatum*), bulltongue arrowhead (*Sagittaria lancifolia*), and cattail (*Typha latifolia*).

Average monthly effluent discharge into the Bayou Chinchuba assimilation wetland from 2006 to 2013 was 1.19 MGD, with average TN and TP concentrations of 14.36 and 3.31 mg/L, respectively (Table 2.8). Mean loading rates for TN and TP were 56.5 and 13.9 g/m^2/year, respectively. This assimilation wetland has higher loading rates than the 15 g/m^2/year for TN and 4 g/m^2/year for TP maximum limits set by the LPDES permit. It is important to note that even though loading rates are much higher than limits set by the LPDES permit, at the Discharge site mean TN concentration of surface water was 7.58 ± 0.95 mg/L and TP concentration was 1.97 ± 0.20 mg/L between 2006 and 2013. By the time surface water reached the Bayou Chinchuba Mid site, concentrations of total nitrogen and total phosphorus were reduced by at least 50% (2.31 ± 0.38 mg TN/L and 0.98 ± 0.14 mg TP/L) from those measured at the Discharge site and concentrations in the Bayou Chinchuba Out site were similar to the Marsh Reference site (Fig. 2.25). Thus, based on these

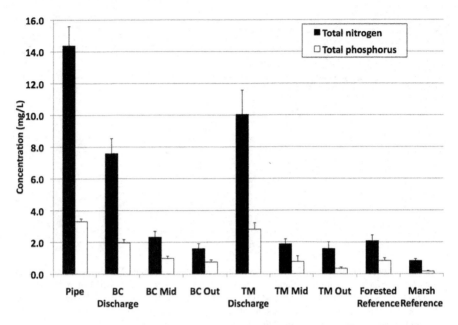

Fig. 2.25 Mean total nitrogen and total phosphorus concentrations at the effluent pipe and in surface water at the Mandeville Bayou Chinchuba (BC) and Tchefuncte Marsh (TM) assimilation wetlands. Data were collected between 2006 and 2013 for Mandeville Bayou Chinchuba and between 2009 and 2013 for Mandeville Tchefuncte Marsh. Error bars are standard error of the mean

data, it appears that the wetland has the capacity to assimilate and reduce nutrients at loading rates much higher than those allowed by the LPDES permit.

Average monthly discharge from 2009 to 2013 into the Tchefuncte Marsh assimilation wetland was 1.44 MGD and average concentrations of TN and TP were 15.52 and 3.02 mg/L, respectively, with mean TN and TP loading rates during this time of 7.48 and 1.46 g/m^2/year, respectively (Table 2.8). Even though the Tchefuncte Marsh and Bayou Chinchuba assimilation wetlands receive effluent from the same source (Mandeville wastewater treatment facility) mean nutrient concentrations are not always the same because they were calculated for different time periods and concentrations of the effluent vary over time.

By the time surface water reached the Mid sites at the Bayou Chinchuba and Tchefuncte Marsh assimilation wetlands, mean total nitrogen concentrations were almost as low as the Reference sites (Fig. 2.25). The Mandeville effluent is 43.2% NH_3, 40.2% NO_x, and 16.6% organic nitrogen. Like the Hammond assimilation wetland, effluent entering these wetlands is high in NO_x and, as stated previously, in anoxic soils NO_x is rapidly removed through denitrification. The high NO_x removal also was seen in the decrease between the point of effluent discharge and the Bayou Chinchuba Discharge and Tchefuncte Marsh Discharge sites. Mean TP concentrations at the Mid, Out, and Reference sites were very similar and generally less than about 1.0 mg/L.

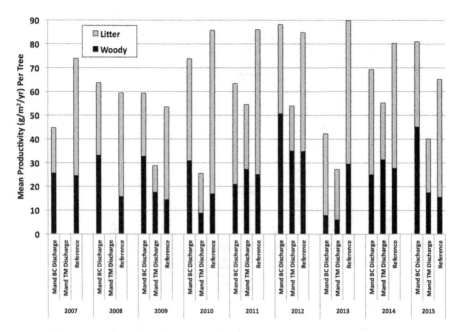

Fig. 2.26 Mean annual litterfall and woody productivity at the Mandeville Bayou Chinchuba (BC) and Tchefuncte Marsh (TM) assimilation wetland

In the Mandeville Bayou Chinchuba assimilation wetland, mean annual NPP was often higher at the Discharge sites than at the Reference site (Fig. 2.26), but results varied from year to year. Monitoring of vegetation biomass did not begin until 2009 at the Mandeville Tchefuncte Marsh wetland. EOSL biomass at the marsh sites (Bayou Chinchuba Out, Tchefuncte Marsh Out, marsh Reference) followed similar trends, but was higher at the Reference site than at the Bayou Chinchuba Out site (Fig. 2.27).

Hammond

During the fall of 2006, the City of Hammond began discharging secondarily-treated effluent into Four Mile Marsh located in the northwest corner of the Joyce wetlands, approximately 11 km southeast of Hammond, Louisiana (Fig. 2.28). The city treatment system has the capacity to treat 8 MGD. Dry weather flow averages about 2.7 MGD but inflow and infiltration can raise discharge as high as 17 MGD (Lane et al. 2015). Influent wastewater is passed through the Hammond wastewater treatment plant headworks and then piped to a three-cell oxidation lagoon. After secondary treatment, effluent is disinfected with chlorine and then de-chlorinated. From 2007 to 2013, average monthly discharge into the assimilation wetland was 3.9 MGD, with average concentrations of TN and TP of 17.91 and 3.64 mg/L,

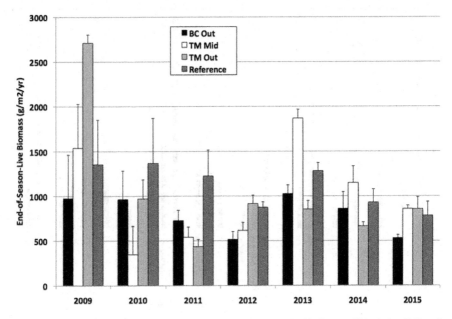

Fig. 2.27 Mean annual end-of-season-live biomass at the Mandeville Bayou Chinchuba (BC) and Tchefuncte Marsh (TM) assimilation wetland sites. Error bars represent standard error of the mean

respectively, and mean TN and TP loading rates of 2.39 and 0.48 g/m²/year, respectively (Table 2.8).

The Hammond assimilation wetland has generated controversy because of marsh deterioration during the winter of 2007–2008. After initiation of discharge in Fall 2006, there was robust marsh growth during the 2007 growing season. During the 2007–2008 winter, about 150 ha of marsh deteriorated. Since that time, there has been partial recovery of the area. Field observations and exclosure experiments indicated that the deterioration was due to herbivory by nutria (Shaffer et al. 2015). Others proposed that the deterioration was due to excessive inundation and nutrient induced decomposition. We address these issues below.

The east Joyce Wetlands (EJW) bordering northwest Lake Pontchartrain have a long history of human induced changes (Lane et al. 2015), such as leveeing of the Mississippi River that eliminated almost all riverine input to the area and segmentation of the east and west Joyce wetlands by the construction of a railroad, U.S. highway 51, and Interstate 55. Dredged drainage canals and associated spoil banks have channeled watershed input around the wetlands, especially South Slough that channelizes most upland runoff north of the assimilation wetlands directly to the I-55 canal. The deep canal associated with I-55 causes both rapid short-circuiting of freshwater runoff to Lakes Maurepas and Pontchartrain and saltwater intrusion deep into fresh and formerly freshwater wetland areas. Increasing salinity has caused widespread loss of freshwater forested wetlands in the area (Shaffer et al. 2009, 2015).

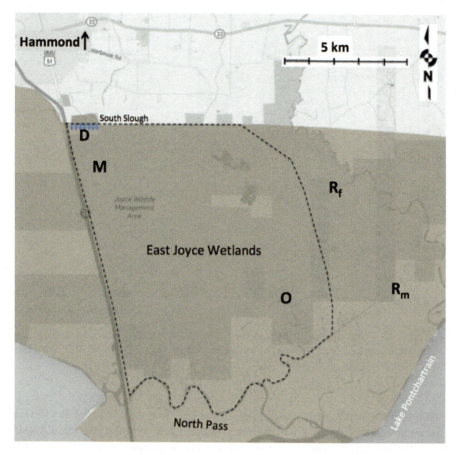

Fig. 2.28 Location of wetland monitoring sites at the Hammond assimilation wetland. R_m Marsh Reference site, R_f Forested Reference site, D Discharge site, M Mid site, O Out site. The *blue lines* show location of the discharge pipe and outlets, and the *black dashed line* the assimilation wetlands

Field measurements and a hydrological model showed that short-circuiting from the wetlands south of South Slough to the I-55 canal was minimal and most flow through the wetlands was to the southeast (Lane et al. 2015). Water levels in the Hammond assimilation wetland were highly variable prior to the beginning of effluent discharge in 2006, with relatively high water levels that did not increase substantially from 2007 through summer 2009 despite the addition of municipal effluent. Post-effluent water levels lacked the variability of the pre-discharge period and were about 20 cm higher from late 2009 until 2014 due to high rainfall in 2009, 2012 and 2013 and high effluent inflow due to significant inflow and infiltration (I&I) into the city collection system. Historical net watershed inputs averaged 2.69 cm/year over the 4 km² area immediately south of the effluent distribution system, compared to 0.38 cm/year for the effluent and 0.13 cm/year for direct precipitation. Salinity increased from north to

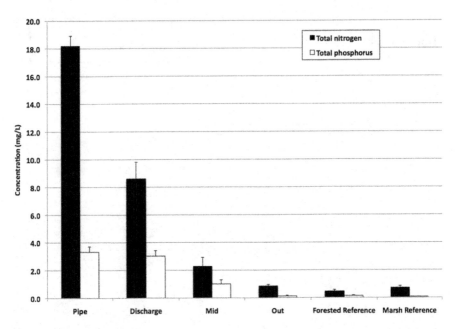

Fig. 2.29 Mean total nitrogen (TN) and total phosphorus (TP) concentrations at the effluent pipe and in surface water at the Hammond assimilation wetland between 2007 and 2013. Error bars represent standard error of the mean

south with strong seasonality, averaging 1.9–2.1 PSU near the lake to 0.4–0.6 PSU in the northwestern EJW. Peak salinities were 4.6–5.1 PSU near the lake and 1.8 PSU in northwestern EJW. There was a significant decrease in salinity beginning in 2010 coinciding with the closure of the Mississippi River Gulf Outlet, high precipitation in the fall and winter of 2009, and in 2012 and 2013, and continuing operation of the assimilation system with high I&I in those years.

The Discharge, Out, and Marsh Reference sites are emergent freshwater wetlands dominated by cutgrass (*Zizaniopsis miliacea*), bulltongue (*Sagittaria lancifolia*), soft rush (*Juncus effusus*) and cattail. The Mid and Forested Reference sites are freshwater forested wetlands dominated by pond cypress (*Taxodium distichum var. nutans*), and further south vegetation transitions to wiregrass (*Spartina patens*) and relict baldcypress forest. The Hammond assimilation wetland is the only assimilation wetland in coastal Louisiana with an herbaceous emergent marsh at the Discharge site; all of the other assimilation wetlands have forested Discharge sites.

Mean TN and TP concentrations in surface water at the Hammond assimilation wetland declined steadily from the Discharge site to the Out site (Fig. 2.29). In particular, mean TN concentration decreased by more than 60% between the Pipe and Discharge site and between the Discharge and Mid sites. Unlike the effluent at the Breaux Bridge, Broussard, and Luling assimilation wetlands, effluent discharged into the Hammond assimilation wetland is highly nitrified. In the Hammond effluent,

TN is 52.6% NO_x, 32.5% NH_3, and 14.9% organic nitrogen. The high percentage of NO_x in the effluent leads to rapid nitrogen removal through denitrification, and is the primary reason why TN decreases rapidly as surface water moves through the wetland; NO_x decreased by almost an order of magnitude within 100 m and to <0.1 mg/L within 700 m.

The forested Reference site is highly degraded and had a much lower stem density than the Mid site until the Reference site was re-located in 2012 (Fig. 2.30). When perennial productivity was normalized for stem density, the Mid site was more productive than the Reference site but litterfall was higher in the Reference site. At the Reference site, as the number of trees declined, the amount of leaf litter produced by each tree increased greatly, most likely due to increased light availability.

At the Hammond assimilation wetland, Shaffer et al. (2015) found a linear decrease occurred in the concentrations of NH_3 and PO_4 from the outfall pipe along a 700-m transect. Inorganic nutrients were essentially non-detectable 600 m from the outfall pipe. In addition, baldcypress seedlings planted where effluent is discharged at the Hammond assimilation wetland had aboveground production that followed a remarkably similar pattern as that of inorganic nutrients, with seedling growth greatest at the outfall pipe and decreasing linearly to 700 m from the discharge pipe. The Mid site is almost 1000 m away from the discharge pipe and nutrient concentrations were low and no differences in growth based on nutrients from the effluent should be expected.

Mean EOSL biomass was higher at the Discharge site than at the Reference site in almost every year monitored (Fig. 2.31).

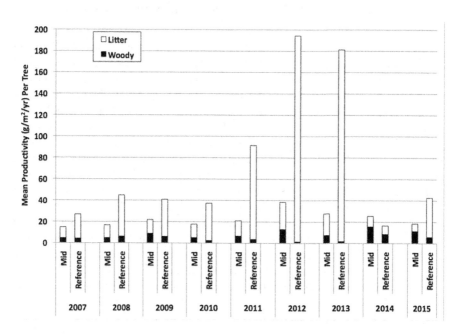

Fig. 2.30 Mean annual litterfall and woody productivity at the Hammond assimilation wetland

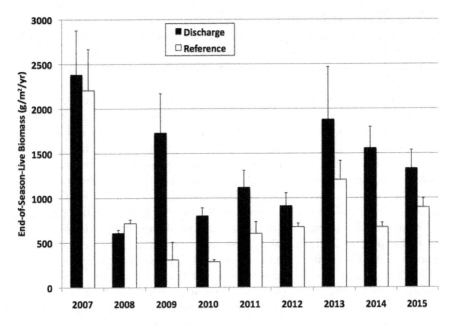

Fig. 2.31 Mean annual end-of-season-live biomass at the Hammond assimilation wetland. Error bars represent standard error of the mean

Adaptive Management at the Hammond Assimilation Wetland

At the Hammond assimilation wetland, immediately following effluent discharge in 2006, there was robust growth of herbaceous vegetation (Shaffer et al. 2015). By late fall 2007, vegetation biomass at the emergent wetland in the immediate vicinity of the effluent discharge began to decline and within months about 150 ha of emergent wetland had converted to open water, degraded marsh, or mudflat. By 2009, however, there had been substantial recovery of the wetland (Shaffer et al. 2015). During spring of 2008, to experimentally determine if the conversion from wetland to open water was caused by nutria herbivory, ten 16-m^2 exclosures were constructed using 2-m wide vinyl-coated crab wire dug approximately 0.4 m into the ground (Shaffer et al. 2015). Ten 16-m^2 paired controls were not enclosed. All of the exclosures and controls were planted with nine individuals of southern cattail (*Typha domingensis*). Cattail displayed nearly 100% cover inside of all ten exclosures within a 3-month period. In contrast, vegetation in all ten controls was completely destroyed within a few days of planting. The control plots were replanted four times, and each time suffered 100% mortality due to nutria herbivory (Shaffer et al. 2015).

The initial enhancement of herbaceous biomass followed by decline and partial recovery has engendered intense controversy and serves as an example of adaptive management to such an event (Shaffer et al. 2015; Lane et al. 2015;

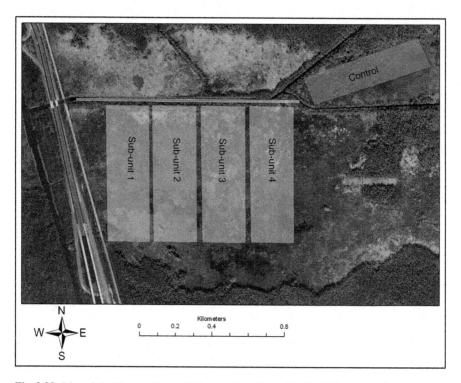

Fig. 2.32 Map of the Hammond assimilation wetland. The *white line* indicates the discharge pipe. Interstate-55 is shown west of the assimilation wetland and Joyce Wildlife Management Area, a baldcypress–water tupelo swamp, is shown south of the marsh

Bodker et al. 2015). The events at Hammond became caught up in a broader controversy of the role of nutrients in coastal wetland health versus the impact of intense herbivory by nutria. In the remainder of this section, we review the results of the monitoring required by the LPDES permit, discuss additional studies carried out to assess causes of wetland deterioration, review actions by the City of Hammond to improve the treatment system, and consider proposals for adaptive management.

To study the impact of effluent addition and marsh deterioration and recovery, a number of studies were conducted at the Hammond assimilation wetland. Four 700-m long subunits were established in the assimilation wetland and a Reference subunit was added at a nearby area isolated from effluent addition (Fig. 2.32). Beginning in winter 2007, and continuing through spring 2008, approximately 6000 baldcypress seedlings grown under different conditions (bare root and potted) were planted in Subunits 1–4 and the Reference subunit (Lundberg et al. 2011). The seedlings were planted from 0 to 700 m from the outfall pipe. Basal diameter growth of the seedlings was monitored over one growing season. Mean basal diameter growth for seedlings in subunits 1–4 ranged from 9.5 ± 0.9 mm to

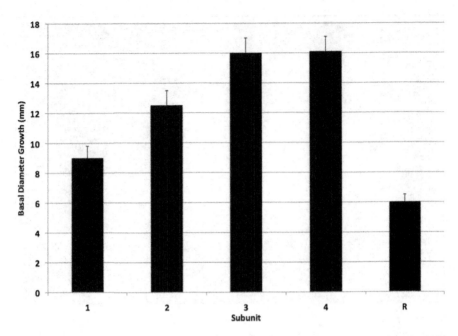

Fig. 2.33 Mean (± SE) relative basal diameter growth of baldcypress seedlings planted within four experimental subunits (1–4) and a reference (R) subunit at the Hammond assimilation wetland

16.1 ± 1.4 mm, with the highest production in Subunits 3 and 4, where most of the effluent was discharged. Mean basal diameter growth for seedlings in the Reference subunit was 6.4 ± 0.9 mm. As mentioned, the planted seedlings were grown under four types of conditions, namely 1-year old 'bare root' seedlings (BR1), 2-year old bare root seedlings (BR2), 5-month old 'Root Production Method' seedlings (RPM5), and 10-month old RPM seedlings grown in 3-gal pots. The RPM seedlings had significantly higher diameter growth than the BR1 seedlings, which had higher growth than the BR2 seedlings (Figs. 2.33 and 2.34; Lundberg et al. 2011).

The Hammond assimilation wetland have been highly effective at improving water quality while providing enormous benefits to wetland health. Once the secondarily-treated effluent reaches 150 m away from the outfall system there is a significant linear decrease in NH_4 and PO_4, declining to undetectable concentrations at 700 m from the outfall system (Fig. 2.35; Lundberg 2008). Growth rates of the baldcypress seedlings serve as surrogates of nutrient assimilation, as they too follow a linear decrease in growth from approximately 150–700 m (Fig. 2.36).

Using the average diameter growth across all distances at the Hammond assimilation wetland and comparing that value with several studies conducted in Manchac/Maurepas ecosystem, indicates a two- to sixfold higher growth rate at the Hammond assimilation wetland (Table 2.9).

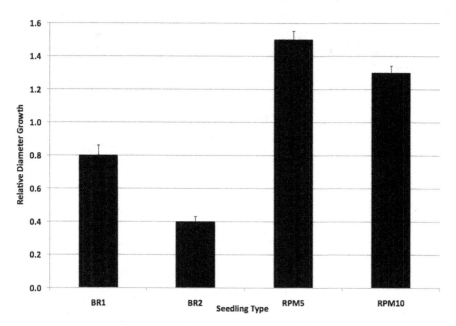

Fig. 2.34 Mean (± SE) relative basal diameter growth of various baldcypress seedling types (*BR1* bare root 1-year olds, *BR2* bare root 2-year olds, *RPM5* root production 5-month olds, *RPM10* root production method 10-month olds)

Summary of the Causes of Wetland Deterioration at the Hammond Assimilation Wetland

In response to the initial wetland deterioration observed at the Hammond assimilation wetland, several studies were conducted to determine the cause of the wetland decline. Shaffer et al. (2015) reviewed a number of hypotheses that were proposed to explain the changes in the assimilation wetland after effluent application began, including herbivory by nutria, excessive nutrients, reductions in above- and belowground biomass, increased soil decomposition due to high nutrient concentrations, prolonged inundation, toxicity, increased pH, and disease. Based on intensive field and mesocosm studies, they concluded that the initial marsh loss was primarily caused by nutria herbivory, and secondarily by waterfowl herbivory, and that significant recovery of the herbaceous vegetation occurred as a result of aggressive nutria control (>2000 eliminated). Marsh destruction due to nutria grazing has been observed frequently in Louisiana (Shaffer et al. 1992; McFalls et al. 2010; Holm et al. 2011) and nutria preferentially graze nutrient-enriched wetland vegetation (Ialeggio and Nyman 2014). After culling of the nutria population, vegetation recovery has been most pronounced near the point of effluent discharge.

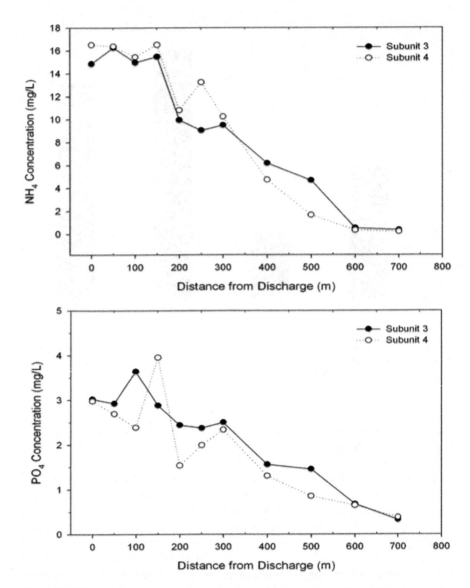

Fig. 2.35 NH_4 and PO_4 concentrations at various distances from effluent discharge at the Hammond assimilation wetland (Lundberg 2008)

Bodker et al. (2015) measured organic matter loss, gas production, and soil strength in experiments with and without added treated municipal effluent and concluded that added nutrients led to a significant increase in decomposition and a decrease in soil strength that was responsible for the wetland deterioration. To test this hypothesis, we determined the annual loading rate for the Hammond assimila-

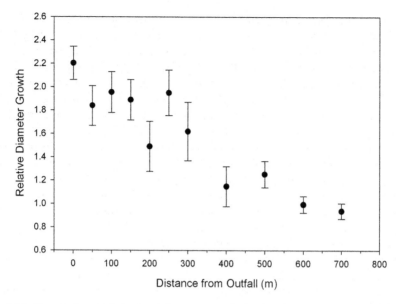

Fig. 2.36 Mean (± 1 s.e.) relative basal diameter growth of RPM baldcypress seedlings at various distances from effluent discharge at the experimental subunits (Lundberg 2008; Shaffer et al. 2015)

Table 2.9 Results of various studies of baldcypress seedling growth within the Maurepas drainage basin

Study	Fertilizer	Diameter (mm)	Height (cm)
Lundberg (2008)	Y	13.48	23.97
	N	6.38	11.32
Beville (2002)	Y	8	–
	N	5	–
Boshart (1997)	Y	4.25	–
	N	4.5	–
Campo (1996)	Y	4.25	–
	N	2.2	–
Forder (1995)	Y	2	–
	N	4	–
Myers et al. (1995)	Y	7.5	9.7
	N	4.1	4.2

Time-release Osmocote 18-6-12 commercial fertilizer was used for all fertilizer treatment studies excluding Lundberg (2008), which used secondarily-treated effluent

tion wetland and calculated how much organic matter could be decomposed if all NO_3 in the effluent stream were denitrified. The equation for denitrification is:

$$C_6H_{12}O_6 + 4NO_3 = 6CO_2 + 2N_2 \quad \text{(Mitsch and Gosselink 2015)}$$

Based on this equation, 1 g of NO_3-N reduced in denitrification results in the oxidation of 1.28 g of carbon, or 2.57 g organic matter assuming a 50% carbon content. If we assume an average NO_3 concentration of 8 mg/L (Shaffer et al. 2015), and discharge of 4.1 MGD (Lane et al. 2015), the total annual nitrogen load would be 44,300 kg-N/year. Using a receiving wetland area of 121 ha, the zone of immediate impact, approximately 94.2 g/m² of soil organic matter could be decomposed by denitrification over an annual period if all NO_3 were reduced to N_2 via denitrification and the organic matter substrate was from soil organic matter. However, this amount is more than compensated for by new vegetative production (Shaffer et al. 2015), generally only 60–70% of available NO_3 is denitrified (Mitsch and Gosselink 2015), and most importantly, most soil organic matter is not a suitable substrate for denitrification while effluent is an excellent source of low molecular weight organic compounds for denitrifying bacteria. We thus reject the hypothesis that nutrients were responsible for the wetland deterioration at the Hammond assimilation wetland. Decomposition rates estimated by Bodker et al. (2015) were equivalent to or less than the first month of litterbag decomposition measured over 15 months that found no differences in decomposition with distance or depth (Shaffer et al. 2015). The decomposition rates based on gas production and stoichiometric calculations were less than 5% and 3.5%, respectively, of the soil organic matter substrate used in the experiments for summer temperatures. The Bodker et al. (2015) gas evolution experiments were carried out at temperatures between 22–35 °C. Decomposition would have been significantly less during winter, and thus could not have been responsible for the marsh deterioration that occurred over a 6-month period in winter 2007–2008. There is an extensive scientific literature on wetlands receiving treated effluent that shows few negative impacts on productivity, accretion, or decomposition (Kadlec and Wallace 2009). We know of no study that reports such rapid deterioration of a wetland over a short period of time due to nutrient enrichment. On the other hand, there are numerous studies showing that nutria herbivory has led to rapid destruction of hundreds of hectares of coastal wetlands in less than a year (e.g., Shaffer et al. 1992; McFalls et al. 2010; Holm et al. 2011).

Over the past several years, the BOD limit of 30 mg/L has been exceeded a number of times at the Hammond wastewater treatment facility, with reported concentrations up to 50 mg/L. The City of Hammond has implemented a number of improvements to better treat wastewater to meet LPDES limits, including:

- Installation of a new aeration system to reduce BOD;
- Upgrading an oxidation ditch system to pre-treat high BOD inputs from a milk processing facility and a medical center and installation of pretreatment systems at these two facilities;
- Installation of a nitrification-denitrification system to reduce NH_3 concentrations;
- An evaluation of toxicity reduction in the effluent; and
- Reduction of inflow and infiltration to the collection system up to 50%.

The City is also exploring other options for improving the overall treatment system. One is the construction of a pipeline that would allow the discharge of treated effluent west of I-55 into a freshwater forested wetland. Pulsing the water east and west would allow the current assimilation wetland to have periods of no flow that would allow better drainage of the wetlands. It also would serve as a buffer to saltwater intrusion in wetlands west of I-55 where there has been widespread loss of freshwater forested wetlands.

Current Issues Concerning Assimilation Wetlands in Coastal Louisiana

Pulsing

In our opinion, all assimilation wetlands should have two independent outfall areas to maximize the pulsing paradigm and allow water levels to subside. Pulsing is currently included at most of the assimilation projects in Louisiana and is accomplished by having at least two discharge outlets so that effluent can be introduced to different parts of the wetland. At the St. Martinville project, the effluent can be discharged into different cells. One of the goals of pulsing is to allow the system to "draw down". The idea is that under natural conditions, most freshwater wetlands in coastal Louisiana generally have no standing water during part of the year. Pulsing has been successful at the three systems that are not impacted by coastal water levels (Breaux Bridge, Broussard, and St. Martinville) but the forested wetland sites that are impacted by coastal water levels (Thibodaux, Luling, Mandeville, Amelia, and Hammond) are permanently flooded to the extent that regeneration cannot occur. Thus, even if discharge to these sites was stopped, there would still be permanent or near permanent flooding. The herbaceous wetlands at Mandeville and Riverbend are tidal and experience regular flooding and draining. The Hammond herbaceous wetland is located about 10 km north of Lake Maurepas and is largely isolated from direct hydrologic exchange with either Lake Maurepas or Lake Pontchartrain. Because of this, there is no daily tidal signal in the assimilation wetland. Water levels are controlled by seasonal water level variability due to changes in lake levels, watershed input, local precipitation, and effluent discharge. The effluent discharge has dampened water level variability in the Hammond assimilation wetland (Lane et al. 2015). As mentioned above, it has been suggested that pulsing to forested wetlands west of I-55 would allow a return to a more normal hydrology as well as buffer saltwater intrusion on both sides of I-55.

Fresh Marshes

The experience at Hammond has raised questions about the use of fresh herbaceous marshes as assimilation wetlands. Although we believe that the marsh deterioration at Hammond was due mainly to nutria herbivory, it has been suggested that adding nutrients to wetlands leads to lower belowground productivity, higher decomposition rates, and loss of soil strength. This is based on considerable discussion over the past several years about the effects of nutrient loading to coastal wetlands (Darby and Turner 2008a, b, c; Day et al. 2013; Davis et al. 2017; Deegan et al. 2012; Graham and Mendelssohn 2014; Morris et al. 2013; Nyman 2014; Swarzenski et al. 2008; Turner 2011; van Zomeren et al. 2012; Snedden et al. 2015). Based on these findings, this is an issue that must be carefully analyzed when considering discharge of treated effluent to fresh marshes. In addition, nutria control needs to be incorporated into management plans.

Global Change

Coastal wetlands in Louisiana will be impacted by increasingly severe climate impacts. These impacts can affect wetland assimilation systems as well as be mitigated by them (Day et al. 2005, 2007, 2016a). These include sea-level rise that is expected to increase by 1–2 m by 2100 (FitzGerald et al. 2008; Pfeffer et al. 2008; Vermeer and Rahmstorf 2009; IPCC 2013; Koop and van Leeuwen 2016; Deconto and Pollard 2016), more category 4 and 5 hurricanes (Emanuel 2005; Webster et al. 2005; Hoyos et al. 2006; Goldenberg et al. 2001; Kaufmann et al. 2011; Mei et al. 2014), drought (IPCC 2007; Shaffer et al. 2015), more erratic weather (Min et al. 2011; Pall et al. 2011; Royal Society 2014) and other factors. The combination of accelerated sea-level rise, more frequent intense hurricanes, and drought will lead to more inundation and salinity stress. For example, the 2000–2001 drought raised salinities above 12 psu in western Lake Pontchartrain (Day et al. 2012) and to 9 psu in the Manchac basin (Shaffer et al. 2009). These conditions will especially impact low salinity to fresh wetlands. By providing fresh water and stimulating vertical accretion, discharge of treated effluent can make low salinity and fresh wetlands more sustainable in the face of these climate extremes.

Decreasing energy availability and higher energy prices will make energy intensive wetland restoration and management more expensive, limit options for restoration and complicate human response to climate change (Day et al. 2007, 2014; Tessler et al. 2015). The implication of future energy scarcity is that the cost of energy will be higher in coming decades (Bentley 2002; Day et al. 2005, 2016a; Hall and Day 2009; Murphy and Hall 2011) and the cost of energy-intensive activities also will increase significantly (Tessler et al. 2015). Advanced wastewater treatment systems are energy intensive and expensive to construct and operate. Wetland assimilation is a low energy approach that can offer a sustainable treatment approach in a time of growing climate impacts and increasing energy costs.

Conclusions

There are ten active assimilation wetlands in coastal Louisiana and another four with permit applications pending. Results of annual monitoring show that nutrient concentrations of surface waters decrease with distance, reaching background levels before water leaves the wetland. While nutrient concentrations decrease, vegetative productivity is enhanced. In degraded forested wetlands being used as assimilation wetlands, baldcypress and water tupelo seedlings are often planted, which thrive in the nutrient rich environment. However, nutria are attracted to vegetation with increased nutrient concentrations, and this introduced species must be monitored and controlled. Pulsing of effluent between two or more sites should be incorporated to prevent prolonged flooding and to encourage seedling development and growth.

Acknowledgements Ecological baseline studies and monitoring at the different sites were funded by the respective communities. JWD, RRL, RGH, JND acknowledge that they carried out both ecological baseline studies and routine monitoring as employees of Comite Resources Inc., which received funding from the communities with assimilation projects. A diversity of funding sources supported additional scientific studies on these wetlands, including EPA, Louisiana Dept. of Environmental Quality, and NOAA. A number of students received M.S. and Ph.D. degrees based on work carried out at these assimilation wetlands, including students from Louisiana State University, Southeastern Louisiana University, Nicholls State University, and Tulane University. Their work is listed in the literature cited. Technical Contribution No. 6573 of the Clemson University Experiment Station. WHC was supported by NIFA/USDA, under project number SC-1700424.

References

Ahel M, Giger W, Koch M (1994) Behaviors of alkylphenol polyethoxylates surfactants in the aquatic environment I. Occurrence and transformation in sewage treatment. Water Res 28:1131–1142

Alexander HD, Dunton KH (2006) Treated wastewater effluent as an alternative freshwater source in a hypersaline salt marsh: impacts on salinity, inorganic nitrogen, and emergent vegetation. J Coast Res 22:377–392

Allen HA, Pezeshki SR, Chambers JL (1996) Interaction of flooding and salinity stress on bald cypress (*Taxodium distichum*). Tree Physiol 16:307–313

Barras JA, Bernier JC, Morton RA (2008) Land area change in coastal Louisiana—a multidecadal perspective (from 1956 to 2006). U.S. Geological Survey Scientific Investigations Map 3019, scale 1:250,000, 14 p

Batt AL, Sungpyo K, Aga DS (2007) Comparison of the occurrence of antibiotics in four full-scale wastewater treatment plants with varying designs and operations. Chemosphere 68(3):428–435

Belmont MA, Ikonomou M, Metcalfe CD (2006) Presence of nonylphenol ethoxylate surfactants in a watershed in Central Mexico and removal from domestic sewage in a treatment wetland. Environ Toxicol Chem 25(1):29–35

Beville SL (2002) The efficacy of a small-scale freshwater diversion for restoration of a swamp in southeastern Louisiana. M.Sc. dissertation, Southeastern Louisiana University, Hammond, LA, 86 p

Bentley RW (2002) Global oil and gas depletion: an overview. Energy Policy 30:189–205

Blahnik T, Day JW (2000) The effects of varied hydraulic and nutrient loading rates on water quality and hydrologic distributions in a natural forested treatment wetland. Wetlands 20:48–61

Bodker JE, Turner RE, Tweel A, Schulz C, Swarzenski C (2015) Nutrient-enhanced decomposition of plant biomass in a freshwater wetland. Aquat Bot 127:44–52

Boesch DF, Josselyn MN, Mehta AJ, Morris JT, Nuttle WK, Simenstad CA, Swift DJP (1994) Scientific assessment of coastal wetland loss, restoration and management. J Coast Res 20:56–65

Boshart WM (1997) The conservation and restoration of a baldcypress swamp: an investigation of macronutrients, competition, and induced vegetation dynamics as related to nutria herbivory. M.Sc. dissertation, Southeastern Louisiana University, Hammond, LA, 67 p

Boumans RM, Day JW (1994) Effects of two Louisiana marsh management plans on water and materials flux and short-term sedimentation. Wetlands 14:247–261

Boyd GR, Palmeri JM, Zhang S, Grimm DA (2004) Pharmaceuticals and personal care products (PPCPs) and endocrine disrupting chemicals (EDCs) in stormwater canals and Bayou St. John in New Orleans, Louisiana, USA. Sci Total Environ 333:137–148

Boyd GR, Reemtsma H, Grimm DA, Mitra S (2003) Pharmaceuticals and personal care products (PPCPs) in surface and treated waters of Louisiana, USA and Ontario, Canada. Sci Total Environ 311(1–3):135–149

Boyt FL, Bayley SE, Zoltek J (1977) Removal of nutrients from treated municipal wastewater by wetland vegetation. J Water Pollut Control Fed 49:789–799

Brantley CG, Day JW, Lane RR, Hyfield E, Day JN, Ko J-Y (2008) Primary production, nutrient dynamics, and accretion of a coastal freshwater forested wetland assimilation system in Louisiana. Ecol Eng 34:7–22

Breaux AM (1992) The use of hydrologically altered wetlands to treat wastewater in coastal Louisiana. Ph.D. dissertation, Department of Oceanography and Coastal Sciences, Louisiana State University, Baton Rouge, LA

Breaux AM, Day JW (1994) Policy considerations for wetland wastewater treatment in the coastal zone: a case study for Louisiana. Coast Manag 22:285–307

Breaux AM, Farber S, Day JW (1995) Using natural coastal wetlands systems for wastewater treatment: an economic benefit analysis. J Environ Manag 44:285–291

Cahoon DR (1994) Recent accretion in two managed marsh impoundments in coastal Louisiana. Ecol Appl 4:166–167

Campbell CJ, Laherrere JH (1998) The end of cheap oil. Sci Am 278:78–83

Campo FM (1996) Restoring a repressed swamp: the relative effects of a saltwater influx on an immature stand of baldcypress (Taxodium distichum (L.) Richard). M.Sc. dissertation, Southeastern Louisiana University, Hammond, LA, 86 p

Cardoch L (2000) Approaches to sustainable management of deltas: Integrating natural systems subsidies with societal needs in the Mississippi and Ebro Deltas. PhD Dissertation. Department of Oceanography and Coastal Sciences, Louisiana State University, Baton Rouge, LA. USA

Chan AW, Zoback MD (2007) The role of hydrocarbon production on land subsidence and fault reactivation in the Louisiana coastal zone. J Coast Res 23:771–786

City of New York Department of Environmental Protection and HydroQual, Inc. (1997) Alternative wastewater disinfection methods. Prepared for New York—New Jersey Harbor Estuary Program Pathogen Work Group, September, 1997, 80 p

Clara M, Strenn B, Saracevic E, Kreuzinger N (2004) Adsorption of bisphenol-A, 17ß-estradiol and 17α-ethinylestradiol to sewage sludge. Chemosphere 56:843–851

Cleaveland MK (2006) Extended chronology of drought in the San Antonio area. Revised Report to the Guadalupe-Blanco River Authority. 26 pp

Conkle JL, Lattao C, Cook RL, White JR (2010) Competition, uptake and release of three fluoroquinolone antibiotics in a Southeastern Louisiana freshwater wetland soil. Chemosphere 80:1353–1359

Conkle JL, White JR, Metcalfe C (2008) Removal of pharmaceutically active compounds by a lagoon-wetland wastewater treatment system in Southeast Louisiana. Chemosphere 73:1741–1748

Conner WH, Day JW Jr (1988) Rising water levels in coastal Louisiana: implications for two coastal forested wetland areas in Louisiana. J Coast Res 4:589–596

Conner WH, Duberstein JA, Day JW Jr, Hutchinson S (2014) Impacts of changing hydrology and hurricanes on forest structure and growth along a flooding/elevation gradient in a South Louisiana forested wetland from 1986 to 2009. Wetlands 34:803–814

Couvillion BR, Barras JA, Steyer GD, Sleavin W, Fischer M, Beck H, Trahan N, Griffin B, Heckman D (2011) Land area change in coastal Louisiana from 1932 to 2010: U.S. Geological Survey Scientific Investigations Map 3164, scale 1:265,000, 12 p

Darby FA, Turner RE (2008a) Below- and aboveground Spartina alterniflora production in a Louisiana salt marsh. Estuaries Coasts 31:223–231

Darby FA, Turner RE (2008b) Effects of eutrophication on salt marsh root and rhizome accumulation. Mar. Ecol Prog Ser 363:63–70

Darby FA, Turner RE (2008c) Below- and aboveground biomass of Spartina alterniflora: response to nutrient addition in a Louisiana salt marsh. Estuaries Coasts 31:326–334

Davis J, Currin C, Morris JT (2017) Impacts of fertilization and tidal inundation on elevation change in microtidal, low relief salt marshes. Estuar Coasts. https://doi.org/10.1007/s12237-017-0251-0

Day JW, Wetphal A, Pratt R, Hyfield E, Rybczyk J, Kemp GP, Day JN, Marx B (2006) Effects of long-term municipal effluent discharge on the nutrient dynamics, productivity, and benthic community structure of a tidal freshwater forested wetland in Louisiana. Ecol Eng 27:242–257

Day JW, Breaux AM, Hesse ID, Rybczyk JM (1992) Wetland wastewater treatment in the Louisiana coastal zone. In: Barataria-Terrebone National Estuary Program: Data Inventory Workshop Proceedings, p 221–238

Day JW, Breaux AM, Feagley S, Kemp P, Courville C (1994) A use attainability analysis of long-term wastewater discharge on the Cypriere Perdue Forested Wetland at Breaux Bridge, LA. Coastal Ecology Institute, Louisiana State University, Baton Rouge, LA

Day JW, Rybczyk J, Pratt R, Westphal A, Blahnik T, Delgado P, Kemp P (1997b) A use attainability analysis of long-term wastewater discharge on the Ramos Forested Wetland at Amelia, LA. Coastal Ecology Institute, Louisiana State University, Baton Rouge, LA

Day JW, Rybczyk J, Pratt R, Sutula M, Westphal A, Blahnik T, Delgado P, Kemp P, Englande AJ, Hu CY, Jin G, Jeng HW (1997a) A use attainability analysis of long-term wastewater discharge to the Poydras-Verret Wetland in St. Bernard Parish, LA. Coastal Ecology Institute, Louisiana State University, Baton Rouge, LA

Day JW Jr, Cardoch L, Rybczyk JM, Kemp GP (1998) Food processor and community development of rural coastal areas through the application of wetland wastewater treatment systems. Coastal Ecology Institute, Louisiana State University, Baton Rouge, LA. National Coastal Resources Research and Development Institute, 43 p

Day JW, Rybczyk JM, Cardoch L, Conner W, Delgado-Sanchez P, Pratt R, Westphal A (1999) A review of recent studies of the ecological and economical aspects of the application of secondarily treated municipal effluent to wetlands in Southern Louisiana. In: Rozas L, Nyman J, Proffitt C, Rabalais N, Turner R (eds) Recent research in Coastal Louisiana. Louisiana Sea Grant College Program, Louisiana State University, Baton Rouge, LA, pp 155–166

Day JW, Ko JK, Rybczyk J, Sabins D, Bean R, Berthelot G, Brantley C, Cardoch L, Conner W, Day JN, Englande AJ, Feagley S, Hyfield E, Lane R, Lindsey J, Mitsch J, Reyes E, Twilley R (2004) The use of wetlands in the Mississippi Delta for wastewater assimilation: a review. Ocean Coast Manag 47:671–691

Day JW Jr, Britsch LD, Hawes SR, Shaffer GP, Reed DJ, Cahoon D (2000) Pattern and process of land loss in the Mississippi Delta: a spatial and temporal analysis of wetland habitat change. Estuaries 23:425–438

Day R, Holz R, Day J (1990) An inventory of wetland impoundments in the coastal zone of Louisiana USA: historical trends. Environ Manag 14:229–240

Day JW, Barras J, Clairain E, Johnston J, Justic D, Kemp GP, Ko J-Y, Lane RR, Mitsch WJ, Steyer G, Templet P, Yañez-Arancibia A (2005) Implications of global climatic change and energy cost and availability for the restoration of the Mississippi Delta. Ecol Eng 24:251–263

Day JW, Boesch DF, Clairain EJ, Kemp GP, Laska SB, Mitsch WJ, Orth K, Mashriqui H, Reed DJ, Shabman L, Simenstad CA, Streever BJ, Twilley RR, Watson CC, Wells JT, Whigham DF (2007) Restoration of the Mississippi delta: lessons from hurricanes Katrina and Rita. Science 315:1679–1684

Day JW, Christian RR, Boesch DM, Yanez-Arancibia A, Morris J, Twilley RR, Naylor L, Schaffner L, Stevenson C (2008) Consequences of climate change on ecogeomorphology of coastal wetlands. Estuar Coasts 31:477–491

Day J, Hunter R, Keim RF, DeLaune R, Shaffer G, Evers E, Reed D, Brantley C, Kemp P, Day J, Hunter M (2012) Ecological response of forested wetlands with and without large-scale Mississippi River input: implications for management. Ecol Eng 46:57–67

Day J, Lane R, Moerschbaecher M, DeLaune R, Mendelssohn I, Baustian J, Twilley R (2013) Vegetation and soil dynamics of a Louisiana estuary receiving pulsed Mississippi River Water following Hurricane Katrina. Estuar Coast Shelf Sci 36:665–682

Day JW, Moerschbaecher M, Pimentel D, Hall C, Yañez-Arancibia A (2014) Sustainability and place: How emerging mega-trends of the 21st century will affect humans and nature at the landscape level. Ecol Eng 65:33–48

Day JW, Agboola J, Chen Z, D'Elia C, Forbes DL, Giosan L, Kemp P, Kuenzer C, Lane RR, Ramachandran R, Syvitski J, Yanez-Arancibia A (2016a) Approaches to defining deltaic sustainability in the 21st century. Estuar Coast Shelf Sci 183:275–291. https://doi.org/10.1016/j.ecss.201606.018

Day JW, Cable J, Lane R, Kemp P (2016b) Sediment deposition at the Caernarvon crevasse during the great Mississippi flood of 1927: implications for coastal restoration. Water 8(2):38. https://doi.org/10.3390/w8020038

DeConto RM, Pollard D (2016) Contribution of Antarctica to past and future sea-level rise. Nature 531:591–597

Deegan LA, Johnson DS, Warren RS, Peterson BJ, Fleeger JW, Fagherazzi S, Wollheim WM (2012) Coastal eutrophication as a driver of salt marsh loss. Nature 490:388–392

Deffeyes KS (2001) Hubbert's peak: the impending world oil shortage. Princeton University Press, Princeton, NJ, 208 p

DeLaune RD, Kongchum M, White JR, Jugsujinda A (2013) Freshwater diversions as an ecosystem management tool for maintaining soil organic matter accretion in coastal marshes. Catena 107:139–144

Effler RS, Shaffer GP, Hoeppner SS, Goyer RA (2007) Ecology of the Maurepas Swamp: effects of salinity, nutrients, and insect defoliation. In: Conner WH, Doyle TW, Krauss KW (eds) Ecology of tidal freshwater forested wetlands. Springer, Dordrecht, Netherlands, pp 349–384

Emanuel K (2005) Increasing destructiveness of tropical cyclones over the past 30 years. Nature 436:686–688

Euliss NH, Gleason RA, Olness A, McDougal RL, Murkin HR, Robart RD, Bourbonniere RA, Warner BG (2006) North American prairie wetlands are important nonforested land-based carbon storage sites. Sci Total Environ 361:179–188

Evers DE, Sasser CE, Gosselink JG, Fuller DA, Visser JM (1998) The impact of vertebrate herbivores on wetland vegetation in Atchafalaya Bay, Louisiana. Estuar Coasts 21:1–13

Ewel KC, Bayley SE (1978) Cypress strand receiving sewage at Waldo. In: Odum HT, Ewel KC (eds) Principal Investigators. Cypress wetlands for water management, recycling, and conservation. Fourth Annual Report to National Science Foundation, p 750–801

FitzGerald DM, Fenster MS, Argow BA, Buynevich IV (2008) Coastal impacts due to sea-level rise. Ann Rev Earth Planet Sci 36:601–647

Forder, DR (1995) Timber harvesting, coppicing and artificial regeneration of a cypress muck swamp. M.Sc. thesis, Southeastern Louisiana University, Hammond, LA, 74 p

Godfrey PJ, Kaynor ER, Pelczarski S, Benforado J (eds) (1985) Ecological considerations in wetlands treatment of municipal wastewaters. Van Nostrand Reinhold, New York, 473 p

Goldenberg SB, Landsea M-NCW, Gray WM (2001) The recent increase in Atlantic hurricane activity: causes and implications. Science 293:474–479

Graham SA, Mendelssohn IA (2014) Coastal wetland stability maintained through counterbalancing accretionary responses to chronic nutrient enrichment. Ecology 95:3271–3283

Hall CAS, Day JW (2009) Revisiting the limits to growth after peak oil. Am Sci 97:230–237

Hesse ID, Day J, Doyle T (1998) Long-term growth enhancement of baldcypress (Taxodium Distichum) from municipal wastewater application. Environ Manag 22:119–127

Holm GO Jr, Peterson EF, Evers DE, Sasser CE (2011) Contrasting the historical stability of wetlands exposed to Mississippi River water. Presentation to Workshop on Response to Louisiana

Marsh Soils and Vegetation to Diversions. Lafayette, LA, Feb 2011, Abstract at http://www.mvd.usace.army.mil/lcast/pdfs/Abstracts.pdf

Horton BP, Rahmstorf S, Engelhart SE, Kemp AC (2014) Expert assessment of sea-level rise by AD 2100 and AD 2300. Quat Sci Rev 84:1–6

Hoyos CD, Agudelo PA, Webster PJ, Curry JA (2006) Deconvolution of the factors contributing to the increase in global hurricane intensity. Science 312:94–97

Hunter RG, Faulkner SP (2001) Denitrification potentials in restored and natural bottomland hardwood wetlands. Soil Sci Soc Am J 65:1865–1872

Hunter RG, Day JW Jr, Lane RR, Lindsey J, Day JN, Hunter MG (2009a) Nutrient removal and loading rate analysis of Louisiana forested wetlands assimilating treated municipal effluent. Environ Manag 44:865–873

Hunter RG, Day JW Jr, Lane RR, Lindsey J, Day JN, Hunter MG (2009b) Impacts of secondarily treated municipal effluent on a freshwater forested wetland after 60 years of discharge. Wetlands 29:363–371

Hypoxia Task Force (2016) Report on point source progress in Hypoxia Task Force States. Mississippi River/Gulf of Mexico Hypoxia Task Force, 80 p

Ialeggio JS, Nyman JA (2014) Nutria grazing preference as a function of fertilization. Wetlands 34:1039–1045

IPCC (2001) Climate change 2001: the Scientific Basis, Contribution of Working Group 1 to the Third Assessment Report, Intergovernmental Panel on Climate Change. Cambridge University Press, Cambridge, UK

IPCC (2007) Climate change 2007: the physical science basis. Contribution of Working Group I to the Fourth Assessment Report of the Intergovernmental Panel on Climate Change. In: Solomon S, Qin D, Manning M, Chen Z, Marquis M, Avery KB, Tignor M, Miller HL (eds) Intergovernmental panel on climate change. Cambridge University Press, Cambridge and New York

IPCC (2013) Climate change 2013: the physical science basis. Contribution of Working Group 1 to the Fifth Assessment Report of the Intergovernmental Panel on Climate Change. In: Stocker TF, Qin D, Plattner GK, Tignor M, Allen SK, Boschung J, Nauels A, Xia Y, Bex V, Midgley PM (eds) Intergovernmental panel on climate change. Cambridge University Press, Cambridge and New York

Izdepski CW, Day JW, Sasser CE, Fry B (2009) Early floating marsh establishment and growth dynamics in a nutrient amended wetland in the lower Mississippi delta. Wetlands 29:1004–1013

Kadlec RH, Wallace S (2009) Treatment wetlands. CRC Press, Boca Raton, FL, 1046 p

Kangas P (2004) Ecological engineering: principles and practice. CRC Press, Boca Raton, FL, 452 p

Kaufmann RF, Kauppi H, Mann ML, Stock JH (2011) Reconciling anthropogenic climate change with observed temperature 1998–2008. Proc Nat Acad Sci U S A 108(29):790–793

Keim RF, Amos JB (2012) Dendrochronological analysis of baldcypress responses to climate and contrasting flood regimes. Can J For Res 42:423–436

Keim RF, Izdepski CW, Day JW (2012) Growth responses of baldcypress to wastewater nutrient additions and changing hydrologic regime. Wetlands 32:95–103

Kesel RH (1988) The decline in the suspended load of the lower Mississippi River and its influence on adjacent wetlands. Environ Geol Water Sci 11:271–281

Kesel RH (1989) The role of the lower Mississippi River in wetland loss in southeastern Louisiana, USA. Environ Geol Water Sci 13:183–193

Ko JY, Day JW, Lane RR, Day J (2004) A comparative evaluation of cost-benefit analysis and embodied energy analysis of tertiary municipal wastewater treatment using forested wetlands in Louisiana. Ecol Econ 49:331–347

Ko JY, Day JW, Lane RL, Hunter R, Sabins D, Pintado KL, Franklin J (2012) Policy adoption of ecosystem services for a sustainable community: a case study of wetland assimilation using natural wetlands in Breaux Bridge, Louisiana. Ecol Eng 38:114–118

Koop SHA, van Leeuwen CJ (2016) The challenges of water, waste and climate change in cities. Environ Dev Sustain 18:1–34

Koplin DW, Furlong ET, Meyer MT, Thurman EM, Zaugg SD, Barber LB, Buxton HT (2002) Pharmaceuticals, hormones, and other organic waste water contaminants in U.S. streams, 1999–2000: a national reconnaissance. Environ Sci Technol 36:1202–1211

Lal R (2004) Soil carbon sequestration to mitigate climate change. Geoderma 123:1–22

Lane R, Day J, Thibodeaux B (1999) Water quality analysis of a freshwater diversion at Caernarvon, Louisiana. Estuaries 2A:327–336

Lane R, Day J, Kemp G, Marx B (2002) Seasonal and spatial water quality changes in the outflow plume of the Atchafalaya River, Louisiana, USA. Estuaries 25(1):30–42

Lane RR, Madden CJ, Day JW, Solet DJ (2010) Hydrologic and nutrient dynamics of a coastal bay and wetland receiving discharge from the Atchafalaya River. Hydrobiologia 658:55–66

Lane RR, Day JW, Justic D, Reyes E, Marx B, Day JN, Hyfield E (2004) Changes in stoichiometric Si, N and P ratios of Mississippi River water diverted through coastal wetlands to the Gulf of Mexico. Estuar Coast Shelf Sci 60:1–10

Lane RR, Day JW, Shaffer GP, Hunter RG, Day JN, Wood WB, Settoon P (2015) Hydrology and water budget analysis of the East Joyce wetlands: Past history and prospects for the future. Ecol Eng 87:34–44

Lane R, Mack S, Day J, Kempka R, Brady L (2017) Carbon sequestration at a forested wetland receiving treated municipal effluent. Wetlands. https://doi.org/10.1007/s13157-017-0920-6

Lemlich SK, Ewel KC (1984) Effects of wastewater disposal on growth rates of cypress trees. J Environ Qual 13(4):602–604

Li Y, Zhu G, Ng WJ, Tan, SK (2014) A review on removing pharmaceutical contaminants from wastewater by constructed wetlands: Design, performance and mechanism. Sci Total Environ 468–469:908–932

Louisiana Department of Environmental Quality (2010) Water quality management plan: volume 3 permitting guidance document for Implementing Louisiana Surface Water Quality Standards. Available via DIALOG. http://www.deq.louisiana.gov/portal/DIVISIONS/WaterPermits/WaterQualityManagementPlanContinuingPlanning.aspx. Accessed 30 July 2014

Louisiana Department of Environmental Quality (2015) Environmental regulatory code: LAC Title 33 environmental quality, Part IX. Water Quality, Baton Rouge, LA

Lundberg CJ (2008) Using secondarily treated sewage effluent to restore the baldcypress—water tupelo swamps of the Lake Pontchartrain Basin: a demonstration study. M.S. dissertation, Southeastern Louisiana University. Hammond, LA, 71 p

Lundberg CJ, Shaffer GP, Wood WB, Day JW Jr (2011) Growth rates of baldcypress (Taxodium distichum) seedlings in a treated effluent assimilation marsh. Ecol Eng 37:549–553

Lytle SA, Grafton BF, Ritchie A, Hill HL (1959) Soil Survey of St. Mary Parish, Louisiana. USDA Soil Conservation Service. In cooperation with Louisiana Agriculture Experiment Station, Franklin, LA

McFalls TB, Keddy PA, Campbell D, Shaffer G (2010) Hurricanes, floods, levees, and nutria: vegetation responses to interacting disturbance and fertility regimes with implications for coastal wetland restoration. J Coast Res 26:901–911

Mcleod E, Chmura GL, Bouillon S, Salm R, Björk M, Duarte CM, Lovelock CE, Schlesinger WH, Silliman BR (2011) A blueprint for blue carbon: toward an improved understanding of the role of vegetated coastal habitats in sequestering CO2. Front Ecol Environ 9:552–560

Meehl GA, Tebaldi C, Teng H, Peterson TC (2007) Current and future U.S. weather extremes and El Nino. Geophys Res Lett 34:L20704. https://doi.org/10.1029/2007GL031027

Meers E, Tack FMG, Tolpe I, Michels E (2008) Application of a full-scale constructed wetland for tertiary treatment of piggery manure: monitoring results. Water Air Soil Pollut 193:15–24

Mei W, Xie S-P, Zhao M (2014) Variability of tropical cyclone track density in the North Atlantic: observations and high-resolution simulations. J Clim 27:4797–4814

Min S-K, Zhang X, Zwiers FW, Hegerl GC (2011) Human contribution to more intense precipitation extremes. Nature 470:378–381

Minor A (2014) Forested freshwater wetland responses to secondarily treated municipal effluent discharge. M.S. Thesis, Nicholls State University, Thibodaux, LA. 143 pp

Mitsch WJ, Jorgensen SE (2003) Ecological engineering and ecosystem restoration. Wiley, New York, 146 p

Mitsch WJ, Day JW, Gilliam JW, Groffman PM, Hey DL, Randall GW, Wang N (2001) Reducing nitrogen loading to the Gulf of Mexico from the Mississippi River basin: strategies to counter a persistent ecological problem. Bioscience 51:373–388

Mitsch WJ, Gosselink JG (2015) Wetlands, 5th ed., John Wiley & Sons, Inc., New York

Murray BC, Pendleton L, Jenkins WA, Sifleet S (2011) Green payments for blue carbon: economic incentives for protecting threatened coastal habitats. Nicholas Institute for Environmental Policy Solutions, Report NI R 11-04

Morris JT, Sundberg K, Hopkinson CS (2013) Salt marsh primary production and its response to relative sea level rise and nutrients in estuaries at Plum Island, Massachusetts, and North Inlet, South Carolina, USA. Oceanography 26:78–84

Morton RA, Bster NA, Krohn MD (2002) Subsurface controls on historical subsidence rates and associated wetland loss in southcentral Louisiana. Gulf Coast Assoc Geolog Soc Trans 52:767–778

Moshiri GA (ed) (1993) Constructed wetlands for water quality improvement. CRC Press, Boca Raton, FL, 632 p

Mossa J (1996) Sediment dynamics in the lowermost Mississippi River. Eng Geol 45:457–479

Murphy DJ, Hall CAS (2011) Energy return on investment, peak oil, and the end of economic growth. Ecolog Econ Rev 1219:52–72

Myers RS, Shaffer GP, Llewellyn DW (1995) Baldcypress (Taxodium distichum (L.) Rich.) restoration in southeast Louisiana: the relative effects of herbivory, flooding, competition, and macronutrients. Wetlands 15:141–148

Nessel JK, Bayley SE (1984) Distribution and dynamics of organic matter and phosphorus in a sewage-enriched cypress swamp. In: Ewel KC, Odum HT (eds) Cypress swamps. University Presses of Florida, Gainesville, FL, pp 262–278

Nyman JA (2014) Integrating successional ecology and the delta lobe cycle in wetland research and restoration. Estuaries Coasts 37:1490–1505

Pall P, Aina T, Stone DA, Stott PA, Nozawa T, Hilberts AGJ, Lohmann D, Allen MR (2011) Anthropogenic greenhouse gas contribution to flood risk in England and Wales in autumn 2000. Nature 470:382–385

Penland S, Ramsey KE, McBride RA, Mestayer JT, Westphal KA (1988). Relative sea level rise and delta plain development in the Terrebonne parish region. Coastal geology technical report. Louisiana Geological Survey, Baton Rouge, LA, 121 p

Pfeffer WT, Harper JT, O'Neel S (2008) Kinematic constraints on glacier contributions to 21st-century sea-level rise. Science 321:1340–1343

Reddy KR, DeLaune RD (2008) Biogeochemistry of wetlands: science and applications. CRC Press, Boca Raton, FL

Richardson CJ, Nichols DS (1985) Ecological analysis of wastewater management criteria in wetland ecosystems. In: Godfrey PJ, Kaynor ER, Pelczarski S (eds) Ecological considerations in wetlands treatment of municipal wastewaters. Van Nostrand Reinhold, New York, pp 351–391

Roberts HH (1997) Dynamic changes of the holocene Mississippi River delta plain: the delta cycle. J Coast Res 13:605–627

Royal Society (2014) Resilience to Extreme Weather, p 122. https://royalsociety.org/topicspolicy/projects/resilience-extreme-weather/

Rybczyk JM, Zhang XW, Day JW, Hesse I, Feagley S (1995) The impact of Hurricane Andrew on tree mortality, litterfall, nutrient flux and water quality in a Louisiana coastal swamp forest. J Coast Res 21:340–353

Rybczyk JM (1997) The use of secondarily treated wastewater effluent for forested wetland restoration in a subsiding coastal zone. Ph.D. dissertation, Louisiana State University, Baton Rouge, LA

Rybczyk JM, Callaway JC, Day JW Jr (1998) A relative elevation model for a subsiding coastal forested wetland receiving wastewater effluent. Ecol Model 112:23–44

Rybczyk JM, Day JW, Hesse ID, Delgado Sanchez P (1996) An overview of forested wetland wastewater treatment projects in the Mississippi River delta region. In: Flynn K (ed) Proceedings of the southern forested wetlands ecology and management conference, Clemson University, Clemson, SC, p 78–82

Rybczyk J, Day J, Conner W (2002) The impact of wastewater effluent on accretion and decomposition in a subsiding forested wetland. Wetlands 22:18–32

Seo DC, Lane RR, DeLaune RD, Day JW (2013) Nitrate removal and nitrate removal velocity in coastal Louisiana freshwater wetlands. Anal Lett 46:1171–1181

Shaffer GP, Sasser CE, Gosselink JG, Rejmanek M (1992) Vegetation dynamics in the emerging Atchafalaya Delta, Louisiana, USA. J Ecol 80:677–687

Shaffer GP, Wood WB, Hoeppner SS, Perkins TE, Zoller J, Kandalepas D (2009) Degradation of baldcypress-water tupelo swamp to marsh and open water in Southeastern Louisiana, U.S.A.: An irreversible trajectory? J Coast Res 54:152–165

Shaffer GP, Day JW, Hunter RG, Lane RR, Lundberg CJ, Wood WB, Hillman ER, Day JN, Strickland E, Kandalepas D (2015) System response, nutria herbivory, and vegetation recovery of a wetland receiving secondarily-treated effluent in coastal Louisiana. Ecol Eng 79:120–131

Shappell NW, Billey LO, Forbes D, Matheny TA, Poach ME, Reddy GB, Hunt PG (2007) Estrogenic activity and steroid hormones in swine wastewater through a lagoon constructed-wetland system. Environ Sci Technol 41(2):444–450

Sifleet S, Pendleton L, Murray BC (2011) State of the Science on Coastal Blue Carbon: a summary for policy makers. Nicholas Institute for Environmental Policy Solutions, Report NI R 11-06

Snedden GA, Cretini K, Patton B (2015) Inundation and salinity impacts to above- and belowground productivity in Spartina patens and Spartina alterniflora in the Mississippi River deltaic plain: implications for using river diversions as restoration tools. Ecol Eng 81:133–139

Stahle DW, Cleaveland MK (1992) Reconstruction and analysis of spring rainfall over the southeastern United-States for the past 1000 years. Bull Am Meteorol Soc 73:1947–1961

Steyer GD, Perez BC, Piazza S, Suir G (2007) Potential Consequences of Saltwater Intrusion Associated with Hurricanes Katrina and Rita. In: Farris GS, Smith GJ, Crane MP, Demas CR, Robbins LL, Lavoie DL (eds) Science and the storms: the USGS response to the hurricanes of 2005. U.S. Geological Survey Circular 1306:283. p 138–147. Available: http://pubs.usgs.gov/circ/1306/

Swarzenski CM, Doyle TW, Frye B, Hargis TG (2008) Biochemical response of organic-rich freshwater marshes in the Louisiana delta plain to chronic river water influx. Biogeochemistry 90:49–63

Tao B, Tian H, Ren W, Yang J, Yang Q, He R, Cai W, Lohrenz S (2014) Increasing Mississippi River discharge throughout the 21st century influenced by changes in climate, land use, and atmospheric CO_2. Geophys Res Lett 41:4978–4986

Tessler ZD, Vörösmarty CJ, Grossberg M, Gladkova I, Aizenman H, Syvitski JPM, Foufoula-Georgiou E (2015) Profiling risk and sustainability in coastal deltas of the world. Science 349:638–643

Turner RE, Swenson EM, Lee JM (1994) A rationale for coastal wetland restoration through spoil bank management in Louisiana, USA. Environ Manag 18:271–282

Turner RE (2011) Beneath the wetland canopy: loss of soil marsh strength with increasing nutrient load. Estuaries Coasts 33:1084–1093

Van Zomeren CM, White JR, DeLaune RD (2012) Fate of nitrate in a vegetated brackish coastal marsh. Soil Sci Soc Am J 76:1919–1927

Verlicchi P, Zambello E (2014) How efficient are constructed wetlands in removing pharmaceuticals from untreated and treated urban wastewaters? a review. Sci Total Environ 470–471:1281–1306

Vermeer M, Rahmstorf S (2009) Global sea level linked to global temperature. Proc Natl Acad Sci 106:21527–21532

Vymazal J (2010) Constructed wetlands for wastewater treatment: five decades of experience. Environ Sci Technol 45:61–69

Webster PJ, Holland GJ, Curry JA, Chang H-R (2005) Changes in tropical cyclone number, duration, and intensity in a warming environment. Science 309:1844–1846

White JR, Belmont MA, Metcalfe CD (2006) Pharmaceutical compounds in wastewater: wetland treatment as a possible solution. Sci World J 6:1731–1736

Williams JR (1999) Addressing global warming and biodiversity through forest restoration and coastal wetlands creation. Sci Total Environ 240:1–9

Zhang X (1995) Use of a natural swamp for wastewater treatment. Ph.D. dissertation, Department of Agronomy, Louisiana State University, Baton Rouge, LA

Zhang X, Feagley SE, Day JW Jr, Conner WH, Hesse ID, Rybczyk JM, Hudnall WH (2000) A water chemistry assessment of wastewater remediation in a natural swamp. J Environ Qual 29:1960–1968

Chapter 3
Recommendations for the Use of Tundra Wetlands for Treatment of Municipal Wastewater in Canada's Far North

Gordon Balch, Jennifer Hayward, Rob Jamieson, Brent Wootton, and Colin N. Yates

Introduction

The treatment of wastewaters in circumpolar regions can vary greatly from one country to another. Gunnarsdóttir et al. (2013) found that wastewater treatment in many Arctic regions is inadequate or completely missing. The harsh climatic conditions, remoteness of Arctic communities and permafrost conditions make the installation and operation of wastewater collection and treatment systems very challenging. Subsequently, many communities in Greenland, Iceland, northern coastal regions of Scandinavia and some parts of Alaska discharge to marine environments with no treatment or at best primary treatment designed to remove suspended material (Huber et al. 2016). The treatment of domestic wastewaters in Canada's Far North and sub-arctic communities is accomplished primarily through the use of long term detention in wastewater stabilization ponds (WSPs); otherwise known as sewage lagoons. Treated effluent exfiltrates from these lagoons, or in some cases where it is intentionally discharged, may flow into shallow tundra lowland wetlands that eventually drain into surface waters. These wetlands containing grasses and sedges are referred to in this chapter as tundra wetland treatment areas (WTAs). This chapter is devoted to summarizing our current understanding regarding the treatment performance of these natural areas, which have not been designed, or engineered. The focus on the potential of these WTAs is distinct from the investigations of other researchers who study the performance of engineered constructed

G. Balch (✉) • B. Wootton
Centre for Alternative Wastewater Treatment, Fleming College, Lindsay, ON, Canada
e-mail: gordon.balch@flemingcollege.ca

J. Hayward • R. Jamieson
Centre for Water Resources Studies, Dalhousie University, Halifax, NS, Canada

C.N. Yates
Ecosim Consulting Inc., St. Catharines, ON, Canada

wetlands within cold climates. Although many of the physical, chemical, and biological processes operative in the treatment of domestic wastewaters are shared by both the WTAs and constructed wetlands, the WTAs are unique.

The studies summarized in this chapter were conducted in Canada's Far North territory of Nunavut. Hamlets in Nunavut are generally small and range in population from about 100 to 2300 people (Statistics Canada 2012). Additional studies were focused in the Northwest Territories, which also has similar remote and small-sized hamlets. The hamlets are typically isolated and in Nunavut accessible only by air, or by ship during the brief summer. Due to the remoteness, communities are dependent on self-supported infrastructure to deliver community services, such as provision of potable water, wastewater treatment, and solid waste disposal (Yates et al. 2013, 2014a; Krkosek et al. 2012). The extreme climate, the logistical challenges of bedrock and/or permafrost, together with the lack of financial and human resources, represent significant impediments to the development and operation of mechanized wastewater treatment infrastructure commonly used in more southern locations of Canada (Hayward et al. 2014; Chouinard et al. 2014a). Rates of per capita residential water use in Nunavut are based on a design standard of 100 L/capita·day (Daley et al. 2014). This water usage rate is approximately one-third of the Canadian average of 274 L/capita·day (Daley et al. 2014). The difference in water usage results in less dilution and higher concentrations of contaminants in municipal wastewater in northern Canada compared to other regions in Canada.

Due to these challenges, most communities are reliant on centralized systems with relatively low maintenance passive wastewater treatment technologies. Passive treatment of wastewater in Northern Canada occurs in most communities during a 3–4 month period extending from the spring freshet in June to the freeze-up in September (Hayward et al. 2014). This ice-free period is termed the treatment season, and represents the period when the water temperatures are high enough to encourage biological treatment processes. Vacuum trucks are used in many communities to collect wastewater from storage tanks within individual dwellings and transport the sewage to lagoons. The lagoons are typically designed as controlled-discharge single-cell retention ponds with storage capacities sized to accommodate the volume of wastewater accumulated over an entire 1-year period (Ragush et al. 2015). Wastewater is stored frozen for a significant portion of the year with scheduled discharge from the lagoons typically occurring at the end of the ice-free treatment season, in August or September for some communities. This is done to maximize the summer treatment potential, and to lower the water levels in the lagoons to increase the winter storage capacity. The wastewater can be discharged manually in a controlled manner from some controlled-designed lagoons, but may at times be discharged early, when the capacities of the WSPs have been reached prematurely. Some lagoons however, experience a continuous discharge of wastewater as it exfiltrates through the berm structure. This discharge can be either intentional by design, or in some cases uncontrolled, as it leaks through the berm.

The WTAs are different from constructed wetlands, as they are typically non-engineered, and generally not intentionally created (Hayward et al. 2014; Chouinard et al. 2014a). As such, they are by nature open and diffuse systems, often with poorly defined boundaries and flow patterns (Kadlec and Wallace 2009). Tundra WTAs are characterized by the natural physical attributes of the landscape such as the topography, soil depth, and drainage area. Despite the similarity to natural wetlands, the tundra WTAs are distinctly different from natural tundra wetlands in their hydrology, vegetation, nutrient availability, and organic loading (Chouinard et al. 2014b; Hayward et al. 2014). Tundra WTAs are not to be confused as being a modification of constructed wetlands which have been engineered to meet specific performance targets. Although many of the physical, chemical and biological treatment processes are similar between tundra WTAs and constructed wetland a direct comparison between tundra WTAs and constructed wetlands in terms of treatment performance is difficult primarily due to the often-unknown hydrology and variable substrates associated with the tundra systems. The use of tundra WTAs appears generally unique to northern Canada and no parallels were found to be in common use within other circumpolar regions and the authors are not aware of any other comparable studies that have investigated the use of tundra wetlands to treat domestic sewage or lagoon effluents. The information presented is intended to characterize the physical, chemical and biological aspects influencing treatment at these sites in an attempt to help regulators and operators to better understand the benefits and limitations of including these tundra WTAs in a wastewater treatment strategy. Despite the differences between constructed wetlands and WTAs, these tundra sites have been shown to improve the effluent quality passing through them (Yates et al. 2012; Hayward et al. 2014; Doku and Heinke 1993, 1995; Dubuc et al. 1986).

There is much interest in Northern Canada among infrastructure managers and regulators who recognize the treatment potential that may be achieved by formally recognizing the role currently offered by many WTAs. This chapter discusses the treatment potential of these unique areas and provides an informed and cautious methodology for their incorporation into treatment strategies that protects both human and environmental health. This chapter offers insight into how WTAs could be optimized to supplement the treatment of domestic wastewaters in the harsh climatic environment of Northern Canada and may provide options for other circumpolar regions.

Regulatory Setting

In Canada, territorial water boards have historically stipulated compliance targets for individual communities for the treated effluents discharged from lagoons. Most water licenses issued by territorial water boards stipulate compliance targets for fecal coliforms, BOD_5, total suspended solids and pH. Currently, a standardized approach for incorporating WTAs in municipal wastewater treatment systems in Nunavut does not exist. Despite this, territorial regulatory provisions are becoming available to incorporate the WTAs into the wastewater treatment. At this point, the Community and Government Services (CGS) department of the Government of

Nunavut (GN) funds wastewater system improvements, the community operates and maintains the facilities, the Nunavut Water Board (NWB) sets the treatment objectives and monitoring requirements, and Indigenous and Northern Affairs Canada (INAC) monitors compliance. These provisions may comprise of compliance sampling and/or monitoring of water quality at the outlets of select tundra WTAs. In order to incorporate a WTA into the facility there may be a requirement for a wetland assessment.

Research Background

The information presented within this chapter draws heavily upon the recent work undertaken by the Centre for Alternative Wastewater Treatment (CAWT), an applied research facility of Fleming College and by the Centre for Water Resources Studies (CWRS) at Dalhousie University. Prior to the work by CAWT and CWRS there was a lack of design criteria and modeling tools that could be used for the development of a standardized design process. The compilation of large comprehensive data sets developed by both the CAWT and CWRS has increased our understanding of the treatment potential of these sites and can now be used towards the development of standardized protocols for the use of tundra WTAs in the Canadian Arctic. Although most of the information presented has been developed from a Northern Canada perspective, it is felt that many aspects of this work may be transferable to other circumpolar regions, and therefore may be of interest to wastewater managers outside of Canada.

Methodologies

Three initiatives were undertaken to study 15 WTAs during the years 2008–2013 (Fig. 3.1). Although the primary goal of the studies was to characterize the ability of these areas to treat domestic wastewaters, each of the studies had a slightly different focus. The first study was conducted by the CAWT (Arctic Summer) in 2008 to examine how treatment performance varied in six WTAs over the course of one Arctic summer extending from June to September. The second study by the CAWT (Interpolated Mapping) took place during the summers of 2009–2011, and investigated the treatment performance of an additional seven wetlands. The focus during this second study was to monitor the change in wastewater concentrations as waters flowed through the wetland, and to present the changes in a series of interpolated maps that visually outlined the changing concentrations of water quality parameters. In the third study the CWRS (Hydrology) characterized the hydrologic regime of three WTAs in order to better understand how hydrology impacts treatment performance. Collectively the studies ameliorate our understanding of the capacity and challenges associated with the use of tundra WTAs in the treatment of domestic wastewaters.

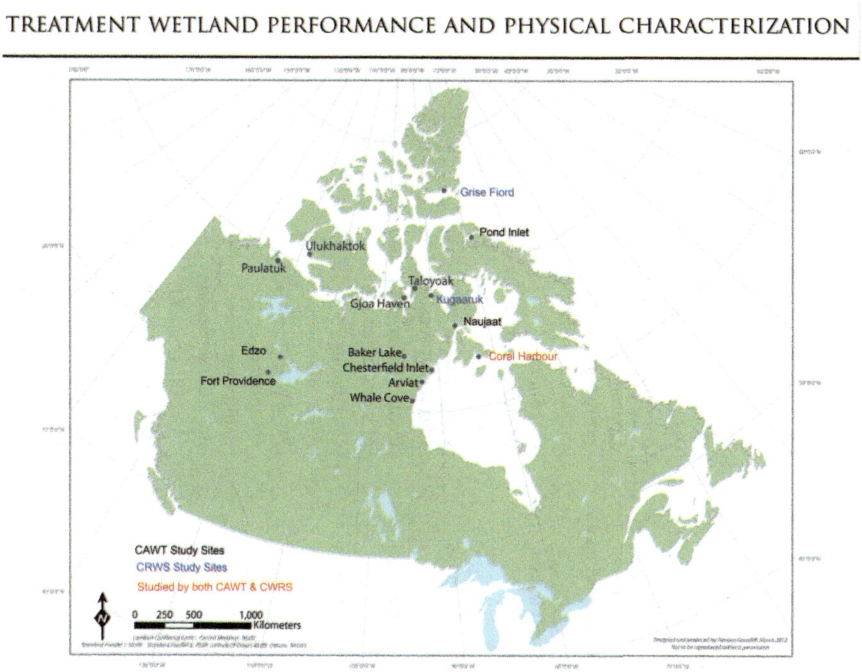

TREATMENT WETLAND PERFORMANCE AND PHYSICAL CHARACTERIZATION

Fig. 3.1 Locations of 15 wetland treatment areas investigated by the CAWT and the CWRS over the period from 2008 to 2013. Figure modified from Yates et al. (2013)

Site Descriptions

Table 3.1 summarizes the attributes of all the WTA sites that were studied as part of the combined efforts of CAWT and CWRS:

Arctic Summer: In the summer of 2008, the CAWT monitored six WTAs over the course of one arctic summer from June 21st to September 24th. The wetlands investigated were associated with the communities of Whale Cove, Arviat, Chesterfield Inlet, Baker Lake, Coral Harbour, and Naujaat; all located in the Kivalliq region of Nunavut, Canada. Details of this study are documented in Yates et al. (2012).

Interpolated Mapping: In the summers of 2009 to 2011, the CAWT visited an additional three WTAs in Nunavut (Pond Inlet, Gjoa Haven, Taloyoak), and four in the Northwest Territories (Paulatuk, Edzo, Fort Providence, Ulukhaktok). The ratio of wetland size to the number of people in the associated community meant that wastewater loading rates varied significantly amongst the study locations; the impact of which will be discussed in later sections. The findings from this study are summarized in Yates et al. (2013), Yates et al. (2014b), and Chouinard et al. (2014b, c).

Hydrology: In 2011 and 2013, CWRS studied the WTAs in the hamlets of Kugaaruk and Grise Fiord, Nunavut. In 2011 and 2012, CWRS also studied the WTA located in Coral Harbour, Nunavut, Canada. The hydraulic loading rates

Table 3.1 Summary table of the characteristics of the study sites that were the focus of research

Study site	Coordinates	Pop.	Approximate daily wastewater production (m³/day)	Pre-treatment	Discharge type	Area (ha)	Measured HRT[a] (days)
Naujaat	66°31′ N, 86°14′ W	850	66	Single-cell lagoon	Continuous exfiltration over the summer season	9.5	n/a
Whale Cove	62°11′ N, 92°35′ W	407	28	Single-cell lagoon (bermed natural lake)	Continuous exfiltration over the summer season	3.7	n/a
Baker Lake	64°19′ N, 96°02′ W	2069	167	Single-cell lagoon	Continuous exfiltration over the summer season	n/a	n/a
Kugaaruk	68° 32′ N, 089° 49′ W	771	76	Single-cell lagoon	Controlled at end of summer season	0.56	>0.4–6
Grise Fiord	76° 25′ N, 082° 54′ W	130	13	Single-cell lagoon	Controlled at end of summer season	0.54	0.01–0.1
Coral Harbour	64° 08′ N, 083° 10′ W	893	95	Single-cell lagoon	Continuous exfiltration over the summer season	1.1–1.5	0.5–14
Paulatuk	69° 21′ N, 124° 4′ W	311	34	Single-cell lagoon (natural lake)	Continuous exfiltration over the summer season	1.5	n/a
Pond Inlet	72° 42′ N, 77° 58′ W	1617	312	Single-cell lagoon	Controlled at end of summer season	0.58	n/a
Edzo	62° 48′ N, 116° 3′ W	320	109	three-celled lagoon	Continuous exfiltration over the summer season	2.1	n/a

Fort Providence	61°21′ N, 117°40′ W	734	n/a	three-celled lagoon with two sludge holding cells	Controlled at end of summer season	0.87	n/a
Gjoa Haven	68°38′ N, 095°52′ W	1197	119	Single-cell lagoon	Continuous exfiltration over the summer season	16.9	n/a
Ulukhatok	70°44′ N, 117°46′ W	402	41	Single-cell lagoon	Continuous exfiltration over the summer season	7.3	est >100
Taloyoak	69°32′ N, 093°32′W	1029	86	Natural lake	Continuous exfiltration over the summer season	6.1	n/a
Arviat	61°05′N, 94°00′W	2318	253	Two-celled lagoon	continuous exfiltration over the summer season	7.8	n/a
Chesterfield Inlet	63°20′ N, 90°42′ W	313	36	Single-cell lagoon (bermed natural pond/depression)	Continuous exfiltration over the summer season	5.5	n/a

n/a not analyzed
[a]*HRT* hydraulic retention time

(HLRs) were variable and uncontrolled onto the WTA, which ranged from 0 to 8 cm/day (0 to 827 m^3/ha/day), with an average of 1 cm/day (97 m^3/ha/day). The direction and spatial extent of flow changed over the duration of the treatment season in Coral Harbour.

Physical Characteristics

The methods used to characterize the physical attributes of the WTAs varied slightly among all three studies. In general, wetland boundaries were approximated by both landscape elevations and vegetation cover. The wetland boundaries were delineated and the sample locations marked using a TopCon 3105 W reflectorless total station in Paulatuk, Gjoa Haven, Ulukhaktok, and Taloyoak. A topographic survey was completed at the Coral Harbour, Grise Fiord and Kugaaruk study sites with a real-time kinematic (RTK) GPS survey, using the methodology described in Hayward (2013) and Hayward et al. (2014). Alternatively, a global positioning system (GPS, Garmin eTrex Vista HCx, and Garmin Montana 650) was used in the remaining wetlands where vegetative understory was too dense to establish a line of sight, or when the required technical personnel were not available.

Digital photographs of 1 m^2 plots were obtained at each of the sample locations (Fig. 3.2). Each photograph was analyzed to record the percent cover of the dominant vegetative groups to provide a semi-qualitative analysis of vegetation in the WTA. ArcMap software was used for the Baker Lake and Chesterfield Inlet sites to gener-

Fig. 3.2 Photograph of vegetative cover in 1 m^2 plot located next to a sampling point equipped with a lysimeter for the sampling of ground water

ate interpolated maps that correlated nutrient parameters with the cover of *Carex aquatilis*; a common sedge found within WTAs. A similar survey was conducted in Coral Harbour in 2011 and 2012 to develop a map of the spatial distribution of the classes of the dominant vegetation, and land cover classes were developed from a three band (RBG) QuickBird satellite image with a spatial resolution of 0.6 m. The maximum likelihood tool in the Image Classification toolbar in ESRI ArcMap 10 was used to perform a supervised image classification.

Hydraulic and Hydrological Characterization

Wastewater flows through the wetland were generally a combination of diffuse surface flow, preferential surface flows, and subsurface flows. Characterization of the hydraulics and hydrology of the Coral Harbour, Kugaaruk, and Grise Fiord sites involved the measurement of the influent and effluent discharges from the WTAs. The measurement of the hydrology of the sites and calculation of the hydraulic parameters is described in greater detail in Hayward et al. (2014) and Hayward (2013).

The hydraulic tracer tests were conducted with Rhodamine WT (RWT) fluorescent dye tracer. Measurement of the tracers was completed *in situ* with an optical fluorescent probe. Analysis of the tracer test data and calculation of the hydraulic parameters is described in detail in Hayward (2013) and Hayward et al. (2014).

In some instances, the volume of potable water was used as a surrogate to estimate the amount of domestic wastewater generated for each community. Summer time averages were tallied to produce an estimated volume of wastewater that could be potentially flowing into each of the study wetlands on a daily basis. Most communities investigated had exfiltration berms where wastewater leaked from the lagoons into the downstream wetlands. In some cases, an assumption was made that the amount leaked to the wetland on a daily basis was similar in magnitude to the amount of raw wastewater added to the lagoon each day.

Treatment Performance

The treatment performance assessments were conducted by collecting water samples at key points in the WTAs. The selection of sampling sites representative of the influent entering the wetland was challenging in systems where the flow of influent from a lagoon berm was diffuse or subsurface. Analysis was performed on water samples for a suite of indicator parameters all according to APHA (2012) protocols. Additional details on the methodology used for the treatment performance assessment are provided in Hayward et al. (2014), Yates et al. (2012) and Hayward (2013).

The total number of sampling visits per wetland varied, as some sites were sampled on a weekly basis, some on a daily basis to get the full spatial distribution of concentrations, and some seasonally on a year-to-year basis.

Factors That Influence Treatment Performance

The following discussion describes six factors that have the potential to influence the treatment performance of the WTAs. These factors include: seasonal influences, pre-treatment, hydraulics and hydrology, influence of discharge technique, vegetation, and spatial distribution of treatment. Findings from the studies are used to highlight how these factors influence performance and why they should be included in future studies of WTAs.

Seasonal Influences

Many biological processes are halted during cold and freezing conditions and in some locations, wastewater seepage onto the wetland can accumulate untreated in a frozen state. During spring freshet, frozen wastewater and snow melt can be released to the wetland in volumes that are higher than average summertime flows (Hayward et al. 2014). All of this can occur during a time when microbial processes are limited and reaction rates are slower due to the cold temperatures. In a likewise manner, the autumn senescence of plants and colder temperatures can also reduce biological treatment, which may be of particular importance to lagoons that routinely perform a late summer discharge of wastewater to the wetland.

The CAWT undertook their first study (i.e. *Arctic Summer*) to investigate how the spring and autumn (shoulder) periods influenced treatment performance. Table 3.2 and Fig. 3.3 should be reviewed together. Table 3.2 summarizes the averaged summer time concentration of wastewater parameters entering the WTAs and percent reductions while Fig. 3.3 is intended to provide a visual representation of how the

Table 3.2 Measured mean concentrations of $CBOD_5$ and TAN in treated wastewater entering 5 tundra wetlands, calculated percent reductions, and estimated loading rates measured over the course of one arctic summer in 2008

	Influent volume (m^3 day^{-1})	Wetland size (ha)	$CBOD_5$			TAN		
			Influent conc. (mg/L)	Loading rate (kg/ha/day)	Effluent % reduction	Influent conc. (mg/L)	Loading rate (kg/ha/day)	Effluent % reduction
Arviat	235	7.8	103	3.1	85	73.2	2.2	85
Chesterfield Inlet	36	5.5	221	1.5	94	39.6	0.26	99
Coral Harbour	96	10	181	1.7	92	21.8	0.21	87
Naujaat	66	9.5	385	2.7	93	70.0	0.49	96
Whale Cove	82	3.7	40.3	0.89	47	9.0	0.20	100

Source: Yates et al. (2012)

concentrations of 5-day carbonaceous biochemical oxygen demand ($CBOD_5$) and total ammonia nitrogen (TAN) changed from week to week. Table 3.2 indicates that relatively good treatment occurred in the WTAs, even though, as shown in Fig. 3.3, the concentration of $CBOD_5$ and TAN entering the wetland varied significantly in four of the five WTAs. The reason for the variations in the influent is unknown. The higher early summer $CBOD_5$ concentrations in Arviat, Coral Harbour, Naujaat, and Whale Cove indicate that these sites may be influenced by the melting of frozen wastewaters. Despite the high and variable concentrations in influent the effluents exiting all wetlands suggested a good level of treatment throughout the study period, even during the spring and autumn shoulder periods. The WTA located at Baker Lake was not included in this analysis, since it consisted of a relatively small

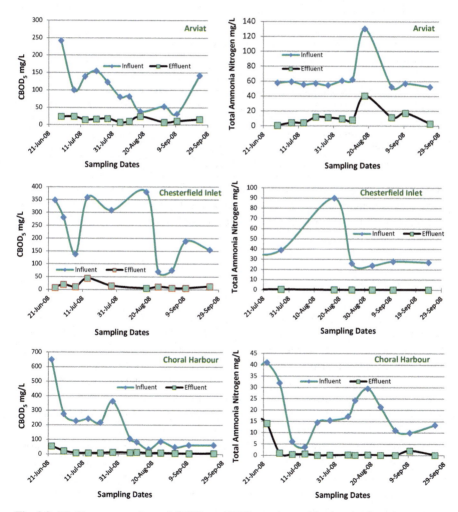

Fig. 3.3 Weekly concentrations of $CBOD_5$ and TAN entering and leaving the 5 tundra wetlands monitored over the course of one arctic summer in 2008. Source: Yates et al. (2012)

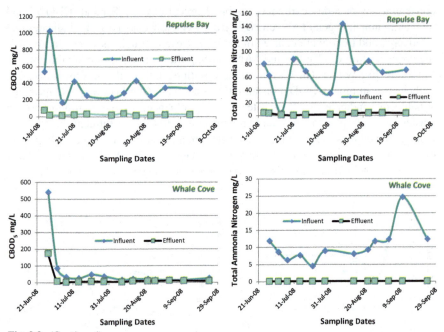

Fig. 3.3 (Continued)

wetland, followed by a series of interconnected ponds. Treated effluent samples were collected at the end of the last pond and for these reasons treatment values do not properly reflect the performance of the wetland alone.

Pre-treatment

Potable water consumption in Nunavut is generally less than one third of the Canadian average (Daley et al. 2014), and because of this, the strength of the wastewaters is relatively high in comparison to most urbanized areas of Canada. Table 3.3 illustrates this strength by summarizing the $CBOD_5$, and total suspended solids (TSS) concentrations observed in some of the raw wastewater samples of four typical Nunavut communities. When compared to literature values, the raw wastewater strengths are within or above the composition of typical medium to high strength wastewaters, as presented in Tchobanoglous et al. (2003).

The strength of the wastewater entering the wetland (i.e. *Arctic Summer*) varied amongst the study sites and depended on whether the raw sewage was discharged directly to the WTA, or if there was some form of pre-treatment. For example, at the time of study, no pre-treatment existed for the communities of Baker Lake, Chesterfield Inlet, or Naujaat. In these communities, raw sewage was generally discharged from the vacuum trucks to a small natural depression directly upstream

Table 3.3 Minimum, maximum and mean concentrations (mg L^{-1}) of CBOD$_5$ and TSS in the lagoons of 4 Nunavut communities

Location	Pre-treatment	n^a	CBOD$_5$ (mg L^{-1})			TSS (mg L^{-1})		
			Min.	Mean	Max.	Min.	Mean	Max.
Pond Inlet	Lagoon	23	436	525	614	272	326	380
Clyde River	Lagoon	15	300	367	434	211	273	335
Kugaaruk	Lagoon and decant cell	8	321	371	421	227	272	317
Grise Fiord	Lagoon	3	504	632	871	380	665	1036

$^a n$ is sample number

of the wetland, which overflowed onto the tundra. In the case of Arviat and Coral Harbour, raw sewage was discharged first to engineered lagoons, which exfiltrated onto the tundra WTAs. The raw sewage generated from Whale Cove was first discharged to a small natural lake with an estimated volume of 15,000 m^3 that significantly diluted the initial strength of the wastewater entering the WTA.

The mean percent reductions in the concentration of wastewater parameters accomplished by these six wetlands over the course of the Arctic summer were relatively high. This was despite the fact that some WTAs wastewaters did not undergo pre-treatment before flowing onto the tundra. The overall trends, excluding the Baker Lake WTA are summarized in Table 3.4.

The concentration of CBOD$_5$ exiting the WTAs averaged from all five sites (direct input + pre-treatment sites) was 18 mg/L, with a range from 14 to 25 mg/L. The WTA at Whale Cove, which is a pre-treatment site, exhibited the poorest percent reduction, although the CBOD$_5$ concentration of 21 mg/L leaving this wetland was comparable to the mean for all five wetlands. The seemingly poor performance of the Whale Cove wetland (e.g., 47% reduction) may be reflective of the fact that pre-treatment removed most of the CBOD$_5$ prior to the wetland resulting in an organic loading of only 0.3 kg/ha/day (i.e. the lowest of all pre-treatment sites), which likely meant that a higher percentage of recalcitrant organics were left for treatment in the Whale Cove WTA. The Arviat WTA, which is also a pre-treatment site had the highest loadings for CBOD$_5$, chemical oxygen demand (COD), TAN, total phosphorus (TP), and TSS, even though the wastewater was pre-treated before being introduced into the wetland. The population of Arviat at the time of study was significantly higher than the populations of both Naujaat and Coral Harbour.

It should be noted that WTAs are not intended to act as a front-end treatment process and should be used exclusively for provision of secondary or tertiary levels of treatment. Doku and Heinke (1995) and Yates et al. (2012) recommended that a minimum of primary treatment should occur prior to effluent discharge into WTAs. Their recommendation should be adopted to ensure that the solids and organic matter loading into the WTAs do not overwhelm the treatment capacity of the wetland. A minimum of primary pre-treatment is recommended for proposed and existing WTAs.

Table 3.4 Water quality parameters for wastewater influent entering tundra wetland treatment areas, and the percent concentration reduction (chemical/physical) or log reduction (microbial), of wastewater constituents in treated effluents leaving the WTAs

	Direct input of raw sewage[a]			Pre-treatment of influent[b]		
	Influent range (mg/L)	Effluent range (reduction)	Loading rate (kg/ha/day)	Influent range (mg/L)	Effluent range reduction	Loading rate (kg/ha/day)
$CBOD_5$	221–385	93–94%	1.6–2.8	40–180	47–92%	0.3–3.1
COD	300–450	79–86%	0.8–1.7	130–310	58–79%	0.6–7.1
TAN	40–90	96–99%	0.07–0.3	9–70	85–100%	0.07–2.2
TP	5.6–9.2	85–92%	0.04–0.07	4–11	80–97%	0.03–0.3
TSS	75–200	86–88%	0.5–1.4	30–90	40–90%	0.2–1.7
E. coli	6–7 log	5 log		3–4 log	2 log	
T. Coliform	7–9 log	5–6 log		5–6 log	2–3 log	

Source: Yates et al. (2012)
[a]Direct input for Chesterfield Inlet and Naujaat WTAs
[b]Pre-treatment of effluent for Arviat, Coral Harbour and Whale Cove WTAs

Hydraulics and Hydrology

The hydraulics and hydrology of the area influence the treatment potential of the tundra WTA (Hayward et al. 2014). The objective of the studies in Kugaaruk and Grise Fiord was to characterize the treatment performance of the wetlands in relation to the unique site-specific hydraulic and hydrological conditions. Figures 3.4 and 3.5 illustrate aspects of the tundra WTAs in Kugaaruk and Grise Fiord, respectively. The surface areas of these WTAs were 0.6 and 0.5 ha, respectively.

The hydraulic loading rate onto the WTA in Kugaaruk ranged from 0.06 to 1 cm/day (6 to 103 m^3/ha/day) during the 4 day decant monitoring. This HLR was within the recommended range from Alberta Environment (2000) of 2.5 cm/day. The middle segment of the wetland had a hydraulic retention time of approximately 5 h. The HRT from the inlet area near the decant cell berm to the outlet was greater than 10 h. The HRT from the decant cell to the outlet was approximately 6 days. These HRTs were shorter than the recommended 14–20 day range for natural wetlands receiving wastewater (Alberta Environment 2000). The watershed area contributing to the WTA was only 2.5 ha (Fig. 3.4a). The WTA is bounded closely along the length by topographic divides. As a result of the topography on site, there is little potential for dilution from external hydrologic contributions.

The hydraulic loading rates (HLRs) onto the WTA in Grise Fiord ranged from 17 to 27 cm/day (1700 to 2730 m^3/ha/day). These HLRs are excessively high compared to the recommended maximum of 2.5 cm/day. During the August 2011 decant, the HRT of the main channel in the WTA was only 20 min. The HRTs in the two smaller channels were 90 and 140 min, respectively. These HRTs are extremely short in comparison to other treatment wetlands. Therefore, very little, if any, treatment would be possible in this WTA due to the short HRTs. There was a defined external hydrologic contribution near the wetland inlet that diluted the effluent

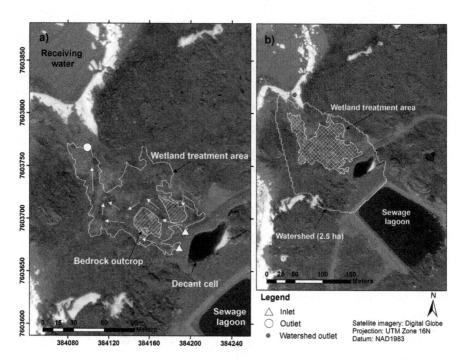

Fig. 3.4 (a) Plan view of the wastewater treatment system in Kugaaruk, with white arrows to denote the direction of effluent flow, and (b) the WTA shown as the cross hatched area and the watershed delineation of the hydrologic contribution to the WTA

stream by 20–40%. The watershed contributing to the WTA was approximately 26.5 ha in area.

Overall, the WTA in Kugaaruk performed well, as the concentrations of effluent $CBOD_5$ and TSS were less than 12 mg/L (Table 3.5). Un-ionized ammonia nitrogen concentrations in the effluent were below 1.25 mg/L during the study periods in Kugaaruk. Overall, the WTA in Kugaaruk provided additional improvement of the effluent water quality prior to discharge into the marine receiving water environment. In contrast, the effluent discharging from the WTA in Grise Fiord (Table 3.6) was of poor quality, with effluent concentrations of 75 mg/L for $CBOD_5$ and 230 mg/L for TSS.

It should be noted that due to logistical challenges with sample collection, only a small sample size was obtained for some parameters. The differences in treatment performance results between the two systems demonstrate the importance of understanding the hydrological and hydraulic context of the individual systems when assessing overall treatment potential. The HRT in the Kugaaruk was long enough to encourage treatment of the effluent prior to release into the marine receiving environment. This contrasted with Grise Fiord, where the HRT of the wetland was too short for any significant treatment.

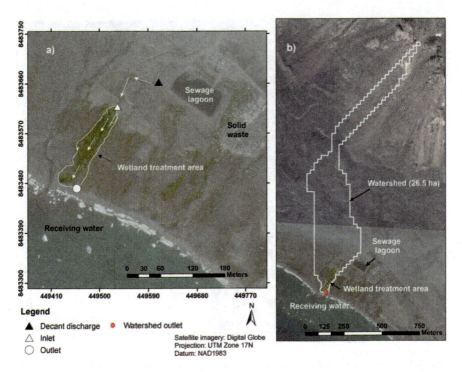

Fig. 3.5 (**a**) Plan view of the wastewater treatment system in Grise Fiord, with white arrows to denote the direction of effluent flow, and (**b**) watershed delineation of the hydrologic contribution to the WTA

Table 3.5 Summary of treatment performance results for the WTA at Kugaaruk, NU

Parameter	Influent		Effluent		
	Mean conc.	Reduction from raw	Mean conc.	Reduction from raw	n
$CBOD_5$ (mg/L)	151	59%	12	97%	2
E. coli (MPN/100 mL)	3.2×10^5	1.8 log	4.23	3.7 log	4
TSS (mg/L)	21	92%	10	97%	4–3
VSS (mg/L)	21	91%	10	96%	4–3
TN (mg/L)	81	38%	55	58%	4
TAN (mg/L)	82	13%	55	41%	4
NH_3-N (mg/L)	0.8	15%	0.4	55%	2
NO_3-N (mg/L)	0.09	−40%	0.15	−121%	2
TP (mg/L)	8.6	24%	4.9	56%	4
Temperature (°C)	5.2		3.0		
DO (mg/L)	2.0		4.9		
pH	7.5		7.4		
Sp. Cond. (µS/cm)	1145		848		

Table 3.6 Summary of treatment performance results for the WTA at Grise Fiord, NU

Parameter	Influent		Effluent		
	Mean conc.	Reduction from raw	Mean conc.	Reduction from raw	n
$CBOD_5$ (mg/L)	194	69%	75	88%	2
E. coli MPN/100 mL	3.0×10^1	4.5 log	1.6×10^2	3.7 log	3
TSS mg/L	140	79%	230	65%	3
TN mg/L	23	72%	8	90%	2–3
TAN mg/L	25	77%	14	88%	13
NH_3-N mg/L	3.2	77%	0.4	88%	13
TP mg/L	3.6	72%	1.2	91%	2–3
Temperature (°C)	8.4		8.7		13
DO mg/L	10.8		11		13
pH	8.5		8		13
Sp. Cond. (µS/cm)	400		323		13

Discharge Techniques and Magnitude

The impacts on treatment performance by specific WTAs that release a large volume of wastewater over a short period of time is poorly understood, but the experience in Grise Fiord (discussed above) indicates that this practice negatively affects performance. In the communities of Edzo, Fort Providence, Gjoa Haven, Pond Inlet and Ulukhaktok, the wastewater entering the associated WTAs was first pre-treated in sewage lagoons. In the communities of Paulatuk and Taloyoak, wastewater was discharged into lake lagoons as the primary treatment method before draining into their downstream WTAs. Fort Providence and Pond Inlet manually discharged effluents from their lagoons; meaning that large volumes of effluent are released to the wetlands in a relatively short period of time. The effluents to the remaining five wetlands were continuously released in smaller volumes as pre-treated exfiltrates through berm walls, or drained from lagoon lakes. The WTAs ranged in size from approximately 0.6 ha in Pond Inlet to 17 ha in Gjoa Haven, as shown in Table 3.7. The treatment efficiency for $CBOD_5$ and TAN typically ranged in the upper 90th percentile, but exceptions to this were observed at Pond Inlet and Fort Providence.

It is also worth noting that the loading rates for $CBOD_5$ and TAN were elevated at both Pond Inlet and Fort Providence. The primary factors contributing to the poor performance at Pond Inlet are believed to be related to the steep slope at the wetland site and the relatively small size of the wetland, which is a vegetated rock filter strip. Although the higher loading rates indicated in Table 3.7 can be attributed to the relatively small wetland size, the actual loading values may even be higher than listed in the table.

Benthic invertebrate studies in the receiving environments were conducted in the communities of: Kugaaruk, Pond Inlet, Grise Fiord, Pangnirtung, and Iqaluit. As part of this work, Krumhansl et al. (2014) found that there were minimal impacts to benthic communities in four out of five of the communities. Impacts were observed

Table 3.7 Measured mean concentrations of CBOD$_5$ and TAN (mg/L) in influent, loading rates (kg/h/day) and reductions in these parameters in the effluents from 7 tundra wetlands

	Influent volume (m³/day)	Wetland size (ha)	cBOD$_5$				TAN			
			Influent conc. (mg/L)	Loading rate (kg/ha/day)	Effluent % Reduction		Influent conc. (mg/L)	Loading rate (kg/ha/day)	Effluent % reduction	
Edzo	109	2.1	26	1.3	92		16.1	0.82	98	
Gjoa Haven	119	17	113	0.80	98		76.4	0.54	99	
Fort Providence	Decanted[a]	0.9	60	5.2	47		26	2.3	31	
Paulatuk	34	1.5	40	0.93	95		3.2	0.07	99	
Pond Inlet	Decanted[a]	0.6	70	12.6	29		75.4	13.5	58	
Taloyoak	86	6.1	80	1.1	86		4.6	0.06	97	
Ulukhaktok	41	7.3	94	0.53	95		9.6	0.05	99	

[a]Loading rates for Fort Providence and Pond Inlet based on daily flow rates of 76 and 104 m³/day. True flow rates are likely higher since wastewaters volumes are stored in lagoons and released quickly during decanting.

at linear distances of less than 225 m from the discharge point of the effluent in all communities, which all had populations of less than 2000 people.

The scale of benthic impacts for each of the study sites was summarized by Krumhansl et al. (2014). Generally, it was suggested that the total volume and duration of the effluent discharge were the most important factors influencing the amount of environmental impact in the receiving environment. Continuous year-round discharge facilities such as those at Pangnirtung and Iqaluit had larger linear distances of impacts than decanted facilities, but it is worth noting that these two facilities did not have WTAs. The larger community of Iqaluit was observed to have significant negative environmental impacts at marine locations over 500 m from the point of discharge. These findings suggested that communities of populations of less than 2000 people, currently have adequate treatment systems to minimize environmental impacts to the receiving environments, due to the relatively low discharge rates (Krumhansl et al. 2014).

Vegetation

Traditionally, the delineation of wetlands has been performed with consideration of the hydrological features, soil conditions and vegetation species observed on the site. Soil depth in most wetlands studied was generally shallow and varied on average 15–30 cm in depth, but sometimes was greater than 1 m. In southern and temperate North American climates, wetland delineation can be done according to the procedure described within the US Army Corps of Engineers Wetland Delineation Manual (Environmental Laboratory 1987). This resource recommends using the National List of plant species that are typically found in wetlands within the United States (Reed 1988). An equivalent resource for tundra wetland treatment areas in an arctic Canada context has not been developed, and therefore, the primary objective of the Coral Harbour study was to identify common vegetation species, and their occurrence and spatial distribution within these wetlands. These tundra WTAs provide an influx of nutrients and water from upstream lagoons. Of particular interest was whether certain tundra vegetation species prefer or thrive in the wetted areas of the tundra WTAs.

Nine distinct classes of land cover and vegetation, of which six are illustrated in Fig. 3.6, were developed for the WTA in Coral Harbour. Land near the effluent flow areas is characterized by wetland species (Classes 1, 2, 5, and 6), the land around the perimeter of the flow areas is mainly a transition area of diverse wetland and upland species (Class 3), while the remaining land is classified as upland which is associated with higher ground and bedrock outcrops (Class 4).

The effluent flow areas for both the spring and post-spring (i.e. mid-summer and fall) conditions are superimposed in Fig. 3.6. *Salix richardsonii* was a dominant species in the spring flow area, while the post-freshet flow area is characterized predominantly by *Bryophyta* spp. and *Hippuris vulgaris*. *Salix richardsonii* was

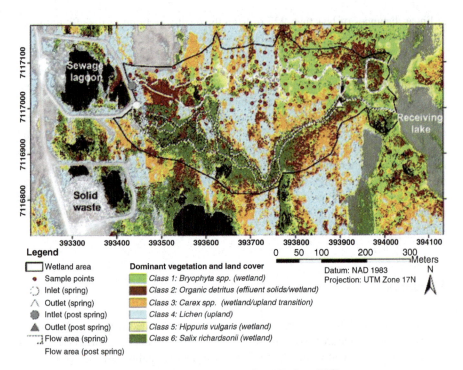

Fig. 3.6 Dominant vegetation and land cover in the Coral Harbour WTA

noted to be on average 35 cm in height in the spring flow area, in comparison to a 21 cm average height in the reference wetland. Tilton and Kadlec (1979) noted an increase in plant biomass near the inlet of a natural treatment wetland receiving municipal wastewater. *Salix* spp. has been demonstrated to function as a tolerant vegetation species in constructed wetlands in Denmark and Sweden for phytoremediation of nutrients and metals in domestic wastewater (Gregersen and Brix 2001; Wittgren and Maehlum 1997).

Organic detritus was distributed along the entire length of the flow path from the inlet to the outlet spring flow area. Whereas, only the upper portion of the post-spring flow path, near the inlet, was dominated by organic detritus from effluent solids. This difference may be explained by the shorter HRT and turbulent flow conditions during spring freshet, which would result in the suspension and settling of solids along the entire length of the flow path. Settling of solids occurred mainly near the inlet during the post-freshet, potentially due to lower hydraulic loading rates and discharge energies. Yates et al. (2012) similarly noted depositions of organic matter in many other Nunavut tundra WTAs.

In total, it is estimated that 60% of the 14 ha Coral Harbour WTA was characterized by vegetation and land cover types indicative of wetland environments (Classes 1, 2, 5, and 6). Approximately 15% of the wetland treatment area was likely composed of vegetation characteristic of the transition area between wetland and upland environments (Class 3). An estimated 19% of the WTA was classified by vegetation and land cover indicative of upland environments (Class 4). The remaining 6% of the wetland treatment area was characterized by surface water in ponds and stream channels, and to a lesser degree, sand from disturbed soil resulting from previous earthworks on-site (Classes 7 and 8).

Overall, the diversity of vegetation species was greatest in the Class 3 wetland/upland transition land type where *Carex* spp. was the most dominant species. The least amount of diversity was observed in Class 2, which was mainly characterized by organic detritus. Class 2 areas were distributed predominantly around the inlet and along the spring flow area. This is in contrast to Class 3 areas, which were distributed primarily around the lower half the post-freshet flow area. The difference in the spatial distribution of classes may suggest that areas where the hydraulic and organic loadings are highest observe a decrease in vegetation diversity. Other authors similarly noted a decrease in vegetation diversity in natural wetlands in response to wastewater addition (Kadlec 1987; Mudroch and Capobianco 1979). Similarly, changes in the community structure of the tundra vegetation in response to nutrient addition have been noted by Yates et al. (2012) and Gough et al. (2002). Additional details on this vegetation classification are provided in Hayward (2013).

It is well-established that within constructed wetlands, but not the tundra WTA featured in this work, much, if not most of the biological treatment can be attributed to microorganisms located within the bed media, or found in association with the root mass of the planted vegetation. Regarding tundra WTAs, however, our understanding of the contribution of native vegetation to biological treatment process is largely unknown. Several factors distinctive to northern cold climate areas may modulate the role that native plants play in the biological treatment of domestic wastewaters. For example, tundra WTAs are generally nutrient poor and because of this, some of the native plants appear to have evolved biological processes that enhance their ability to more efficiently utilize nutrient sources (Yates et al. 2016). Other studies, such as those conducted by Stein and Hook (2005) and Allen et al. (2002) provide interesting insights suggesting that under cold conditions and periods of plant dormancy, internal oxygen consumption rates of the plant decrease, which in turn favors conditions that allow roots to leak unneeded oxygen to the surrounding root zone. They surmised that during summer time, the microbial pathway in wetland soils is often characterized as anaerobic but can shift to predominantly aerobic respiration and oxygen dependent processes during cold periods in which there is plant dormancy. The current understanding of the role that plants may play in an arctic environment is based primarily on our knowledge of temperate regions, meaning we are left to extrapolate the general principles of plant ecology to arctic conditions. The reliability of these assumptions still needs study through better ground-truthing.

Arctic-specific investigations have included studies by Hobbie (2007) who commented on the longevity of Arctic species, and Woo and Young (2003) who explored the response to long-term ecosystem stressors. Numerous other studies have documented short-term response relationships between particular plant species, nutrients and Arctic herbivores (Cadieux et al. 2005; Ngai and Jefferies 2004; Tolvanen et al. 2004). What is largely unknown is the interaction between the plant species or communities with the surrounding environment, such as the influx of nutrients into these systems from municipal wastewater, or the influence of natural nutrient additions from snow geese (*Chen caerulescens*). Kotanen (2002) noted that rapid responses to fertilization generally only occur in freshwater species, and only when nutrients are added at levels much greater than that of natural background concentrations. For example, Pineau (1999) observed that the within-season growth response of sedge fen species was significant only when inorganic N was added at 20 times the natural rate. Cornelissen et al. (2001) and Press et al. (1998) saw that plant communities shifted from moss-lichen to graminoid communities (i.e. grass and sedges) only when there was a long-term presence of increased nutrients. Hobbie et al. (2005) suggested that high levels of nutrients may be toxic to mosses and lichens and could account for the shift to grasses and sedges. They further suggested that other changes in environmental conditions, such as shading/moisture, may also influence the shift in community structure.

In general, it is well understood that the growing seasons for plants are relatively short in northern latitudes, and yet the number of daylight hours in summer in these regions leads to increased photosynthesis and evapotranspiration that may enhance the overall treatment potential provided by plants. Although not well documented, there is anecdotal information to suggest that the ratio of root mass to soils in shallow tundra regions may be greater than experienced in constructed wetlands in southern latitudes that typically have much deeper bed profiles.

Some plants, in particular *Carex aquatilis*, showed a moderate correlation between plant occurrence and ammonia (measured as TAN) in many of the studied WTAs. *C. aquatilis* stands generally dominated the wetland where wastewater concentrations were high, typically anoxic, and rich in organic matter and unmineralized nutrients. In these locations *C. aquatilis* formed near mono-culture stands particularly at the influence of wastewater into the tundra WTA. In respect to these observations, nutrient concentrations were compared to the dominant plant species in an attempt to determine if the plant community was re-ordering itself in response to the high nutrient loading originating from wastewater inputs. A spatial correlation was performed using *C. aquatilis* as an indicator species. Findings from this exercise indicated a moderate relationship between stands of *C. aquatilis* and ammonia at the Chesterfield Inlet WTA. This relationship was verified via principle component (PCA) analysis and findings are described more fully in Yates (2012). Applied laboratory studies have been conducted by the CAWT to further investigate the relationship between *C. aquatilis* to ammonia and other nitrogen species. In these studies, *C. aquatilis* plants were obtained from the Baker Lake WTA and were transported to the laboratory to be grown in an environmental chamber under simulated summer and autumn Arctic climatic conditions. These experiments demonstrated that *C. aquatilis* was responsible for the loss of ammonia in simulated wastewater, supposedly via direct uptake. These results indicate that *C. aquatilis* may be selecting the first available source of inorganic-N (Yates et al. 2016).

Further away from the input of wastewater, plant species richness increased, which suggested that over time the plant species diversity had decreased in impacted areas as a response to the addition of nutrients and possibly other environmental variables that had been altered by the presence of wastewater, such as the hydrologic regime. Such re-organization of vegetation communities is not unexpected. Kadlec and Bevis (2009) observed great shifts in the community, at the Houghton Lake treatment wetland in Michigan, where partially treated wastewater was being discharged in a natural wetland. They observed *Typha* spp. displacing the original plant community which was likely negatively impacted by the newly altered hydrologic nutrient regime, which was more favourable for *Typha* sp. A similar response likely occurred in the Chesterfield Inlet wetland, where *C. aquatilis*, a nitrophilic and hydrophilic species, was becoming dominant.

It is well understood that many abiotic factors (temperature and nutrients) strongly influence plant communities in the Arctic (Chapin and Shaver 1985; Hobbie 2007), because of the oligotrophic conditions present in most Arctic systems. The extreme environment has allowed some species to evolve in very nutrient limited conditions, often resulting in low biomass production. Studies of abiotic factors in Arctic plant communities by Chapin and Shaver (1985), and Chapin et al. (1995) have shown that competition within a plant community is primarily driven by nutrient availability in the system, and many Arctic plant species have been shown to respond rapidly to the addition of nutrients. Although temperature does not directly affect plants in the Arctic (Chapin 1983), it indirectly influences the plant community through nutrient cycling and nutrient availability (Hobbie 2007; Nadelhoffer et al. 1991). Jonasson and Shaver (1999) suggest that nutrient pools entering from external sources or in vegetative material present in Arctic wetland systems are small, while organically fixed nutrients in the soil are large but are often unavailable for plant uptake.

Fertilization studies in various Arctic and alpine systems have also been used to demonstrate how communities can rapidly respond to increased nutrients and changes in environmental conditions (e.g. light and moisture), often imitating conditions which are expected with a changing climate Gough et al. (2002). Hobbie et al. (2005) showed how biomass rapidly increased in *Betula nana* in Arctic tundra with the addition of N and P over 2 years. However, these responses have been more variable than changes attributed nutrient availability alone (Hobbie 2007).

Changes in polar systems from climate change have been shown to change nutrient uptake in simulated environments (Wasley et al. 2006), as temperature directly influences nutrient input from N_2 fixation (Ju and Chen 2008). Despite this empirical evidence of regional climate change, little is understood with respect to how or to what extent plant communities respond in natural Arctic tundra wetlands when the system experiences regular nutrient loading on a landscape scale from thawing nutrient pools in the permafrost. Small scale fertilization studies (i.e., addition of N, and P) in Arctic wet meadows have shown a rapid positive association to increase in plant biomass to specific nutrients generally when added to the system (Gough et al. 2002; Hobbie et al. 2005). However, most studies in upland tundra environments showed that plants responded to the addition of N rather than P (Gebauer et al. 1995). Contrastingly, Ngai and Jefferies (2004) demonstrated that in freshwater marshes the addition of P from the feces of geese was more important than N. It should be noted that these studies are generally small on a spatial scale, 5×20 m or

2.5 × 2.5 m (Cornelissen et al. 2001; Hobbie et al. 2005), and their applicability at a spatially landscape scale is unknown.

As a result of receiving wastewater for long periods of time (i.e., decades), these WTAs provide a ready-made environment to test the observations of nutrient response by plant communities and individual species at a landscape scale. From pre-study observations of the Chesterfield Inlet WTA, *Carex aquatilis* was observed to dominate many portions of the WTA. Mono-culture stands were most prevalent near of the locations where wastewaters entered the WTA. *Carex aquatilis* is often associated with freshwater wetlands (Aiken 2007), and is known to be nitrophilous and maintains a high concentration of nitrogen in its above ground tissue (Murray 1991). It is also a common species with circumpolar distribution, commonly found along rivers, pond edges, and wet meadows (Hulten 1968; Porsild and Cody 1980). *C. aquatilis* also has great ecotypic differentiation in size and phenology, respiration, photosynthesis, and nutrient absorption across regions, and even in micro habitat (Chapin and Chapin 1981). Muskoxen (*Ovibos moschatus*) regularly feed on stands of this species, fertilizing it with feces and urine. Raillard (1992) showed that *C. aquatilis* may be responding to the presence of more nutrients from muskoxen feces and urine promoting greater *C. aquatilis* stands on Ellesmere Island, Nunavut.

Collectively these studies indicate that plants may have a more significant role in the treatment of wastewater in arctic treatment wetlands than what has been generally observed in more temperate regions. The information gathered during these investigations may have implications concerning how manipulations to vegetation cover could be used to more effectively manage the treatment potential of WTAs.

Spatial Representation of Treatment

Performance was assessed by monitoring a series of sample locations situated within each of the WTA study sites. From this information, concentration gradients for individual parameters was interpolated and visually presented in maps that provided a "snap shot" of treatment. A sample of the data generated for the Paulatuk WTA is used for illustrative purposes below. Figure 3.7 provides an overview of the WTA and lake lagoon where raw wastewater is first introduced into the system and is intended to help orientate the reader to the interpolated mapping illustrated in Fig. 3.8. Wastewater flows from the lake lagoon through the WTA, and empties into the Arctic Ocean. Figure 3.8 provides data on the interpolated mapping for $CBOD_5$, TAN, Nitrate and Total Coliforms. This type of mapping can be used to illustrate generalized trends that existed at the time of sampling. For example, the concentration of $CBOD_5$ and Total Coliforms decrease significantly before exiting the wetland and in the case of nitrogen, it is evident that TAN was being converted to nitrate.

Fig. 3.7 A Google Earth image of the wastewater treatment area located in Paulatuk, NU, and the lake lagoon that receives the hauled truck raw wastewater

The interpolated maps provide a visual illustration of the locations in the wetland where concentration of a specific water quality parameter is the greatest and the least, which gives an indication of treatment as the wastewater progresses through the WTA.

Performance Models

In general, a performance model can be used for a variety of purposes when assessing and or designing a WTA. Examples include:

1. Generating estimates of the treatment performance of a WTA under the expected range of operational conditions;
2. Assessing and optimizing the effects of operational changes on treatment performance;
3. Comparing options for system upgrades when the modeled effluent quality does not meet requirements.

The choice of modeling approach depends on whether the WTA is characterized by surface or subsurface flow. The approach for surface flow wetlands also differs slightly between new proposed WTAs and existing WTAs.

The range of predictive tools for constructed wetlands vary from simple scaling factors to elaborate 2-dimensional and 3-dimensional process based computer models. Readers are encouraged to review Chouinard et al. (2014a) for a more

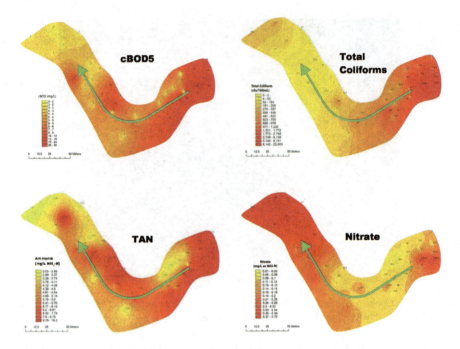

Fig. 3.8 CBOD$_5$, total coliform bacteria, total ammonia nitrogen (TAN), and nitrate concentrations (mg/L) in the Paulatuk wetland. Arrows indicate flow direction. *Dark red* coloration indicates high concentrations while *light yellow* coloration indicates low concentrations. Modified from Yates et al. (2013)

comprehensive discussion on predictive tools for assessing wetland performance. Most of the predictive tools are used during the design of constructed wetlands to determine the wetland's aerial size needed to accommodate a defined wastewater flow of a known strength in order to achieve a desired level of treatment. Sizing tools can be generally categorized under the following headings:

- Scaling factors (sometimes also called rule of thumb);
- Regression equations and loading charts;
- Ideal chemical reactor models (e.g., $k - C^*$ model); and
- Variable—order, mechanistic or compartmental models (e.g., SubWet 2.0) and sophisticated 2D and 3D models (e.g., HYDRUS, WASP, TABS-2).

In general terms, the scaling factors contain the greatest amount of uncertainty and thus are often used primarily as a "first-cut" estimate of wetland size. The variable—order and compartmental models can provide the most precise measurements, but their use is often hampered by the need for a large data set of site specific information, which often does not exist or is not easily obtained. Without the calibration of these models to the specific conditions of the site, the results can be quite inaccurate.

In tundra WTAs, it is much more difficult to apply predictive tools since each wetland is unique and many of the characteristics in terms of hydrology, wetland size, and hydraulic retention times require comprehensive site-specific studies. The CAWT has however, had reasonable success in applying the SubWet 2.0 computer model. The SubWet model is a software program package used to simulate the treatment of wastewater in subsurface horizontal flow artificial wetlands. SubWet can be used to allow managers to predict the impact to treatment efficiency based on different scenarios involving an alteration to the HRT, aerial loading rates, and the desired level of influent treatment. The model can be used to predict treatment performance anticipated from alterations to the size of the treatment area that could be increased through the construction of infiltration/dispersion ditches and structures that divert flow to other parts of the wetland that are not currently involved in treatment of the influent. The procedure used to calibrate SubWet 2.0 to site conditions has been outlined by Chouinard et al. (2014c). In brief, the calibration is achieved by comparing measured wastewater effluent concentrations exiting the wetland site against the simulated concentrations generated by the SubWet 2.0 model and adjusting rate constants of the model to bring model simulations closer to measured values.

Chouinard et al. (2014b) presented an analysis of five different hypothetical scenarios to demonstrate how SubWet 2.0 can provide Arctic municipal wastewater managers with a tool to adapt to changing treatment conditions, as well as predicting the impact to treatment when wetland systems are altered. The simulated scenarios illustrate how a reduction in wetland size, or increased wastewater loadings, as well as temperature changes can impact treatment potentials. For example, SubWet 2.0 has the ability to simulate how the release of wastewaters from lagoon systems could impact treatment performance in the downstream wetlands; thus providing a predictive tool to help identify management options for lagoon operators. Readers are encouraged to review Jørgensen and Gromiec (2011) for insight into the process equations used in the SubWet model.

Hayward and Jamieson (2015) used a different approach to model wetland performance. They applied a modified Tanks-in-Series (TIS) mathematical model to simulate the Coral Harbour wetland and refine treatment rate constants. The modified TIS model is based on a conventional TIS chemical reactor model, which was modified to account for the external hydrologic contributions from the watershed that is cumulatively added along the length of the wetland. A simple TIS model can be created in a software program, such as Microsoft Excel. The model represents the wetland hydraulically by a series of completely mixed tanks with equivalent HRTs. Hayward and Jamieson (2015) specify the general mass balance for each tank in the modified TIS model. The methodology for the use of this type of model for the prediction of treatment performance of a given wetland is detailed in Hayward and Jamieson (2015) and Jamieson and Hayward (2016). Advantages of this type of model is that is relatively simple to construct and parameterize, and it is currently accepted as an adequate method to assess treatment wetland performance (Kadlec and Wallace 2009). However, the hydraulic parameterization data requires site-specific data collection programs for tundra wetlands due to the intersystem variability observed between sites.

Best Practices for Design and Assessment

Tundra WTAs present a low cost and low maintenance option for secondary and tertiary levels of treatment in Nunavut. The uncertainties with their use will likely be reduced with comprehensive site-specific datasets to verify whether adequate treatment is provided. The ongoing monitoring programs will be costly; however, these studies will, in many cases, still be significantly more economical than conventional wastewater infrastructure with large associated capital and O&M costs. The WTAs may also be less prone to the occasional operational failure observed with WWTP technology in northern communities as discussed in Johnson et al. (2014). As a result of the intersystem variability between study sites, it is recommended that site-specific studies are conducted on WTAs that are going to be used intentionally as part of a treatment system.

Proposed Assessment Framework

It is recommended that a standardized assessment framework be applied to adequately assess treatment performance in WTAs. A proposed standardized assessment framework is outlined in detailed in Fig. 3.9. In summary, the initial step in the assessment process is the desktop mapping to inform the site-specific study proposal which outlines the fieldwork to be conducted on site. The proposal then undergoes review by the applicable regulatory bodies. After approval, a public consultation program is initiated in conjunction with a site-specific data collection program. The site-specific data collection program characterizes the physical and biological environments, hydraulics, hydrology, hydrogeology, and assesses the treatment performance and biogeochemistry of the site. The data collected from the site-specific study is used to inform the performance model for the site to estimate performance of the WTA. If the modeling results do not meet the regulatory requirements, prior to discharge into the receiving environment, then options for upgrades should be assessed and re-modeled, as an iterative process. If the modeling results meet the regulatory requirements, then a long-term monitoring plan is established and implemented. The findings from the site-specific study are submitted to and assessed by the applicable regulatory bodies. Ongoing annual monitoring and annual data reviews are recommended to verify that adequate treatment is provided after approval of the design and operation of the system.

3 Recommendations for the Use of Tundra Wetlands for Treatment of Municipal... 111

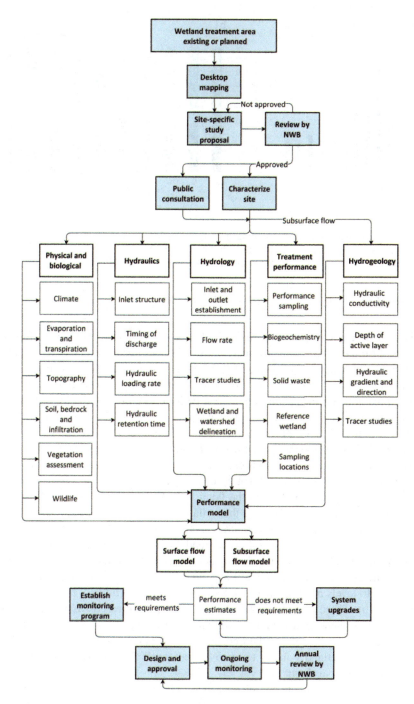

Fig. 3.9 Proposed standardized framework to assess the WTAs

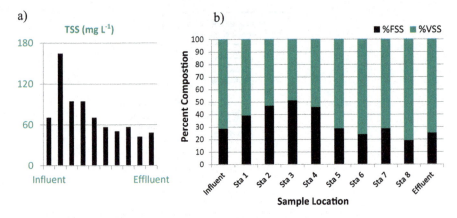

Fig. 3.10 Concentration and composition of TSS in wastewater samples collected from the Pond Inlet tundra WTA. Graph (**a**) illustrates the concentration (mg/L) of TSS at sampling locations with the wetland. Graph (**b**) illustrates the percent composition of FSS to VSS at these sites

Inadequacies of Performance Parameters

The wastewater compliance parameter TSS is often applied to treated effluents released from engineered treatment systems. TSS does not always provide a reliable measure of treatment performance for WTAs as some wetlands can generate TSS within the wetland that does not originated from the wastewater. Results demonstrated that a better indication of treatment performance in WTAs is to examine the percent composition of volatile suspended solids (VSS) to fixed suspended solids (FSS) within TSS. VSS represents the organic fraction of TSS that can be oxidized while FSS represent the inorganic fraction. Monitoring the change in percent composition between these two fractions can provide an indication of treatment efficiency. For example, Fig. 3.10 indicates that the concentration of TSS (Fig. 3.10a) decreased as the wastewater passes through the Pond Inlet wetland while the overall percent composition of VSS to FSS exiting the wetland (Fig. 3.10b) remained relatively stable, which suggested that physical filtration may be the primary method of removal.

Figure 3.11a illustrates a different trend in the composition of TSS at the Ulukhaktok WTA. At this site, the organic fraction decreases significantly as the wastewater passes through the wetland while the overall concentration of TSS increases to levels much greater than found in the original influent entering the site (Fig. 3.11b). These two trends suggest that the TSS increased from an inorganic fraction being generated within the wetland. Therefore, TSS alone would suggest that the treatment efficiency of the wetland for this parameter would be low. In reality, it appears the wetland's ability to remove the organic fraction associated with the original influent was high.

In regard to the monitoring of microbial indicators, waterfowl have been suspected to be a source of fecal indicator bacteria to wetland treatment areas (Hayward et al. 2014; Yates et al. 2012). During periods when migratory waterfowl are present

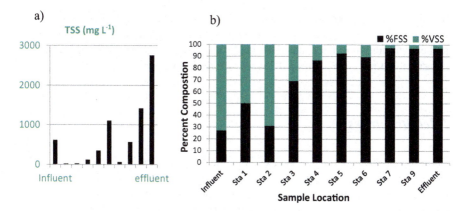

Fig. 3.11 Concentration and composition of TSS in wastewater samples collected from the Ulukhaktok tundra treatment wetland. Graph (**a**) illustrates the concentration of TSS (mg/L) at associated sampling locations with the wetland. Graph (**b**) illustrates the percent composition of fixed suspended solids (FSS) to volatile suspended solids (VSS) at these sites

within a WTA, it is recommended that fecal indicator bacteria (e.g., *E. coli*) be exempt from performance assessments.

Timing and Frequency of Monitoring

The seasonal variations in hydraulic, hydrological, hydrogeological and biogeochemical characteristics affect the treatment performance of tundra WTAs. For this reason, it is recommended that site-specific studies are conducted at key times during the treatment season. For existing WTAs which undergo manual discharge at scheduled times during the treatment season, the timing of the data collection should be centered on anticipated discharge events to ensure the treatment profile is captured at these times. For WTAs which discharge continually over the treatment season, the timing of the data collection should capture both the spring freshet, and post-freshet discharge conditions. Furthermore, the post-freshet conditions should be characterized during the middle, and at the end of the treatment season. For proposed WTAs, site-specific data should be collected during the periods when wastewaters discharges to the proposed WTA are anticipated.

The WTA should be sampled at a frequency capable of capturing changes in hydraulic loading rates and changes in seasonal conditions. If discharge into a WTA occurs during the spring freshet, then collection of a minimum of three performance samples, and flow measurements are recommended, at the inlet and outlet, during the spring freshet and post-spring freshet conditions, respectively (i.e., six total). If the discharge into a WTA is less than 2 weeks in length, then the collection of a minimum of three samples and flow measurements are at the beginning, middle, and end of the decant, is recommended. If the discharge into a WTA is greater than 2 weeks in length, then the collection of a minimum of three samples and flow measurements at the beginning, middle, and end of the decant, and one additional sample and flow measurement per each additional 2 weeks of discharge, is recommended.

Inlet Structures

There is opportunity for new types of inlet structures to be used for design of new WTAs. Flow control is important to be able to optimize the hydraulic loading onto the wetland. Inlet structures which disperse the influent over a wide area improve the hydraulic efficiency of the WTAs. Low flow inlet structures with deep zones, for settling would encourage settling of solids, and enhance vegetation growth down gradient of the structure, and discourage re-suspension.

Flow Rates

The influent and effluent flow rates and variability are important datasets for assessments of the treatment potential of the tundra WTAs. The findings from Hayward et al. (2014) and Hayward and Jamieson (2015) suggest that it is important to characterize the external hydrologic contributions to enable proper assessment of the treatment potential. To accomplish this, flow gauging points may be established at locations along the effluent flow path where water from external hydrologic contributions are suspected to enter the wetland. The tracer study may be helpful to discern where external hydrologic contributions may be entering the wetland.

Loading Rates

Doku and Heinke (1993) recommended an HLR for WTAs of between 1 and 2 cm/day (100 and 200 m^3/ha/day). Based on recommendations in Kadlec and Knight (1996), Alberta Environment (2000) recommended HLRs ranging from 0.2 cm/day (20 m^3/ha/day) for secondary treated effluent to 0.5 cm/day (50 m^3/ha/day) for nitrified secondary effluent, for natural treatment wetlands. In cases where BOD$_5$, TSS, and TP are reduced in the pretreatment step, an HLR of 2.5 cm/day (250 m^3/ha/day) was recommended (Alberta Environment 2000). HLRs at other northern Canadian WTAs were estimated at 0.43 cm/day (43 m^3/ha/day) for a 7 ha wetland in Teslin, Yukon; 0.63 cm/day (63 m^3/ha/day) for a 32 ha wetland in Hay River, NWT; and 1.25 cm/day (125 m^3/ha/day) for a 6 ha wetland in Haines Junction, Yukon (Doku and Heinke 1993; Doku and Heinke 1995). The hydraulic loading rate may change over the treatment season and this variation should be quantified. A maximum HLR of 2.5 cm/day is recommended for WTAs in Nunavut. There may be WTAs that work better at lower HLRs that may be optimized based on the long-term monitoring program data.

Results suggest that treatment performance may be negatively impacted when organic loading rates exceed 1–2 kg/ha/day $CBOD_5$, and or 1 kg/ha/day TAN. These higher loading rates were associated with wetlands less than 1 ha in size, or during periods when effluents were being intentionally decanted from discharged in large volumes.

Hydraulic Retention Time

A nominal HRT of 14–20 days is recommended for a natural treatment wetland (Alberta Environment 2000; Kadlec and Knight 1996). For an existing WTA, the HRT can be determined by conducting a tracer test. Seasonal variation in hydraulic loading rates may present the requisite for multiple tracer tests to characterize the range of HRTs. Methodologies for conducting tracer studies are detailed in Jamieson and Hayward (2016). Treatment performance may be improved by designing and construction of flow retention structures within WTAs, such as diversion berms and flow control structures to increase the HRT.

Timing of Discharge

Currently, there are existing systems that decant in an uncontrolled manner during the spring freshet. Ideally, exfiltration or manual discharge of effluent into WTAs should occur during the middle of the summer or late summer before plants become senescent and temperatures cool. This will allow for time for the vegetation to become productive, for water temperatures to rise, and to avoid high flows from melting water. Discharge of wastewater into WTAs when melt flows are high during the spring freshet should be avoided. If possible, it is beneficial to extend the discharge period out over a long time frame at a low hydraulic loading rate.

Emerging Areas of Research

This chapter demonstrated that there is a breadth of research demonstrating the estimated treatment performance of WTA in Arctic regions. Despite this new understanding of the important function of WTAs in northern Canadian wastewater management, there are still emerging areas for research to help better inform their use, including: (1) monitoring of the performance of augmented or altered WTAs; (2) studies of the fate and transport of emerging biological contaminants and chemical contaminants; and (3) characterization of the impact of climate change on the performance and ecology of the WTAs.

Recommendations have been presented for the beneficial management practices (BMPs) of these WTAs based on findings summarized within this chapter. There have been no studies which have assessed the performance and vegetation recovery of WTAs which have undergone alteration in attempt to improve hydraulic efficiency. Baseline data monitoring for a WTA before and after alteration for improved hydraulics would be useful data to verify the effectiveness of the recommended BMPs.

Gunnarsdóttir et al. (2013) indicated that further research is needed in the areas of fate and transport of emerging biological contaminants and pharmaceuticals and personal care products (PPCPs) associated with municipal wastewater in arctic regions. Work has been initiated by Chaves-Barquero et al. (2016) to characterize the behavior of PPCPs and antibiotic resistance genes (ARGs) associated with municipal wastewater in the Cambridge Bay sewage lagoon and WTA. These authors concluded that the levels of PPCPs and ARGs in the receiving environment would not pose a significant risk to the aquatic environment at that particular site.

Another area for future research is to study the performance of northern wastewater treatment systems and WTAs in light of climate change. The Intergovernmental Panel on Climate Change (IPCC) suggested that in high latitude environments characteristic of the Arctic; the precipitation is projected to increase; higher average ambient air temperatures are projected to increase the rate of evaporation from freshwater lakes and reservoirs (including sewage lagoons); and evapotranspiration from the tundra landscape (Intergovernmental Panel on Climate Change (IPCC) 2014). The long-term impacts of these anticipated changes of the climate on the performance and risk of the municipal wastewater treatment systems in the Canadian arctic are unknown. Preliminary work estimates that impact of climate change is uncertain. The most important factor driving the performance of the WTAs is the HRT, and increased precipitation can theoretically shorten the HRT, and vice versa. Warmer air temperatures could theoretically extend the treatment season and facilitate an increase in biological treatment. Furthermore, warmer temperatures could lead to increased algae growth associated with the wastewater, which can have implications for eutrophication, un-ionized ammonia, and TSS concentrations.

Conclusions

Tundra WTAs are a valuable resource that with the appropriate management procedure, can be an important part of the treatment strategy in Northern municipal wastewater systems. Published literature to date has demonstrated that water quality improvements are observed in the tundra WTAs. The site-specific research programs conducted by CAWT and CWRS focused on identifying the specific elements required to develop an adequate understanding of the WTAs from a risk and engineering stand-point. Design recommendations, and a proposed framework for conducting site-specific studies have been developed which provides a standardized approach and tools to assess the treatment performance expected from WTAs.

Due to the variability and uncertainty associated with the use of tundra WTAs, it would be ill-advised to extrapolate data between sites, and expect similar performance results. The assessment frameworks that have been generated from this research cover many aspects that affect the complex functioning of WTAs, therefore highlighting that a robust data collection component is required. This includes the recommendation that comprehensive site-specific assessments of WTAs are essential to reduce the risks associated with the use of this type of treatment process.

The modified TIS model and SubWet 2.0 models have been shown to be effective tools for the assessment of expected treatment performance of WTAs at a range of operational conditions. The wetland designer must understand the particular limitations of the selected model chosen to predict treatment and the required input data. If particular WTAs do not meet the targeted treatment objectives according to the model results, then modification options for WTA may be explored at the modeling stage.

Finally, there is a recommendation for the long-term monitoring for each WTA that will build a database for the performance of the systems in the Canada's Far North. This database will refine parameters that are used for modeling these systems, as well as, optimize other operational features of the WTAs.

Acknowledgments The authors would like to thank the Canadian Government and their contributions to the International Polar Year program along with Environment Canada, the RBC Blue Water Foundation and the Community and Government Services division of the Government of Nunavut, the Canadian Water Network, and NSERC for funding support that made this work possible. The authors would also like to thank Jamal Shirley and the Nunavut Research Institute for support with the Northern Water Quality laboratory.

References

Aiken SG (2007) Flora of the Canadian arctic archipelago. NRC Research Press: Canadian Museum of Nature, Ottawa

Alberta Environment (2000) Guidelines for the approval and design of natural and constructed treatment wetlands for water quality improvements. No. t/518.Municipal Program Development Branch, Environmental Sciences Division

Allen WC, Hook PB, Biederman JA, Stein OR (2002) Temperature and wetland plant species effects on wastewater treatment and root zone oxidation. J Environ Qual 31:1010–1016

APHA (2012) Standard methods for the examination of water and wastewater. American Public Health Association, Washington, DC

Cadieux MC, Gauthier G, Hughes RJ (2005) Feeding ecology of Canada geese (*branta canadensis* interior) in sub-arctic inland tundra during brood-rearing. Auk 122(1):144–157

Chapin FS (1983) Direct and indirect effects of temperature on arctic plants. Polar Biol 2(1):47–52

Chapin FS, Shaver GR (1985) Individualistic growth-response of tundra plant-species to environmental manipulations in the field. Ecology 66(2):564–576

Chapin FS, Shaver GR, Giblin AE, Nadelhoffer KJ, Laundre JA (1995) Responses of arctic tundra to experimental and observed changes in climate. Ecology 76(3):694–711

Chapin FS, Chapin MC (1981) Ecotypic differentiation of growth processes in *Carex aquatilis* along latitudinal and local gradients. Ecology 62(4):1000–1009

Chaves-Barquero LG, Luong KH, Mundy CJ, Knapp CW, Hanson ML, Wong CS (2016) The release of wastewater contaminants in the Arctic: a case study from Cambridge Bay, Nunavut, Canada. Environ Pollut 218:542–550

Chouinard A, Balch GC, Jørgensen SE, Wootton BC, Anderson BC, Yates CN (2014a) Tundra wetlands: the treatment of municipal wastewaters—RBC Blue Water Project: performance and predictive tools (manual only). Centre for Alternative Wastewater Treatment, Fleming College, Lindsay, ON, Canada

Chouinard A, Yates CN, Balch GC, Jørgensen SE, Wootton BC, Anderson BC (2014b) Management of tundra wastewater treatment wetlands within a lagoon/wetland hybridized treatment system using the SubWet 2.0 wetland model. Water 6(3):439–454

Chouinard A, Yates CN, Balch GC, Wootton BC, Anderson B, Jørgensen SE (2014c) SubWet 2.0. Modelling the performance of treatment wetlands in a cold climate. In: Jørgensen SE, Chang N-B, Xu F (eds) Ecological modelling and engineering of lakes and wetlands. Elsevier, New York, NY

Cornelissen JHC, Callaghan TV, Alatalo JM, Michelsen A, Graglia E, Hartley AE (2001) Global change and arctic ecosystems: is lichen decline a function of increases in vascular plant biomass? J Ecol 89(6):984–994

Daley K, Castleden H, Jamieson R, Furgal C, Ell L (2014) Municipal water quantities and health in Nunavut households: an exploratory case study in Coral Harbour, Nunavut, Canada. Int J Circumpolar Health 73:15–28

Doku IA, Heinke GW (1995) Potential for greater use of wetlands for waste treatment in northern Canada. J Cold Reg Eng 9(2):75–88

Doku IA, Heinke GW (1993) The potential for use of wetlands for wastewater treatment in the northwest territories. Report prepared for the Department of Municipal and Community Affairs. Government of the Northwest Territories, Yellowknife, NT, Canada

Dubuc Y, Janneteau P, Labonte R, Roy C, Briere F (1986) Domestic waste-water treatment by peatlands in a northern climate—a water-quality study. Water Resour Bull 22(2):297–303

Environmental Laboratory (1987) Corps of Engineers wetlands delineation manual. Technical Report Y-87-1. U.S. Army Engineer Water-ways Experiment Station, Vicksburg, MS

Gebauer RLE, Reynolds JF, Tenhunen JD (1995) Growth and allocation of the arctic sedges *Erophorum angustifolium* and *E. vaginatum*: effects of variable soil oxygen and nutrient availability. Oecologia 104(3):330–339

Gough L, Wookey PA, Shaver GR (2002) Dry heath arctic tundra responses to long-term nutrient and light manipulation. Arct Antarct Alp Res 34(2):211–218

Gregersen P, Brix H (2001) Zero-discharge of nutrients and water in a willow dominated constructed wetland. Water Sci Technol 44:407–412

Gunnarsdóttir R, Jenssen PD, Jensen PE, Villumsen A, Kallenborn R (2013) A review of wastewater handling in the Arctic with special reference to pharmaceuticals and personal care products (PPCPs) and microbial pollution. Ecol Eng 50:76–85

Hayward J (2013) Treatment performance assessment and modeling of a natural tundra wetland receiving municipal wastewater. MASc dissertation, Dalhousie University, Halifax, NS, Canada

Hayward J, Jamieson R (2015) Derivation of treatment rate constants for an arctic tundra wetland receiving primary treated municipal wastewater. Ecol Eng 82:165–174

Hayward J, Jamieson R, Boutilier L, Goulden T, Lam B (2014) Treatment performance assessment and hydrological characterization of an arctic tundra wetland receiving primary treated municipal wastewater. Ecol Eng 73:786–797

Hobbie SE (2007) Arctic ecology. In: Pugnaire FI, Valladares F (eds) Functional plant ecology, 2nd edn. CRC Press, New York, pp 369–388

Hobbie SE, Gough L, Shaver GR (2005) Species compositional differences on different-aged glacial landscapes drive contrasting responses of tundra to nutrient addition. J Ecol 93(4):770–782

Huber S, Remberger M, Kaj L, Schlabach M, Jörudsdóttir H, Vester J, Arnórsson M, Mortensen I, Schwartson R, Dam M (2016) A first screening and risk assessment of pharmaceuticals and additives in personal care products in waste water, sludge, recipient water and sediments from Faroe Islands, Iceland and Greenland. Sci Total Environ 562:13–25

Hulten E (1968) Flora of Alaska and neighboring territories. A manual of the vascular plants. Stanford University Press, Stanford, CA

Intergovernmental Panel on Climate Change (IPCC) (2014) Fifth assessment report (AR5). http://www.ipcc.ch/report/ar5/. Accessed 8 Mar 2016

Jamieson R, Hayward J (2016) Guidelines for the design and assessment of tundra wetland treatment areas in Nunavut. Centre for Water Resources Studies, Dalhousie University. Technical report prepared for the Community and Government Services department of the Government of Nunavut. Halifax, Nova Scotia, Canada

Johnson K, Prosko G, Lycon D (2014) The challenge with mechanical wastewater systems in the Far North. Conference proceeding paper at: Western Canada Water Conference and Exhibition, Regina, SK, Canada, 23–26 September 2014

Jonasson S, Shaver GR (1999) Within-stand nutrient cycling in arctic and boreal wetlands. Ecology 80(7):2139–2150

Jørgensen SE, Gromiec MJ (2011) Mathematical models in biological waste water treatment—Chapter 7.6. In: Jørgensen SE, Fath BD (eds) Fundamentals of ecological modelling, Applications in environmental management and research, vol 23, 4th edn. Elsevier, Amsterdam, The Netherlands, pp 1–414

Ju W, Chen JM (2008) Simulating the effects of past changes in climate, atmospheric composition, and fire disturbances on soil carbon in Canada's forests and wetlands. Biogeochem. Cycles 22(3):GB3010

Kadlec RH, Wallace S (2009) Treatment wetlands. CRC Press, Boca Raton, FL

Kadlec RH, Bevis FB (2009) Wastewater treatment at the Houghton Lake wetland: vegetation response. Ecol Eng 35(9):1312–1332

Kadlec RH, Knight RL (1996) Treatment wetlands. Lewis Publishers, Boca Raton, FL

Kadlec RH (1987) Northern natural wetland water treatment systems. In: Reddy KR, Smith WH (eds) Aquatic plants for water treatment and resource recovery. Magnolia Publishing, Orlando, FL, pp 83–98

Kotanen PM (2002) Fates of added nitrogen in freshwater arctic wetlands grazed by snow geese: the role of mosses. Arct Antarct Alp Res 34(2):219–225

Krkosek WH, Ragush C, Boutilier L, Sinclair A, Krumhansl K, Gagnon GA, Lam B (2012) Treatment performance of wastewater stabilization ponds in Canada's Far North. Cold Regions Engineering:612–622

Krumhansl K, Krkosek W, Greenwood M, Ragush C, Schmidt J, Grant J, Barrell J, Lu L, Lam B, Gagnon G, Jamieson R (2014) Assessment of arctic community wastewater impacts on marine benthic invertebrates. Environ Sci Technol 49(2):760–766

Mudroch A, Capobianco JA (1979) Effects of treated effluent on a natural marsh. J Water Pollut Control Fed 51:2243–2256

Murray JL (1991) Biomass allocation and nutrient pool in major muskoxen-grazed communities in sverdrup pass, 79°N, Ellesmere island, N.W.T., Canada. M.Sc. dissertation, University of Toronto, Toronto, ON, Canada

Nadelhoffer KJ, Giblin AE, Shaver GR, Laundre JA (1991) Effects of temperature and substrate quality on element mineralization in six arctic soils. Ecology 72(1):242–253

Ngai JT, Jefferies RL (2004) Nutrient limitation of plant growth and forage quality in arctic coastal marshes. J Ecol 92(6):1001–1010

Pineau C (1999) Facteurs limitant la croissance des plantes graminoides et des mousses dans les polygones de tourbe utilisespar la grande oie des neiges. M.Sc. dissertation, Université de Laval, Quebec City, QC, Canada, 72 p

Porsild AE, Cody WI (1980) Vascular plants of continental Northwest Territories, National Museum of Natural Sciences, Ottawa, Canada

Press MC, Potter JA, Burke MJW, Callaghan TV, Lee JA (1998) Responses of a subarctic dwarf shrub heath community to simulated environmental change. J Ecol 86(2):315–327

Ragush CM, Schmidt JJ, Krkosek WH (2015) Performance of municipal waste stabilization ponds in the Canadian Arctic. Ecol Eng 83:413–421

Raillard MC (1992) Influence of muskox grazing on plant communities of sverdrup pass (79°N), Ellesmere island, N.W.T. Canada. Ph.D. dissertation, University of Toronto, Toronto, ON, Canada, 262 p

Reed PB (1988) National list of plant species that occur in wetlands. Biological report 88 (24). National Wetlands Inventory (U.S.), U.S. Fish and Wildlife Service, U.S. Department of the Interior, Fish and Wildlife Service. Washington, DC

Statistics Canada (2012) Nunavut (Code 62) and Canada (Code 01) (table). Census Profile. 2011 Census. Statistics Canada Catalogue no. 98–316-XWE. Ottawa. Released October 24, 2012. Available online at: http://www.12.statcan.gc.ca/census-recensement/2011/dp-pd/prof/index.cfm?Lang=E. Accessed 6 Aug 2016

Stein OR, Hook PB (2005) Temperature, plants and oxygen: how does season affect constructed wetland performance? J Environ Sci Health 40:1331–1342

Tchobanoglous G, Burton FL, Stensel HD (2003) In: Metcalf & Eddy, Inc. (ed) Wastewater engineering: treatment, disposal, and reuse, 4th edn. McGraw-Hill, New York, NY

Tilton DL, Kadlec RH (1979) The utilization of a fresh-water wetland for nutrient removal from secondarily treated waste water effluent. J Environ Qual 8:328–334

Tolvanen A, Alatalo JM, Henry GHR (2004) Resource allocation patterns in a forb and a sedge in two arctic environments—short-term response to herbivory. Nord J Bot 22(6):741–747

Wasley J, Robinson SA, Lovelock CE, Popp M (2006) Climate change manipulations show antarctic flora is more strongly affected by elevated nutrients than water. Glob Chang Biol 12(9):1800–1812

Wittgren HB, Maehlum T (1997) Wastewater treatment wetlands in cold climates. Water Sci Technol 35(5):45–53

Woo MK, Young KL (2003) Hydrogeomorphology of patchy wetlands in the high arctic, polar desert environment. Wetlands 23(2):291–309

Yates CN, Varickanickal J, Cousins S, Wootton BC (2016) Testing the ability to enhance nitrogen removal at cold temperatures with *C. aquatilis* in a horizontal flow wetland system. Ecol Eng 94:344–351

Yates CN (2012) Developing an understanding for wastewater treatment in remote communities of Nunavut, Canada: investigating the performance, planning, practice and function of tundra and constructed treatment wetlands. Ph.D. dissertation, University of Waterloo, Waterloo, ON, Canada

Yates CN, Wootton BC, Balch GC (2014b) Framing the need for application of ecological engineering in Arctic environments. In: Jørgensen SE, Chang N-B, Fuliu X (eds) Ecological modelling and engineering of lakes and wetlands. Elsevier, New York, NY

Yates CN, Balch GC, Wootton BC, Jørgensen SE (2014a) Practical aspects, logistical challenges, and regulatory considerations for modelling and managing treatment wetlands in the Arctic. In: Jørgensen SE, Chang N-B, Fuliu X (eds) Ecological modelling and engineering of lakes and wetlands. Elsevier, New York, NY

Yates CN, Wootton BC, Murphy SD (2012) Performance assessment of Arctic tundra municipal wastewater treatment wetlands through an Arctic summer. Ecol Eng 44:160–173

Yates CN, Wootton BC, Jorgensen SE, Murphy SD (2013) Wastewater treatment: wetlands use in Arctic regions. In: Encyclopedia of environmental management. Taylor and Francis, New York, pp 2662–2674

Chapter 4
The Long-Term Use of Treatment Wetlands for Total Phosphorus Removal: Can Performance Be Rejuvenated with Adaptive Management?

John R. White, Mark Sees, and Mike Jerauld

Introduction

Free water surface water flow constructed wetlands have been utilized as a convenient and relatively cost effective method for treatment of wastewater streams for over 100 years all around the globe (Sundaravadivel 2001). Treatment wetlands utilize a number of natural processes in order to process wastewater improving the resultant water quality. These processes include infiltration, sorption, settling, coupled reduction-oxidation processes (redox), microbial degradation, algal and macrophyte uptake, and photolysis among others (Reddy and DeLaune 2008). In many locations, wastewater treatment plants are linked to free water surface treatment wetlands, with the wetlands performing the "polishing" phase of wastewater treatment which includes nutrient removal, primarily nitrogen (N) and phosphorus (P) as well as increasing dissolved oxygen and stabilizing pH before discharge to the receiving water body. These treatment systems were designed to provide nutrient removal capacity based on the areal extent of the wetland and calculated rates of organic matter accretion, a process which eventually decreases the volume or accommodation space of the wetland over time (Wang et al. 2006).

Removal of phosphorus in constructed wetlands can be mediated through abiotic or biotic mechanisms (White et al. 2006). Low water velocities and the matrix of submerged plant stems allows settling and trapping of P associated with parti-

---

J.R. White (✉)
Department of Oceanography and Coastal Sciences, Louisiana State University, Baton Rouge, LA, USA
e-mail: jrwhite@lsu.edu

M. Sees
City of Orlando Wastewater Division, Orlando, FL, USA

M. Jerauld
DB Environmental, Loxahatchee, FL, USA

cles. Chemical reactions with Fe, Al, Ca and Mg can all lead to the removal of soluble reactive phosphorus (SRP) given supportive conditions for precipitation/ sorption reactions (Reddy et al. 1998; Richardson 1985). Additionally, plant uptake can be a significant removal mechanism of P in wetland systems (White et al. 2004). The P contained in the organic matter is deposited in the system and stored as peat which continues to accrete year by year. Unlike N, there are no significant volatilization reactions for P and hence all the P removed from the wastewater remains in the wetland soil, increasing in mass and concentration over the years (Reddy et al. 1998).

There are two major problems related to the continual accretion of P in the constructed wetland soil related to treatment effectiveness. The high organic matter accretion rates seen in treatment wetlands leads to the eventual decrease in the volume or holding capacity for the wastewater in the wetland. When the volume capacity of a wetland decreases, it can be expected that the surface water travel time increases (Wang et al. 2006). This decrease in residence time, coincident with the decrease in wetland volume, leads to increases in the velocity by which the wastewater moves through the system which is inversely related to treatment effectiveness. Additionally, organic matter accretion is not uniform throughout the wetland and preferential flow paths can develop increasing the velocity of the wastewater in the front end of the system (Wang et al. 2006). The second factor is related to the release of previously stored phosphorus from the soil, back to the water column (Reddy et al. 2011). The treatment performance declines because the soils are now oversaturated with P and they begin releasing significant P into the water column thereby increasing water column concentrations (Bostic et al. 2010). This release of previously stored phosphorus is sometimes referred as the "internal load" or in natural systems impacted by nutrient loading also referred to as "legacy phosphorus load" (Reddy et al. 2011). Treatment effectiveness of the vast majority of constructed wetland systems is defined and measured through the decrease in the influent concentration to the outflow concentrations. Hence, any internal process that increases concentration, as in the case of release of P from the soil, will lead to a decline in rated performance of the treatment system.

This chapter documents the sustained, multi-decadal success of the Orlando Easterly Wetlands in reducing P concentrations in tertiary-treated wastewater effluent before discharge to a sensitive riparian ecosystem, and outlines management actions that have helped maintain the viability of the system. Direct and ancillary benefits to the operator, City of Orlando, Florida, are reviewed, along with some unique challenges presented by this wetland.

Study Site

The Orlando Easterly Wetlands (OEW) is a 485 ha wetland treatment system designed and constructed in the mid-1980s to polish excess nutrients from municipal reclaimed wastewater effluent (Fig. 4.1). The OEW, located east of Orlando, FL

Fig. 4.1 The three flow paths for the Orlando Easterly Treatment Wetland

is one of the oldest and largest constructed wetlands for treatment of municipal wastewater in the United States (Lindstrom and White 2011). The wetland was originally designed to polish up to 37.1 ft^3/s of effluent from the Iron Bridge wastewater treatment plant.

The OEW was situated on an existing cattle ranch within the riparian zone of the St. Johns River, FL, USA. The area had been converted to pasture land shortly after the turn of the twentieth century through the construction of a series of ditches to improve drainage. The OEW was originally designed with 17 treatment cells oriented across the site containing three independent flow trains (northern, central and southern) all converging to a single discharge point which emptied into a canal which leading to the St. Johns River. To create the wetlands, approximately 2.1 million aquatic macrophytes were planted along with 160,000 trees, primarily cypress. In 2003, an alternate discharge point was constructed along with the installation of an internal berm, creating an 18th treatment cell to help improve hydraulic control and to facilitate isolating various parts of the flow path to carry out renovation activities.

The wastewater effluent arrives at the wetland by underground pipe to a single point where it can be split into three different longitudinal flow paths (Fig. 4.1). The resultant hydraulic detention time for the northern, central and southern flow pathways, determined in 2001, were 18.1, 37.6 and 54.6 days, respectively from a chemical tracer study (Wang et al. 2006). Greater detail on the plant communities and site can be found in Wang et al. (2006) and Malecki-Brown et al. (2010).

Table 4.1 Maximum allowable total P discharge concentration (mg/L) targets for the Orlando Easterly Treatment Wetlands

Regulatory agency	Averaging period	Threshold	Notes
FL Dept. of Environ. Protection	Daily	0.40 mg/L	Specified under a NPDES operating permit
FL Dept. of Environ. Protection	Weekly	0.30 mg/L	Specified under a NPDES operating permit
FL Dept. of Environ. Protection	Monthly	0.20 mg/L	Specified under a NPDES operating permit
St Johns River Water Management District	Semi-annually	0.07 mg/L	Discharge concentrations specified by a "Modified License Agreement"
St Johns River Water Management District	Semi-annually	0.09 mg/L	Incremental penalties assessed if semi-annual averages are above threshold

Performance Criteria and History

In evaluating the OEW's performance, it is perhaps helpful to understand the regulatory thresholds pertaining to P that the system must meet. Table 4.1 documents the time-dependent criteria which include daily thresholds of 0.4 mg/L as well as monthly averages of 0.2 g/L for total P which had been set by the Florida Department of Environment Protection. Additionally, the OEW falls within the purview of the Saint John River Water Management District, 1 of 5 water management districts within the state of Florida whose borders are based on watershed boundaries. This additional agency has negotiated a semi-annual discharge limit of 0.07 mg/L, an agreement reached in order to allow the OEW to increase discharge loads into the adjacent St. Johns River. In general, the target level of nutrient removal from wastestreams is not related to the best achievable or true capacity of systems, but is primarily driven through legislation or rule-making based upon the capacity of the receiving water body to assimilate the nutrient load without becoming impaired or losing ecological integrity.

Twenty-two years (1993–2015) of flow and chemistry data were compiled (Tables 4.2 and 4.3; Fig. 4.2). We have delineated the record into four non-contiguous periods of distinct operational conditions which are evident in the time series of inflow and outflow TP concentrations and areal mass loads (1993–1998, characterized by low to moderate inflow concentrations and consistent, low outflow concentrations; 1999–2008 plus 2014–2015, marked by relatively high inflow concentrations and loads and winter-time spikes in outflow TP concentrations; 2009–2013, which had a wide range of inflow concentrations but very low outflow concentrations. Summary values for selected operational and performance parameters for the whole data record and each of these periods are given in Table 4.2. Overall, the OEW has removed 74% of the TP loaded to the system, on a flow weighted basis. On an annual basis over the entire 22 year record, the wetland has reduced the concentration of TP from a mean inflow concentration of 0.27 mg/L down to a mean discharge

Table 4.2 Summary inflow/outflow characteristics for the Orlando Easterly Treatment Wetlands

Period	Flow (MDG) In	Out	HRT cm/day	FWMC (mg/L) In	Out	% Reduction	TP load (g/m²/year) In	Out	TP settling rate (m/year)
1993–2015	**14.8**	**17.2**	**1.2**	**0.23**	**0.06**	**75%**	**1.1**	**0.3**	**6.3**
1993–1998	14.1	12.8	1.2	0.16	0.05	68%	0.7	0.2	5.0
1999–2008	15.8	19.5	1.4	0.28	0.07	74%	1.4	0.5	6.6
2009–2013	14.6	17.8	1.2	0.19	0.03	83%	0.8	0.2	7.9
2014–2015	12.0	17.6	0.9	0.37	0.07	81%	1.3	0.4	5.5

Data are presented as means as well as broken down into various time periods. *MDG* millions of US gallons per day, *HRT* hydraulic retention time, *FWMC* flow weighted mean concentration, *TP* total phosphorus

concentration of 0.06 mg TP/L (Table 4.3). While this level of performance clearly meets or exceeds the annual discharge criterion for the treatment wetland, the additional discharge criteria based on much smaller time increments (daily or monthly) also need to be considered for this system.

Beginning as early as 1999, winter outflow TP concentrations from the OEW unexpectedly increased over historical levels, resulting in a series of winter time outflow TP spikes in concentration (Fig. 4.3). These excursions are the largest detriment to water quality performance, and occasionally threatened to exceed the shorter term outflow concentration targets (Table 4.1).

Wintertime Peaks in TP Concentration

To quantify the presence and magnitude of the winter peaks, we calculated a "spike-factor" (the ratio of the average wintertime [Jan–Mar] outflow TP concentration compared to the average 9-month [Apr–Dec] outflow TP concentration) for each calendar year from 1993 to 2015. The calculated spike-factors for each data year 2000–2008, and 2014–2015 exceeded 1.5 (range: 1.6–3.6), distinguishably greater than the range of 0.9–1.3 for the better performing years during 1993–1998 and for more recent years (i.e. 2009–2013). These wintertime spikes, distinguishable in Fig. 4.2 and isolated in Fig. 4.3, are of great concern to OEW managers since they are the greatest threat to compliance with regulatory outflow concentration targets. The monthly average outflow TP concentration approached or exceeded the discharge limit of 0.2 mg/L during winter months in years when the spike factor was greater than 1.5. The inflow/

Table 4.3 Annual performance for loading and removal of total phosphorus by the Orlando Easterly Treatment Wetland, including a lifetime average of 70.54% reduction in concentration

Year	Influent flow (mdg)	Influent conc. (mg/L)	Loading in (lbs/d)	Effluent flow (mdg)	Effluent conc. (mg/L)	Loading out (lbs/d)	Percent reduction from influent to effluent (%)
1987							
1988	10.00	0.57	47.54	10.68	0.10	8.91	81.3
1989	13.33	0.72	80.04	10.68	0.08	7.13	91.1
1990	13.28	0.41	45.41	10.68	0.09	8.02	82.3
1991	12.90	0.230	24.74	13.40	0.090	10.06	59.4
1992	12.77	0.240	25.56	11.60	0.060	5.80	77.3
1993	12.63	0.180	18.96	10.00	0.060	5.00	73.6
1994	12.42	0.200	20.72	12.52	0.050	5.22	74.8
1995	15.12	0.180	22.70	8.83	0.050	3.68	83.8
1996	15.63	0.120	15.64	16.34	0.050	6.81	56.4
1997	15.22	0.140	17.77	16.67	0.040	5.56	68.7
1998	14.22	0.136	16.13	13.93	0.054	6.27	61.1
1999	17.20	0.320	45.90	19.43	0.060	9.72	78.8
2000	17.45	0.300	43.66	13.69	0.060	6.85	84.3
2001	17.86	0.240	35.75	16.76	0.070	9.78	72.6
2002	16.59	0.235	32.51	22.51	0.075	14.08	56.7
2003	17.36	0.207	29.97	24.87	0.070	14.52	51.6
2004	17.20	0.240	34.48	21.48	0.060	10.75	68.8
2005	18.27	0.401	61.10	25.25	0.088	18.53	69.7
2006	12.68	0.333	35.22	17.63	0.129	18.97	46.1
2007	12.33	0.276	28.38	14.62	0.066	8.05	71.6
2008	12.17	0.210	21.31	15.06	0.062	7.79	63.5
2009	14.14	0.153	18.04	15.39	0.041	5.26	70.8
2010	15.29	0.134	17.09	18.96	0.041	6.48	62.1
2011	15.17	0.159	20.12	17.18	0.036	5.16	74.4
2012	14.33	0.194	23.19	18.11	0.036	5.44	76.5
2013	14.16	0.285	33.66	19.18	0.037	5.92	82.4
2014	14.32	0.464	55.41	18.55	0.079	12.22	77.9
2015	11.80	0.247	24.31	16.57	0.075	10.36	57.4
Average	**14.49**	**0.27**	**31.98**	**16.09**	**0.06**	**8.66**	**70.54**

mdg millions of US gallons per day, *lbs/d* pounds per day

outflow TP concentration and load time series suggest that the 2000–2008 winter outflow spikes occurred contemporaneously, with generally elevated inflow concentrations and loads during that period (Fig. 4.4a, b).

Early efforts identified significant performance issues in the front end of the wetland flow train (Martinez and Wise 2003; DB Environmental 2004). It was recognized that the treatment wetland had been in existence for almost 15 years at that point, and was perhaps in need of some adaptive management to help

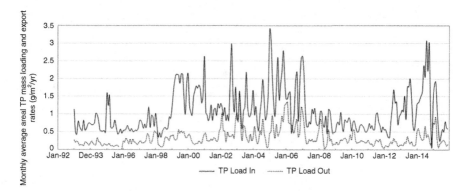

Fig. 4.2 Monthly average areal TP mass loading and export rates.

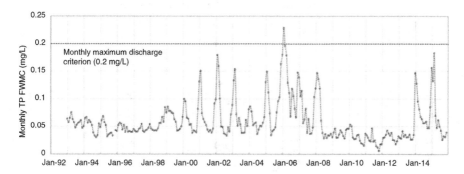

Fig. 4.3 Monthly average outflow TP concentrations. Note winter P spikes beginning around 1999 ending in 2009 and beginning again in 2014

restore the effective removal capacity. Since the OEW was one of the largest and longest-operating treatment wetlands receiving municipal wastewater, it was generally unknown what management activities would be both effectual in the long term, as well as cost effective. Therefore, the City of Orlando began a methodical research program to determine the cause of these seasonally-linked TP spikes in outflow concentration and to evaluate potential management options to maximize the removal of P from wastewaters during the winter months. The City of Orlando recognized this opportunity as one that would provide important and critical management information to the many other large constructed wetland systems worldwide as they begin to age and also experience eventual treatment decline. The following summarizes these research efforts with appropriate citations to the scientific literature for a more detailed account of each measure.

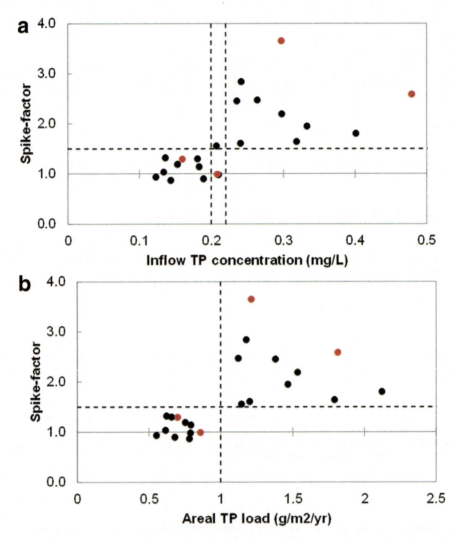

Fig. 4.4 The spike-factor (the ratio of the [Jan–Mar]:[Apr–Dec] average outflow TP concentrations) for each year 1994–2015 with respect to (**a**) the annual average inflow TP concentration and (**b**) the annual areal inflow TP load in the preceding year. Data years 2012–2015 indicated in *red*

Review of Adaptive Management Options for Maintaining P Removal Performance

Preliminary Laboratory Investigations

The use of chemical amendments in wastewater treatment is well established and is employed at the wastewater treatment plant which feeds the OEW. Therefore, it seemed a natural extension of testing and modifying a known treatment technology at the treatment wetland. A laboratory study was conducted on intact cores collected from the front cells, the most P enriched, of the OEW in order to test several chemical amendments, as well as the potential of organic soil removal or soil consolidation through drying on the resultant release of P. Experimental details of this study can be found in Lindstrom and White (2011). Briefly, triplicate intact cores were collected from the first cells to receive the wastewater and were subjected to following treatments; drydown (consolidation), organic soil removal, powdered calcium carbonate amendment, liquid aluminum sulfate (alum) amendment and a continuously flooded control. The 49-day drydown event, while consolidating the organic soil, had no effect on reducing P flux from the soil upon rewetting when compared with the control. The calcium carbonate treatment also showed no significant difference in P flux rate from the control. However, cores with the organic soil removed had flux rates that were 90% lower than the control (Lindstrom and White 2011). Clearly, removal of the P-saturated wetland soil decreased release back to the water column. In addition, the aluminum sulfate additions showed a negative flux rate, in that it not only intercepted any P released from the sediment but it also took the ambient wastewater P out of the water column. Results from this laboratory study suggested two potential management options; (1) treatment with aluminum sulfate during these high P winter spikes in water column concentrations or (2) remove the organic wetland soil which has accreted over the sandy, pasture soil during the previous years of operation. Each of these potential management options has environmental and economic considerations. For example, with soil removal (de-mucking) , there is the cost of removal and perhaps a more significant issue of disposal. Given the cost of this option, there was concern about how long any treatment benefit would last. It was also unclear whether it would it be necessary to de-muck the entire wetland or just a section. While the use of alum is a proven technology in wastewater treatment plants, there were concerns on how the low-pH amendment might affect microbial processes in the soil as well as potential impacts to macrophytes in the system.

Field-Scale Trial Implementation of Adaptive Management Options

Soil Removal or De-mucking

Muck removal is the process of scraping the organic soil which has accreted over time down to the native soil and this procedure had proven effective in renovating wetland soils and P removal capacity in other systems (e.g., Reddy et al. 2007). Those successes, combined with the local evidence provided by the lab study described above, recommended physical muck removal as a viable method to rejuvenate the OEW's ability to polish excess P from inflow wastewater. Therefore, the city initiated a large scale muck removal project in the cells near the wetland inflow in 2001. A bromide and lithium tracer study and an internal surface water quality campaign were conducted both prior and post muck removal to determine the effects of the soil removal on the characteristics of flow, total wetland volume and reduction of internal loading of P from the soil to the water column. Details of the tracer and water quality studies can be found in Wang et al. (2006). In short, the average volume of the cells increased in holding capacity from 230×10^3 m^3 in 2001 before the de-mucking to 347×10^3 m^3 in 2003 after the soil removal for an increased accommodation space or volume of 50%. The water total P concentrations determined from 31 stations distributed across the muck removal project area decreased from 0.463 mg P/L in 2001 to 0.048 mg P/L post project, a concentration decrease of ~90%. Therefore, two significant benefits were seen as a result of the soil removal, the increase in volume will increase the hydraulic retention time of the wastewater improving P removal and the removal of the internal load of P (the soil) prevents the re-release of P back into the water column which negatively affects treatment performance. Significant management drawbacks to soil removal are the issues of significant cost of removal and disposal of dredged material as well as the need to remove sections of the treatment wetland from operation for as long as 6 months.

Since 2001, the City of Orlando has periodically renovated various treatment cells within the OEW. Figure 4.5 documents the various de-mucking projects that have occurred to date. Muck removal efforts typically target specific areas of the wetland where organic matter accretion rates are highest (front end cells) as well as areas proximal to the outflow region, where release of P from the soil would exert the greatest influence on the outflow concentrations. The wetlands managers utilize data on water quality performance, muck depth observations and changes in vegetation composition to decide which cells are to be renovated.

The demucking procedure for the wetland cells occurs in several steps. First, all flow entering a cell or cells is rerouted and the standing water within the cell is drained. Rim ditches are installed using excavators to collect remaining surface water and large hydraulic pumps are used to de-water the cell (Fig. 4.6). Next, dewatered muck and organic debris (in the case of the OEW, 30–45 cm of material) is pushed into large linear rows typically 20–30 cm high, allowing the muck to drain and dry (Fig. 4.7). Once significant water has been removed from the muck, it is physically removed and landfilled on various upland portions of the OEW property. The treatment cell is then graded, leveled and rehydrated (Fig. 4.8). Various replanting schemes have been attempted throughout the various cells. Some of the cells have been planted with giant bulrush (*Schoenoplectus californicus*); another was allowed to regrow with cattails (*Typha spp.*). Another cell was replanted with submerged aquatic species such as Illinois pondweed (*Potamogeton illinoensis*) and southern naiad (*Najas quadalupensis*) while other cells that have been renovated were allowed to naturally recruit vegetation species. Eventually, flow is restored to the cell as wastewater is loaded to the system (Fig. 4.9).

4 The Long-Term Use of Treatment Wetlands for Total Phosphorus Removal

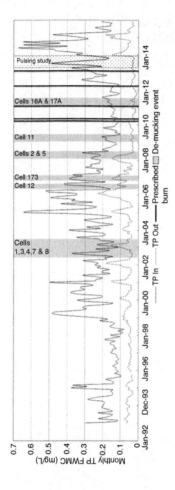

Fig. 4.5 Monthly average inflow and outflow TP concentrations with de-mucking and prescribed burn events superimposed on the water quality time series

Fig. 4.6 Excavators install ditches to help dewater wetland muck before windrowing and muck removal

Fig. 4.7 De-mucking scene of the northern flow train shows dewatering of rows of muck, prior to removal

Fig. 4.8 Crews plant vegetation in the now de-mucked cells

Fig. 4.9 The northern flow train back in operation after de-mucking operation

Aluminum Sulfate (Alum) Additions

As previously mentioned, most wastewater treatment plants employ alum as part of the treatment process and the city of Orlando is no exception. The previously described lab study suggested that the use of alum in the flow path during the winter season spike periods might be beneficial in reducing both the surface water concentrations as well as sequestering P released from the sediment and senescent vegetation. This alternative was investigated in a mesocosm scale experiment as a potential management tool for wetland systems which cannot either afford the cost of de-mucking or allow the cessation of wetland treatment for a considerable period while the soil removal occurs. There is an issue with pH in the use of alum as aluminum sulfate quickly dissociates in water to form Al hydroxides (Beecroft et al. 1995; Lind 2003). In doing so, Al^{3+} combines with the OH^-, thereby decreases the pH as there is an increase in H^+ ions produced by the separation of the water molecule into the two component parts. The alum becomes a flocculent material and settles to the bottom in low flow environments. This method of removing P has been widely adopted in lakes for a number of years with variable results on long term water quality (Welch and Cooke 1999; Berkowitz et al. 2005, 2006). Since lakes generally have a much larger volume of water to sediment ratio compared to wetlands, there is generally a more moderate and ephemeral pH effect. It was unknown at the time as to the effects of continual or repeated alum additions of surface water and soil pH in shallow, vegetated wetland systems.

The research on alum additions was delineated into three distinct components. These included the effects of alum additions on (1) water column characteristics (Malecki-Brown and White 2009); (2) biogeochemical processes of the wetland soil (Malecki-Brown et al. 2007) and (3) macrophyte and submerged aquatic vegetation (Malecki-Brown et al. 2010). Overall, the alum treatment had 40% greater removal of SRP across the various vegetative treatments (emergent and submerged), however up to three times more particulate P was discharged from the alum mesocosms primarily as alum floc material (Malecki-Brown and White 2009). This release could potentially be an issue for the use of alum in areas in close proximity to the discharge points of any treatment wetlands, if discharge criteria include total suspended solids, as in the case of this treatment wetland, and/or the discharge permit is based on total P. Biogeochemical measures for two emergent vegetation treatments (cattail and bulrush) showed significantly lower soil microbial biomass P as well as lower potentially mineralizable P (PMP) rates (Malecki-Brown et al. 2007). The former represents the size of the overall soil microbial pool while the PMP assay represents overall heterotrophic microbial activity. Finally, the most significant effect of alum on macrophyte growth and nutrient uptake were constrained to only the submerged aquatic vegetation, which had 50 times greater tissue Al compared to the control, while the emergent plants showed a more modest increase in Al tissue concentration ranging from 2 to 4 times greater than the control. Therefore, it is possible that continued use of alum can affect the soil pH and possibly lead to aluminum toxicity in SAV (Kochian 1995). Consequently, more research would need to be done to determine longer term effects of alum dosing on various species of macrophytes.

Fire as a Management Tool for Removing Organic Matter

The rationale for using fire as a management tool in this constructed wetland was to decrease the amount of macrophyte tissue which is deposited in the wetland each year, preventing or, at least, retarding the rate of organic peat accretion and restoring more even flow across the cells. It should be noted that burning the organic matter will release P and consequently, the cells may need to remain closed post-burn until P is re-assimilated by the new growth. The history of prescribed burns is documented in Fig. 4.5. A large rate of peat accretion is the primary factor in the loss of accommodation space or wetland volume for wastewater treatment as well as a source of P which could be released back into the water column leading to a decrease in nutrient removal performance. Duplicate cells, side by side, of similar size and vegetative makeup were selected for the study and were fed surface water from the same cell. Details of this study can be found in White et al. (2008). In review, the P removal performance (inflow and outflow sampling) of the replicate burn and control cells were monitored for over 1 month prior to the burn. Surface water loading was ceased to both cells 1 week prior to the fire to lower water levels and expose the detrital material. The single burn event removed almost 100% of the standing dead and/or surface detrital material while reducing the live plants to dead material in the process (Figs. 4.10 and 4.11). The burn triggered a release of SRP concentrations 3.7 times higher than the control cell for 23 days after which, there was no significant difference in SRP water column concentration between the burned and the control (no burn). The total P was approximately three times higher in the burn compared to the control cell, but returned to control levels within 17 days. Limited sampling, from days 50 to 90 days post burn, showed that SRP and TP values were always lower in water from the burn cells than from the control, however too few sampling points were taken to run definitive statistical tests ($n = 4$). The fire did slow potential peat accretion by removing significant amounts of detrital material while having a minimal and potentially a positive impact on water quality. Longer term studies are needed to determine the effectiveness of burns on nutrient removal since the regrowth of the plants, immediately after the burn, could potentially skew the water quality data in the direction of higher removal rates.

Continued Management Activities

The wetland has been continually undergoing renovation activities since the original de-mucking in 2001 of the front cells in the northern flow train (Fig. 4.7). The cells along the front-end of the central and southern flow paths have been de-mucked

Fig. 4.10 A prescribed fire is used as a management tool to remove excess standing dead detritus

Fig. 4.11 Cattail-dominated marsh treatment cell undergoing a prescribed fire at the Orlando Easterly Wetlands

between 2006 and 2009. De-mucking of several back-end cells started in 2006 and finished in 2011. The rationale for removing the wetland soil in the front-end cells was due to high accretion rates, creating short circuiting and consequently poor performance as previously described. The rationale for the much removal of the backend cells or cells closest to the outfall was due to little change in treatment as wastewater entered this area and the potential for increasing concentrations as the water moves to the outfall through internal loading of P from the soil to the water column. The de-mucking events are shown on the monthly TP performance record, the last one completed in 2011 (Fig. 4.5).

Challenges Ahead

Increased Loads and Variability from the Iron Bridge Wastewater Treatment Plant (WWTP)

The peak OEW inflow volumes are expected, on average, to increase in the future as the population served by the WWTP grows. Also, there is a highly seasonal demand for treated wastewater for reuse in the summer for irrigation of golf courses, etc., which could alternatively decrease the flow to the wetland during this time of the year. A recent modeling exercise commissioned by the City estimated that outflow concentrations under a hypothetical 40 MGD at an inflow concentration of 0.15 mg/L would be approximately 0.066 mg/L, which is similar to the 1993–2013 average outflow concentration (DB Environmental 2004). This result suggests that the wetland could effectively treat increased effluent volumes, winter-time P spikes notwithstanding (see following section). In addition, a pulsing test in 2012–2013 demonstrated that the wetland handled peak loading pulses quite well, but performance deteriorated during periods of stagnation. Wetland managers may have to ensure minimum maintenance water deliveries to the wetland during period of high reuse demand to prevent diminished performance.

Wintertime P Spikes

The wintertime P spikes, as mentioned previously, are one of the biggest potential challenges for maintaining permit TP outflow compliance of the OEW. While the P spikes decreased in the period after 2008, suggesting that adaptive management had improved the situation, the P spikes again appeared in 2014 and 2015. The inflow/outflow TP concentration and load time series (Fig. 4.2) suggest that the 2000–2008 and 2014–2015 winter outflow spikes occurred contemporaneously with generally elevated inflow concentrations and loads during those periods. Indeed, the presence/absence of wintertime spikes may have been more closely related to the annual inflow TP load of the preceding year

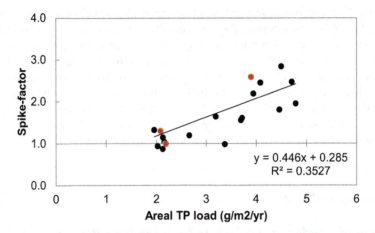

Fig. 4.12 The spike-factor for each year 1996–2015 with respect to the cumulative TP loading during the preceding 3 years. Data years 2012–2015 indicated in *red*

(Fig. 4.12). The data distribution suggests that the OEW may be a threshold-based system with elevated winter concentrations (spike-factor > 1.5) manifested after every year in which the TP load exceeded 1.0 g/m/year. Although a mechanistic explanation for the manifestation of the winter P spikes has not been determined, the apparently strict association with elevated TP loads may constrain the boundaries of future operation of the OEW.

Conclusions

In reviewing the long term water quality record of the OEW, it is evident that the period from 2009 to 2013 shows the wetland performing well, with the absence of previously observed period wintertime spikes. This level of performance is similar to earlier operating periods, which suggests that this 25-year-old wetland continues to remove P effectively. Intensive adaptive management of the OEW has aided in the rejuvenation and perhaps extended the capacity of the OEW to remove nutrients from the city's wastewater into the future. The most notable management activity is the de-mucking which has been shown to increase the wetland volume, increasing retention time and hence the useful lifespan of the wetland. The return of the wintertime P spikes in 2014–2015, concomitant with significant increased P loading, suggests there may be a upper limit at which this system can treat P in order to meet permit discharge criteria that must be met on a monthly, or smaller time-scale. However, on an annual basis, the wetland has been remarkably efficient at removing 70% of load TP over its lifespan.

Acknowledgements The authors would like to acknowledge the Water Reclamation Division of the City of Orlando for monetary and logistical support over the many years of collaboration, in working to identify and evaluate management options for water quality improvement. Efforts of the following graduate student researchers are also acknowledged: Lisa (Gardner) Chambers, Susan Lindstrom, Lynette Malecki-Brown and Carrie Miner.

References

Beecroft JR, Koether MC, van Loon GW (1995) The chemical nature of precipitates formed in solutions of partially neutralized aluminum sulphate. Water Res 29:1461–1464

Berkowitz J, Anderson MA, Amrhein C (2006) Influence of aging on phosphorus sorption to alum floc in lake water. Water Res 40:911–916

Berkowitz J, Anderson MA, Graham RC (2005) Laboratory investigation of aluminum solubility and solid-phase properties following alum treatment of lake waters. Water Res 39:3918–3928

Bostic EM, White JR, Corstanje R, Reddy KR (2010) Redistribution of wetland soil phosphorus ten years after the conclusion of nutrient loading. Soil Sci Soc Am J 74:1808–1815

DB Environmental (2004) Preliminary design and pilot development of sediment management protocols to enhance the long-term performance of the city of Orlando's Easterly Wetlands Treatment System. Final Report. Prepared for Post, Buckley, Smith and Jernigan, Orlando, FL

Kochian LV (1995) Cellular mechanisms of aluminum toxicity and resistance in plants. Annu Rev Plant Physiol 46:237–260

Lind CB (2003) Alum chemistry, storage, and handling in lake treatment applications. Paper presented at the 12th Proceedings of North American Lake Management Society Southeastern Lakes Management Conference, Orlando, FL, 2–5 June 2003

Lindstrom SM, White JR (2011) Reducing phosphorus flux from organic soils in surface flow treatment wetlands. Chemosphere 85:625–629

Malecki-Brown LM, White JR, Brix H (2010) Alum application to improve water quality in a municipal wastewater treatment wetland: effects on macrophyte growth and nutrient uptake. Chemosphere 79:186–192

Malecki-Brown LM, White JR (2009) Effect of aluminum-containing amendments on phosphorus sequestration of wastewater treatment wetland soil. Soil Sci Soc Am J 73:852–861

Malecki-Brown LM, White JR, Reddy KR (2007) Soil biogeochemical characteristics influenced by alum application in a municipal wastewater treatment wetland. J Environ Qual 36:1904–1913

Martinez CJ, Wise WR (2003) Hydraulic analysis of the Orlando Easterly Wetland. J Environ Eng 129(6):553–560

Reddy KR, Wang Y, DeBusk WF, Fisher MM, Newman S (1998) Forms of soil phosphorus in selected hydrologic units of the Florida Everglades. Soil Sci Soc Am J 62:1134–1147

Reddy KR, DeLaune RD (2008) Biogeochemistry of wetlands: science and applications, 1st edn. CRC Press, Boca Raton, FL

Reddy KR, Newman S, Osborne TZ, White JRC, Fitz C (2011) Phosphorus cycling in the Greater Everglades ecosystem: legacy phosphorus implications for management and restoration. Crit Rev Environ Sci Technol 41:149–186

Reddy KR, Fisher MM, Wang Y, White JR, James RT (2007) Potential effects of sediment dredging on internal phosphorus loading in a shallow, subtropical lake. Lake and Reservoir Management. 23:27–38.

Richardson CJ (1985) Mechanisms controlling phosphorus retention capacity in freshwater wetlands. Science 228:1424–1427

Sundaravadivel M (2001) Constructed wetlands for wastewater treatment. Crit Rev Environ Sci Technol 31:351–409

Wang H, Jawitz JW, White JR, Martinez CJ, Sees MD (2006) Rejuvenating the largest municipal treatment wetland in Florida. Ecol Eng 26:132–146

Welch EB, Cooke GD (1999) Effectiveness and longevity of phosphorus inactivation with alum. J Lake Reserv Manage 15:5–27

White JR, Reddy KR, Moustafa MZ (2004) Influence of hydrologic regime and vegetation on phosphorus retention in Everglades stormwater treatment area wetlands. Hydrol Process 18:343–355

White JR, Reddy KR, Newman JM (2006) Hydrologic and vegetation effects on water column phosphorus in wetland mesocosms. Soil Sci Soc Am J 70:1242–1251

White JR, Gardner LM, Sees M, Corstanje R (2008) The short-term effects of prescribed burning on biomass removal and the release of nitrogen and phosphorus in a treatment wetland. J Environ Qual 37:2386–2391

Chapter 5
An Investment Strategy for Reducing Disaster Risks and Coastal Pollution Using Nature Based Solutions

Ravishankar Thupalli and Tariq A. Deen

Introduction

Disasters are results of exposure to natural hazards, and the severity of a disaster depends on the impacts of the hazard on society and the environment. The intensity of the impact depends on the choices we make for our lives and for our environment. These choices are on how we produce our food, where and how we build our homes, type of governance we have, the way our financial systems function and also on what we teach in schools. Human populations have tended to concentrate along coastal areas because of the numerous benefits these areas provide to human activities and interests. Unfortunately, these areas, many of which are located in economically less developed countries, are also extremely vulnerable to natural hazards like tsunamis, tropical cyclones, and other environmental disturbances. As such, there is a need to take pre-emptive measures to protect coastal communities from environmental disturbance. Disaster Risk Reduction (DDR) is a form of emergency management used around the world to protect communities against natural hazards, which has been defined by The United Nations Office for Disaster Risk Reduction (UNISDR) as a *"concept and practice of reducing disaster risks through systematic efforts to analyse and reduce the causal factors of disasters including through reduced exposure to hazards, lessened vulnerability of people and property, wise management of land and the environment, and improved*

R. Thupalli (✉)
International Mangrove Management Specialist, 206-Madhura Apartments, Kakinada, India

Center for South and South East Asian Studies, School of Political and International Studies, University of Madras, Chennai, India
e-mail: rthupalli@hotmail.com

T.A. Deen
United Nations University—Institute for Water, Environment and Health, Hamilton, ON, Canada

© Springer International Publishing AG 2018
N. Nagabhatla, C.D. Metcalfe (eds.), *Multifunctional Wetlands*, Environmental Contamination Remediation and Management, DOI 10.1007/978-3-319-67416-2_5

preparedness for adverse events" (UNISDR 2009). Coastal communities have many options for disaster risk reduction, including hard engineering structures like dykes and seawalls, soft engineering solutions like beach re-nourishment, and policy-based approaches like population resettlement. Regardless of which option is decided upon, there needs to be a Global Disaster Risk Reduction strategy that communities can follow to ensure they remain resilient to natural hazards, especially as the effects of climate change will result in stronger and more frequent environmental and weather disturbances.

Nature-based Solution (NbS) take advantage of the various mechanisms that nature provides to protect communities from the destructive forces of natural hazards and other environmental changes. The Intergovernmental Platform on Biodiversity and Ecosystem Services (IPBES 2013), which is a leading global authority on assessing the state of biodiversity and ecosystem services, as well as providing scientific information to policy makers, includes the benefits of nature within their conceptual framework (Fig. 5.1). In this chapter, we present evidence for the adoption by communities of the NbS approach for coastal protection, with a particular emphasis placed on mangroves. Mangrove ecosystems

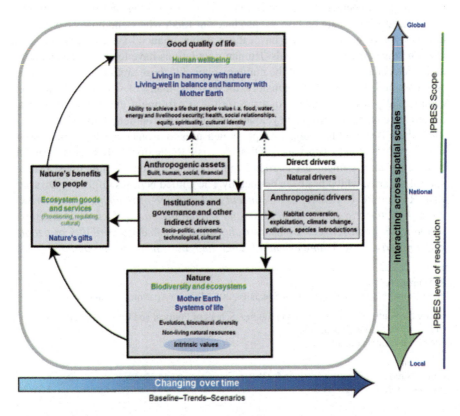

Fig. 5.1 IPBES conceptual framework. Source: IPBES (2013)

serve as a bio-shield for communities and villages against cyclones, storms, tsunamis and other catastrophic events (Spalding et al. 2014), and are cheaper to implement than hard engineering options like dolos or artificial reefs, and less intrusive than forced population resettlement.

Nature Based Solutions

NbS as a New Approach

The International Union for Conservation of Nature (IUCN) defines NbS as *"actions to protect, sustainably manage and restore natural or modified ecosystem that address societal challenges effectively and adaptively, simultaneously providing human well-being and biodiversity benefits"* (Cohen-Shacham et al. 2016). The term "Nature-based Solutions" was first proposed in the Millennium Ecosystem Assessment report entitled, *Ecosystem and Human Well-Being: Synthesis* (MEA 2005), and this concept has since been adopted by many international institutions, including the World Bank, the IUCN, and the International Council for Local Environmental Initiatives (ICLEI), as well as regional institutions like the European Commission, the European Environmental Agency, and the Asian Development Bank (ADB). Instead of being viewed as a single protective solution, NbS should be thought of as a collection of different concepts related to DDR and to the protection of the environment and human populations (Table 5.1).

"Bio-shields" related to coastal vegetation are a form of ecosystem-based disaster risk reduction or ecosystem-based adaptation (EbA) approaches that protect ecosystems and livelihoods from tsunamis, storms and cyclones by absorbing strong wave action and reducing wind speed. Bio-shields are viewed as attractive options, compared to hard engineering structures because they are lower in cost, and they are less intrusive on the tourism industry because they do not affect the aesthetic appeal of coastal landscapes. Additionally, bio-shields have ecological benefits such as protection from soil erosion by reducing the wind velocity, reduction of water loss due to evaporation of the soil moisture, providing habitats for aquatic life and reducing the impacts of coastal pollution. They can also provide socio-economic benefits for local communities, since, depending on the species of vegetation used for the bio-shield, these ecosystems can provide fruits, fodder for livestock and wood resources for fire or construction. Also, the inclusion of the local community in the development and implementation phase of the bio-shield ensures a sense of ownership and post-plantation care and management. Bio-shields, depending on their size and species, might also contribute to national carbon sequestration goals (Donato et al. 2011), as outlined in a country's Intended Nationally Determined Contribution (INDC) or Nationally Determined Contribution (NDC).

Table 5.1 Different concepts related to NbS and their examples

NbS approach	Sub-category	Examples
Ecosystem restoration approaches	• Ecological restoration • Ecological engineering • Forest landscape restoration	This approach focuses on the recovery of degraded, damaged or destroyed ecosystem (c.f. Mitsch 2012; Lewis 2005)
Issue-specific ecosystem-related approaches	• Ecosystem-based adaptation (EbA) • Ecosystem-based mitigation • Climate adaptation services • Ecosystem-based disaster risk reduction	Ecosystem-related approaches vary based on their objective, but they can range from wetland and floodplains for flood control to forest restoration and peat conservation for carbon sequestration (c.f. Baig et al. 2016; Doswald and Osti 2011; Olivier et al. 2012)
Infrastructure-related approaches	• Natural infrastructure • Green infrastructure	Commonly found in cities, examples include roof gardens and urban parks, among others (c.f. Ozment et al. 2015; Roth 2013)
Ecosystem-based management approaches	• Integrated coastal zone management • Integration water resource management	Both ecosystem-based management and ecosystem protection incorporates aspects of conservation, and environmental and ecosystem management (c.f. Arkema et al. 2006; Leslie and McLeod 2007)
Ecosystem protection approaches	• Area-based conservation approaches including area management	

Source: Cohen-Shacham et al. (2016)

Bio-shields Planting Procedure

Although coastal vegetation has been used as bio-shields for some time now, there have been certain technical lapses in planning and managing these bio-shields. For example, a mistake that is commonly made is to plant bio-shields at the high tide line. This has serious implication for the ecology of the coastal areas, and can even affect wildlife since sandy beaches are used as nesting grounds for sea turtles and vegetation can therefore prevent them from nesting. With this in mind, listed below are some factors that should be considered when planning bio-shield projects, as described by the Massachusetts Office of Coastal Zone Management in their fact sheet entitled, *Planting Vegetation to Reduce Erosion and Storm Damage* (MOCZM 2014).

1. *Plant selection*
   Plant selection is dependent on site-specific conditions like wind, soil type and quality, soil moisture, changes in topography (i.e., shifting sand), frequency of storms, and exposure to waves. It is recommended that native, salt-tolerant species be planted since they are more adapted to the conditions of the region and will therefore require less maintenance to grow and thrive. If the objective of

the bio-shield is erosion control, then vegetation with deep root system should be planted. Another factor to consider when selecting plant species is the time of planting. Table 5.2 lists several vegetation species that can be used as bio-shields in tropical coastal ecosystems, along with a description of other uses that can be made of the plants.

2. *Protecting plants*

   Vegetation is most vulnerable when a root system has not been established, so protective measures need to be taken to ensure that the root systems are able to develop. One way this can be achieved is through the use of natural fiber blankets, whereby blankets are placed on the ground to keep the soil in place and prevent the young roots from becoming exposed. Alternatively, fences can be made from natural fiber blankets to protect the vegetation from wind. These techniques can be used together to result in more effective results.

3. *Invasive plants*

   Invasive species, and especially those plant species that thrive at the expense of native species, should never be planted in coastal areas. If there are invasive species along coastal areas, they should be removed and replaced with appropriate native plants. Replacing invasive species may take years and require close monitoring and management. It should be noted that removing and replacing invasive species might result in temporary destabilization of sand banks. In some cases, the entire root system of the invasive species will need to be removed, but if this approach is not needed, the invasive plants can be cut down to ground level so that the remaining root system will retain soil stability. In some cases, invasive species can be removed with the use of small amounts of herbicide, which removes the stems but leaves the roots still in the soil. Targeted used of herbicide reduces overspray which might negatively affect native species, soil or groundwater quality. Invasive species can be removed by hand to minimize the use of heavy machinery that might destroy native species. Proper scheduling is needed when removing invasive species and planting native species, because the soil will be exposed after removal (regardless of which method is used), making it vulnerable to wind and water erosion.

4. *Slope stability*

   For bio-shield projects along steep banks, the stability of the slope is an important factor. If the slope of a bank is steep on the upper portion relative to the lower portion, then the bank is likely unstable and vulnerable to slumping or collapse. This is possible even if the bank is planted with erosion-control vegetation. A solution is to fill in the lower part of the bank with similar bank or beach soil so that the upper portion of the bank is less steep. However, if the new slope extends to the high tide line, then the fill will be eroded away. If this is the case, then sediment can be taken from the upper portion of the slope to match the grade of the lower portion of the slope. If a slope is not properly stabilized, even with a bio-shield it is possible that the slope will be eroded and lost during storms or other extreme events.

   Once the site has been prepared and the species of vegetation for the bio-shield have been selected, steps should be taken to ensure the vegetation grows

Table 5.2 Plant species recommended for bio-shields and their uses

Plant species (English), family	Description	Uses	Planting method
Anacardium occidentale L. (Cashew nut tree) ANACARDIACEAE	Trees, evergreen up to 6 m tall. The canopy has a spread area of about 5–10 m. Leaves simple, alternate, obovate-elliptic, glabrous, base attenuate, apex obtuse. Flowers yellow with pink streaks, fragrant in terminal panicles. Nuts reniform, seated on fleshy pedicel *Flowering and fruiting*: February-May	Fleshy pedicel and roasted kernel edible	Pits of 0.3 m³ should be dug and the nursery-raised saplings should be planted during the commencement of the rainy season. Either 8 m × 8 m or 10 m × 10 m spacing should be followed for this species. Causality replacement and proper care should be taken of the plantation. The trees will start bearing fruits from fifth year onwards
Azadirachta indica. Juss. (Neem tree) MELIACEAE	Trees, up to 12 m tall. Leaves simple pinnate; leaflets lanceolate, serrate, base oblique, apex acuminate. Flowers white, fragrant in axillary panicles. Drupes 1-seeded, yellow *Flowering and Fruiting*: February; July	Tender leaves and inflorescence along with jaggery (*Saccharum officinarum*) consumed as a vegetable. Tender twigs used as toothbrush. Leaves fumigated as a mosquito repellant. Leaf bits put into granaries as an insect repellent. Wood used for house building. Leaf twigs and branches used in religious rituals and ceremonies	The nursery raised saplings as well as seeds could be used for raising plantation. The pits measuring 0.3 m³ should be dug at an interval of 5 m × 5 m. The saplings along with the soil should be planted. Watering during the initial stages of planting and in summer for the first 2 years is a must and helps in successful establishment of the plantation

(continued)

Table 5.2 (continued)

Plant species (English), family	Description	Uses	Planting method
Bambusa arundinacea (Retz.) Roxb. (Spiny or Thorny bamboo) POACEAE	A long thorny bamboo, up to 40 m tall; green or purplish green when young, turning to golden yellow when it matures *Flowering and Fruiting*: Once in 30 or 45 or 60 years	Poles used in house construction, basket and mat weaving; highly useful in cottage industries and handicrafts. Poles used by fishermen in fishing	Pits of 60 cm × 30 cm × 30 cm dimension should be dug at an interval of 8 m × 8 m or 10 m × 7 m. These saplings should be transplanted in the pits during the rainy season. Adding ammonium sulphate or calcium ammonium nitrate (200 g) and super phosphate (200 g) would enhance the growth of the saplings. Intercropping could be done with a row of Ipil Ipil/subabul or Eucalyptus in the middle
Bixa orellina L. (Lipstick tree, Saffron) BIXACEAE	Shrubs or small trees. Leaves simple. Flowers white or purplish in color. Fruits in capsules, reddish brown in color. *Flowering and Fruiting*: September–November; December–February	Seeds as a source of natural dye. Used in dye industries	The nursery-raised saplings should be planted at a spacing of 4 m × 4 m. Pits of 0.30 m³ should be dug and the sapling along with the mud should be planted. Watering should be provided during summer months during the first year
Borassus flabellifer L. (Borassus Palm tree) ARECACEAE	Trees, dioecious, up to 20 m tall; trunk greyish-black. Leaves palmatifid, base sheathing. Peduncles sheathed with spathes. Fruits subglobose, black when ripe *Flowering and Fruiting*: February; May	Toddy tapped from the inflorescence. Boiled primary root, tender kernel and fruit pulp edible. Trunks from 50 to 60 years old trees used for house building. Leaves used for thatching, making baskets, mats and umbrellas. Fiber from petiole used for making ropes	Direct sowing of seeds in the early monsoon season could help in establishment of plantation. It requires very little attention. It can be cultivated on every type of wasteland

(continued)

Table 5.2 (continued)

Plant species (English), family	Description	Uses	Planting method
Cassia fistula L. (Indian Laburnum) CAESALPINIACEAE	Trees, up to 5 m tall; bark rough, dark brown. Leaves pinnate; leaflets opposite, ovate or ovate-oblong, base cuneate, apex acute. Flowers yellow, in axillary lax racemes. Fruits indehiscent, terete, brownish-black *Flowering and Fruiting*: April–September	Inflorescence used as vegetable. And also kept along with unripened mangoes for quick ripening. Bark used for extraction of dye. Wood used for making agricultural implements	Planting is done by either direct sowing or through nursery-raised saplings or stump planting. Root suckers could also be used for regeneration. Saplings should be in the nursery for at least 1–2 years and the sapling height should be 20–30 cm while planting. Pits of 0.30 m^3 should be dug and the sapling along with the mud should be planted with a spacing of 6 × 6 m. The seedlings are sensitive to weeds and hence weeding is very important. Roots suckers could also be used for regeneration
Casuarina equisetifolia Forst. (Horse tail tree) CASUARINACEAE	Tall trees up to 40–60 ft, straight trunk, rough and furrowed in older tree. Leaves in finely branched whorls of 6–8. Fruits grey or yellowish-brown, woody and cone-like *Flowering* (twice a year): February–April and September–October *Fruiting*: June and July–December	Poles used in scaffolding, fuel and construction material. Fishermen use them as fishing poles. A good bio-shield plant	Small pits of 0.3 m^3 should be dug and the sapling should be planted at 1 × 1 m or 2 × 2 m interval. Intercropping with groundnut or pulses is normally practiced. Irrigation is required in the first year
Clerodendrum serratum (L.) (Moon Beetle Killer) VERBENACEAE	Shrubs, up to 2 m tall; stems 4-angled. Leaves oblong-elliptic, coarsely serrate, apex acute. Flowers bluish-purple, in long pyramidal panicles. Drupes broadly obovoid, black *Flowering and Fruiting*: May–September	Roots as well as the leaf twigs boiled in water and the water used for bathing for rheumatic pains	Small pits of 0.2 m^3 should be dug and the sapling should be planted at 1 × 1 m interval

(continued)

Table 5.2 (continued)

Plant species (English), family	Description	Uses	Planting method
Cocos nucifera L. (Coconut palm) PALMAE	Tall trees, up to 40–80 ft. Leaves up to 15 ft. Long. Fruit green or yellowish. *Flowering and Fruiting*: Throughout the year	Trunks used in house construction. Leaves used for thatching. Toddy obtained. Fruit edible and is a source of cooking oil. Coir used in micro-enterprises	One cubic meter pits with an interval of 7–9 m should be dug and the dugout soil should be mixed with organic manure. The sapling is planted and mulched. Manuring and watering are important for sustainable yield
Hibiscus tiliaceus L. (Coast cotton tree) MALVACEAE	Trees up to 4 m tall; stems much branched, glabrous, close to ground level. Leaves, orbicular crenulate, stellate beneath, acute or acuminate at apex, cordate at base; stipules 2–3 cm long, subulate. Flowers 7–10 cm across, campanulate, bright yellow with crimson eye in the center, turning bright purple when old, solitary or rarely two, on terminal peduncles; bracteoles 5–6, lanceolate. Capsules 3–5 cm across, ovoid, closely tomentose, splitting into 5 mericarps. Seeds black with pale dots *Flowering and Fruiting*: June–July	Bark uses as fiber. A good ornamental plant used in shelterbelts	Seeds and cuttings could be used for raising planting material

(continued)

Table 5.2 (continued)

Plant species (English), family	Description	Uses	Planting method
Pongamia pinnata L. (Beach tree) FABACEAE	Trees, up to 5 m tall; bark soft, greyish-green. Leaves imparipinnate; leaflets opposite, ovate-oblong, entire, base rounded or acute, apex acuminate. Flowers white or pale rose, in axillary racemes. Pods obliquely oblong, compressed, 1-seeded *Flowering and Fruiting*: March–August	Seed oil warmed and applied for skin diseases. Seed oil widely used for bio-fuel	The trees are grown in variety of soils ranging from sandy to black cotton soil. But they establish very well in properly-drained alluvium soils. The seeds could be directly sown or nursery raised saplings could be used to raise the plantation. One-year-old saplings should be planted in 0.3 m^3 pits with an interval of 5 m × 5 m
Salvadora persica L. (Tooth Brush Tree) SALVADORACEAE	Much branched, evergreen shrub or small tree. Leaf: elliptic ovate and slightly succulent. Flower: greenish white or greenish yellow. Fruits: red when ripe	Grows in wide range of soils; stem used as tooth brush, leaves used for asthma and cough. Fruits sweet and edible	The saplings should be planted at an interval of 5 m × 5 m. The pits should be dug for 0.3 m^3
Sapindus emarginatus Vahl (Soap nut) SAPINDACEAE	Trees, up to 10 m tall. Leaves paripinnate; leaflets coriaceous, elliptic obovate or oblong, entire, apex emarginate. Flowers brownish-yellow, in terminal panicles. Drupes ovoid, 3-lobed *Flowering and Fruiting*: September–March	Fruit juice mixed with water, used as hair wash; fruits sold in market	Pits measuring about 0.30 m^3 should be dug at an interval of 6 m × 6 m. The saplings should be planted during the onset of the monsoon and if required watering should be done in summer for the first year

(continued)

Table 5.2 (continued)

Plant species (English), family	Description	Uses	Planting method
Thespesia populneoides (Roxb.) Kostel. (Indian Tulip tree) MALVACEAE	Trees, 3–6 m tall, young twigs covered with bronze-colored lepidotes. Leaves deltoid to cordate or subcordate; stipules early caducous. Flowers yellow, red in center, axially, solitary, recurved in fruits. Capsules 3–4 cm across, globes, exude deep yellow latex when young, mature fruits dehiscing apically into two distinct layers *Flowering and Fruiting*: June–July	Fruits and flowers yield yellow dye, which are useful for coloring the cloths	Nursery raised saplings grown for 6–8 months and stem cuttings could be planted. The pits should be dug with the dimension of about 0.30 m^3. Saplings should be planted at an interval of 5 m × 5 m
Vitex negundo L. (Chinese chaste tree) VERBENACEAE	Shrubs, up to 3 m tall; bark thin, grey. Leaves 3–5 foliolate; leaflets elliptic-lanceolate or lanceolate, entire, glabrous above, white tomentose beneath, base acute, apex acuminate. Flowers blush-purple, in terminal panicles. Drupes subglobose, black when ripe *Flowering and Fruiting*: Throughout the year. Common; along hedges and waste places	Leaf twigs put in hot water and taken bath for rheumatic pains	The stem cuttings and the root suckers could be used as planting material for raising the plantations

and becomes an effective barrier. The seedlings should be transported from the nursery to the planting site in light wooden or wire trays. The bag of soil that holds seedlings should not be broken in transit. The usual spacing for all plants in bio-shields is 2.5 m × 2.5 m, but this can change, based on the vegetation species. Pits for the seedlings should be wide and deep enough to fit the bag of soil that holds the seedlings. Pitting should be dug at the same time of planting and not before, since the sand will collapse over time. Additionally, the collar of the seedling should be in the same position during planting as it was while in the nursery, the centre of the pit should be slightly elevated, and planting should be done in the morning. Organic manure can be applied while filling the pit along with the soil. However, while applying inorganic fertilizers, care should be taken to avoid direct contact of the salts with the plant, especially the root portion and watering should be done immediately after adding inorganic fertilizer. Once the seedling has been planted, water should be added immediately if there is no rain, and should continue for 4 days in the event of no rainfall. The entire plantation should be monitored for dead or dying seedlings, which should be replaced with healthy seedlings.

Bio-shield Development

The location of bio-shields should be more or less perpendicular to the main wind direction, and the number of rows in a bio-shield is largely dependent on the velocity of the wind. The higher the velocity of the wind, the broader the bio-shield should be. Conventionally, a bio-shield should have 10–50 rows of vegetation, with first and last rows planted mainly consisting of shrubs and the rows in between planted with a combination of tall and medium-sized trees. A triangular method or "V" shaped method with 1-m spacing should be maintained between tree/shrubs. The spacing may be decided based on the habit of the species used and the geographical locality of the bio-shield.

Resilience

Conceptual Background

A key concept in NbS and DRR is *resilience*, which was initially related to the ability of an ecosystem following a disturbance to return to its original state, or absorb the disturbance and reorganize into a new state while still retaining its original structure and function (Holling 1973). Later, the concept of resilience was adopted by other disciplines, including anthropology (Vayda and McCay 1975) and ecological economics (Perrings et al. 1992). Within the context of NbS and DDR, resilience is defined as the *"ability of a system to reduce, prevent, anticipate, absorb and adapt, or recover from the effects of a hazardous event in a timely and efficient manner, including through ensuring the preservation, restoration, or improvement of its essential basic structures and functions"* (CEB 2013). The *system* in this definition refers to a social-ecological system (SES), referring to the interactions that occur between human and biophysical systems (Gallopín 2006).

Assessing Resilience

There are a number of ways to assess the resilience of communities. Commonly used terms in DRR include *adaptation, adaptive capacity, probability, interval frequency, sensitivity* and many more. All of these indicators can be used to assess resilience, either individually or collectively. In fact, many of the concepts in resilience are interlinked and proportional to each other. With this in mind, the relationship between *vulnerability, exposure*, and *hazard* is used to determine the *risk* (Fig. 5.2). Hazard refers to any disturbance or stress that a community may experience. This could include storms, tsunami, sea level rise, coastal erosion, pollution, or others. Exposure refers to the extent to which people and property are physically

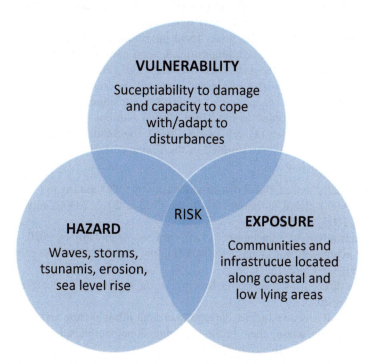

Fig. 5.2 The interaction of three factors (vulnerability, exposure, and hazard) that influence risk. Adapted from Spalding et al. (2014)

exposed to a hazard. Exposure is usually measured in relation to where a community is situated. For example, communities located closer to the shoreline are more exposed to tsunamis than communities further from the shoreline. Exposure changes, based on the hazard that is threatening the community. Vulnerability is a measure of how susceptible a community is to the effect of a hazard. Vulnerability is increased by a communities' ability to reduce or mitigate the effects of a disturbance. The intersection of these three indicators indicates the risk to which a community will be impacted by a hazard.

Both exposure and vulnerability can be tackled, for example if mitigation options like NbS are utilized along the shoreline. These approaches allow communities to reduce the impact of hazards, thereby reducing their vulnerability. In contrast, exposure can be reduced though resettlement or land-use planning. If communities take pre-emptive actions to reduce vulnerability and/or exposure, they can increase their resilience.

Sendai Framework

The Sendai Framework is the successor to the 2005–2015 Hyogo Framework for Action entitled, *Building the Resilience of Nations and Communities to Disasters*. It is a non-binding disaster risk reduction agreement that asserts that it is the responsibility of the State, local governments, the private sector and other stakeholders to protect communities against natural hazards. The outcome of the Framework is *"the substantial reduction of disaster risk and losses in lives, livelihoods and health and in the economic, physical, social, cultural and environmental assets of persons, businesses, communities and countries"* (UNISDR 2015). Adopted at the Third United Nations World Conference on Disaster Risk Reduction (March 14–18, 2015), the Sendai Framework covers a period from 2015 to 2030. Outlined in the Framework (UNISDR 2015) are its seven goals, which are:

- Substantially reduce global disaster mortality by 2030, aiming to lower average per 100,000 global mortality rate in the decade 2020–2030 compared to the period 2005–2015.
- Substantially reduce the number of affected people globally by 2030, aiming to lower average global figure per 100,000 in the decade 2020–2030 compared to the period 2005–2015.
- Reduce direct disaster economic loss in relation to global gross domestic product (GDP) by 2030.
- Substantially reduce disaster damage to critical infrastructure and disruption of basic services, among them health and educational facilities, including through developing their resilience by 2030.
- Substantially increase the number of countries with national and local disaster risk reduction strategies by 2020.

- Substantially enhance international cooperation to developing countries through adequate and sustainable support to complement their national actions for implementation of this Framework by 2030.
- Substantially increase the availability of and access to multi-hazard early warning systems and disaster risk information and assessments to the people by 2030.

Mangroves

Mangroves as Bio-shields

Mangroves are trees, shrubs, palms, or ground ferns that grow in saline or brackish waters along coastal marine environments and estuarine margins. The term "mangrove" can also refer to a forest of several mangrove species growing in these regions. These ecosystems are sometimes called "mangrove forests" or "tidal forests" to differentiate them from the vegetation growing inside them (Selvam et al. 2006). Mangroves grow in tropical and subtropical regions, approximately between 30°N and 30°S (Fig. 5.3), and are typically located in environmental conditions with low wave energy where seedlings will not be uprooted, and where there is low salinity, high temperature, no tidal extremes, high sedimentation and muddy anaerobic soils, and areas regularly flushed by tidal water (Giri et al. 2011; Selvam et al. 2006).

Because of their extensive root system and dense growth (Fig. 5.4), mangrove forests are an effective NbS to protect coastal communities from natural hazards and environmental changes. Table 5.3 lists the special properties required to develop effective DRR strategies using mangrove ecosystems. Mangroves have a long history of protecting coastal communities. For instance, the role of mangroves as a bio-shield in lessening the impact of tsunami waves was documented during the 2004 Indian Ocean Tsunami (EJF 2006). The study observed 18 hamlets along coastal India that were struck by the 2004 tsunami and found that of the three hamlets that had no loss of lives, two of them were located behind mangrove forests, as well as three of the four hamlets that had the lowest death toll (Kathiresan and

Fig. 5.3 Global distributions of mangroves, with areas of extensive growth shown as darker green areas. Figure taken from Giri et al. (2011)

Fig. 5.4 Mangrove vegetation from coastal zone in Colombia, showing the complex root system. Photo taken by C. D. Metcalfe

Rajendran 2005). The study also found that mangroves helped reduce the death toll in Indonesia, West Bengal (India), Bangladesh, and Thailand (Kathiresan and Rajendran 2005). Other studies reinforce the idea that mangroves act as effective coastal protection solutions. Das and Vincent (2009) concluded that "villages with wider mangroves between them and the coast experienced significantly fewer deaths than ones with narrower or no mangroves" following the 1999 super-cyclone in Orissa, India. Danielsen et al. (2005) examined satellite images of 12 villages on the southeast coast of India that were behind areas of dense tree vegetation, open tree vegetation, or no trees and found that villages behind dense or open mangrove forests were less damage than those that had no protection from trees.

As with many other NbS, mangroves are beneficial investments for local communities because of their low cost in comparison to hard engineering structures. Marshy area mangrove species suitable for growth in the prevailing edaphic conditions can be maintained through consultation with local communities. This allows for a sense of ownership by the local community, resulting in long-term sustainability, compared to structures like dykes and seawalls that require more maintenance and technical knowledge.

Mangroves for Pollution Control

Another aspect of mangroves that supports their use as a DRR strategy is their potential to treat wastewater and remove chemical contaminants. The capacity of mangroves to mitigate coastal zone pollution is a major benefit to maintaining these natural systems. In studies with an "old system" of *Avicennia marina* mangroves, and a "new system" of *Rhizophora* spp., *A. marina*, *Bruguiera cylindrica* and *Ceriops tagal,* Boonsong et al. (2003) showed that there was significant removal in the new system of mangroves of total suspended solids (TSS), PO_4-P, and total phosphorous (TP). Retention of these pollutants by mangroves is influenced by soil properties like texture, organic matter, pH, salinity and redox potential (Boonsong et al. 2003). These findings are consistent with other research conducted on the potential of mangroves for pollution control, including studies on removals of pollutants from domestic wastewater (Wu et al. 2008; Yang et al. 2008) and a study on improvements to water quality by mangroves in areas impacted by aquaculture (Peng et al. 2009). In a mesocosm study, Tam and Wong (1997) evaluated the potential for mangroves to retain metals using constructed tide tanks with young *Kandelia candel* plants irrigated with wastewater. Their results showed that most metals (e.g. zinc, cadmium, lead, nickel) present in the wastewater were retained in the soils, with little being taken up by the mangrove plants. On the other hand, Banerjee et al. (2018) describe in a chapter in this book, the significant degree of accumulation of metals (i.e. zinc, copper, lead) in the roots and vegetation of mangroves in a polluted coastal region of India. Also, because of the ability of mangroves to reduce soil erosion (Spalding et al. 2014), these natural systems reduce the negative effects of soil erosion on water quality (Sthiannopkao et al. 2006; Issaka and Ashraf 2017).

These theoretical arguments and empirical evidence about mangrove ecosystems role as coastal bio-shields and as a natural controller of pollution present a good case for evaluating the value of protecting and enhancing mangrove systems as a long-term strategy for DRR at local, regional and global scales. Table 5.3 lists the characteristics of mangroves that are essential for providing protection from waves, storm surge, tsunamis, erosion and sea level rise.

Mangrove Loss and Rehabilitation

In addition to their role in disaster risk reduction and mitigation of coastal pollution, mangroves provide a number of other co-benefits, such as providing food, timber, fuel and medicine, as well as being a habitat for coastal fish, birds and other fauna (Ravishankar and Ramasubramanian 2004; UNEP 2007). It is estimated that the economic value of 1 km^2 of mangrove forest can range between US$ 200,000 to US$ 900,000 (UNEP 2007). Unfortunately, mangrove forests have been in decline in recent years, due largely to conversion of mangrove land to agriculture, aquaculture, tourism, urban development and overexploitation (Tanaka et al. 2009). The Millennium Ecosystem Assessment report estimates that approximately 35% of mangrove forests were lost between 1980 and 2005 (MEA 2005). Additionally, of the 59 mangrove species listed on the IUCN's Red List of Threatened Species, 20 are listed as experiencing a population decline, with the fate of another 29 species being unknown, 11 as being stable, and 1 as increasing. It is believed that 100% of mangrove forests may be lost to extinction within 100 years if the current annual rate of 1–2% loss continues (Duke et al. 2007). Figure 5.5 below shows the global distribution of areas where mangrove species are threatened.

Table 5.3 Mangrove properties that increase coastal protection from natural, including waves, storm surge, tsunami, erosion and sea level rise

		Hazards				
		Waves	Storm surge	Tsunami	Erosion	Sea level rise
Mangrove forest properties	Width	Mangrove forests need to be hundreds of meters in width to significantly reduce waves. Wave height is reduced by 13–66% per 100 m	Mangrove forests need to be hundreds of meters in width to significantly reduce wind and storm surges Thousands of meters of mangroves are needed to reduce flooding impact (storm surge heights is reduce 5–50 cm/km)	Mangrove forests need to be hundreds of meters to reduce flood depth by 5–30%. However mangroves do not provide a secure defense against tsunamis	Mangrove forests need to be of significant width to maintain sediment stability and encourage soil build-up	
	Structure	A denser aerial root system and branches will help reduce wave strength	Open channels and lagoons allow free passage, while dense aerial root systems and canopies obstruct flow		Complex aerial roots systems will help slow water flow, allowing sediment to settle thereby reducing erosion	
	Tree size	Young and small mangroves can be effective	Smaller trees and shrubs may be overtopped by tsunamis and very large storm surges		Young trees can allow for soil to build up, however the more biomass in the coil the better	
	Link to other ecosystem	Sand dunes, barrier islands, saltmarshes, seagrasses and coral reefs can all play an additional role in reducing waves				Allow room for landward retreat of the mangroves
	Underpinning factors	Healthy mangroves are essential for all aspects of coastal protection. Healthy mangroves require: sufficient sediment and fresh water supply and connections with other ecosystems. Conversely, pollution, subsidence (due to deep groundwater/oil extraction or oxidation upon conversion) and unsustainable use negatively affects mangroves				

Source: Spalding et al. (2014)

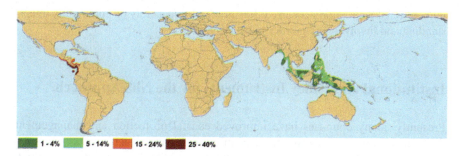

Fig. 5.5 Global map of threatened (critically endangered, endangered, and vulnerable) mangrove forests. Threatened mangrove forests are particularly concentrated in south-east Asia and Central America. Source: Polidoro et al. (2010)

Not only does the loss and degradation of mangrove forests and associated ecosystem services affect the livelihoods of coastal communities, but the absence of mangroves also places these communities at a greater risk of natural hazards and environmental change, especially in light of the predicted effect of climate change on the frequency and intensity of natural hazards. Also, because of the role of mangroves as a blue carbon sink, experts estimate that as much as an additional 1.02 billion tons of carbon dioxide per year will be emitted to the atmosphere because of the degradation of coastal ecosystems, with the loss of mangroves contributing to approximately half of these added emissions (Pendleton et al. 2012). To mitigate these losses, Lewis et al. (2006) identified five steps to restore mangroves:

1. Understand the autecology (individual species ecology) of the mangrove species at the site; in particular the patterns of reproduction, propagule distribution, and successful seedling establishment.
2. Understand the normal hydrologic patterns that control the distribution and successful establishment and growth of targeted mangrove species.
3. Assess modifications of the original mangrove environment that currently prevent natural secondary succession (recovery after damage).
4. Design the restoration program to restore appropriate hydrology and, if possible, utilize natural volunteer mangrove propagule recruitment for plant establishment.
5. Only utilize actual planting of propagules, collected seedlings, or cultivated seedlings after determining that natural recruitment will not provide the quantity of successfully established seedlings, rate of stabilization, or rate of growth of saplings established as objectives for the restoration project.

In addition to the bio-physical factors, socio-economic conditions also need to be taken into account when restoring degraded mangrove ecosystem. If people initially removed the mangroves, then there is a good chance that they will be removed again. Therefore, the root causes of this situation need to be explored and addressed. Alternative economic activities need to be presented if the mangrove ecosystems were over-exploited, and/or educational programmes need to be provided if unintentional degradation occurred. Restoration projects rely on community understanding and involvement in order to be successful.

Institutional and Policy Instruments for the NbS Approach

Presently, many countries have improved coastal protection and management through conservation or restoration of soft engineering structures. For example, programs in Vietnam have restored or protected 12,000 ha of mangroves, which has improved coastal protection as well as increased national carbon storage, enhanced biodiversity, and reduced dike maintenance costs from the previous $7 million annually (IUCN France 2016). However, certain countries still lack effective coastal protection and management. For instance, Ghana has been experiencing coastal erosion at a rate of 1.5–2 m/year of coastline due to a number of reasons, including inadequate coastal management (Anim et al. 2013). One way to address national and regional "disaster risk reduction deficits" is through a bottom-up approach to NbS. Local communities are the first to be affected by natural disasters and environmental changes, so community ownership is important for ensuring that strategies are long-lasting and effective. In order to achieve community ownership, we propose an institutional framework (Fig. 5.6) that can be used by planners, policy makers, and project managers. This framework focuses on rules and guidelines to get community and local stakeholders involved

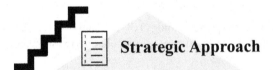

Strategic Approach

Step 1: Community Organization and Mobilization for Mainstreaming
Step 2: Formation of new or strengthening of existing Local Institution – EDC/VSS
Step 3: Situation Analysis for identifying issues related to DRR
Step 4: Training and Capacity Building in DRR
Step 5: Preparation of Microplan for Implementing DRR plan and programme
Step 6: Joint implementation of action plans
Step 7: Monitoring and Evaluation
Step 8: Process Documentation and Replication of Best Practices of mainstreaming

Fig. 5.6 Step-by-Step Strategic Approach for developing soft structures involving stakeholders and local community (adapted from MSSRF Project on Coastal Wetlands: Mangrove Conservation and Management 1996–2004, India)

Fig. 5.7 The Gazi Mangrove Project in Kenya; a community led mangrove project funded by UNDP-GEF. Photo by N. Nagabhatla

in the planning, implementation, and monitoring phases of soft engineering initiatives along coastal regions.

Step 1: Community Organization and Mobilization for Mainstreaming

Community involvement is a prerequisite for addressing and implementing DRR strategies, and can be achieved through proper community organization and mobilization exercises. Figure 5.7 illustrates a community-based project in Kenya to enhance mangrove ecosystems. Additionally, along with community involvement, the inclusion of other stakeholders is important, including government agencies, NGOs, academic and research institutions.

When natural hazards occur, they not only affect the coastal ecosystem (like mangroves) but also the livelihoods of communities. Therefore, with the involvement of local stakeholders, a community should be organized enough to face the aftermath of any natural hazard and be prepared to make collective decisions and take actions to effectively implement a DRR strategy to address biodiversity and livelihoods issues, thereby minimizing the impacts of any disaster.

Step 2: Formation of New or Strengthening of Existing Local Institutions—EDC/VSS/BMC

Prior to the formation of local level institutions, the community and other stakeholders should be educated through awareness programs on the concepts and benefits of investing in and creating soft engineering structures as a DRR strategy, as well as the community's roles and responsibilities in implementing the strategies. All over the world, there are community institutions implementing and benefitting from community forestry, such as the REDD+ programs (Reduced Emissions form Deforestation and forest Degradation in developing countries), as well as other socio-economic activities. These community and grass-roots level institutions, such as Eco Development Committees (EDC), Vana Samrakshana Samithis (VSS), Forest Conservation Committees (FCC) and Biodiversity Management Committees (BMC) need to be formed and strengthened in order to implement DRR action plans. It should be ensured that all the members of a community irrespective of socio-economic class and gender are represented in these institutions so that DRR is addressed transparently and democratically to avoid disruption of communal harmony. At least 33% representation by women is required as a strategy for mainstreaming gender.

Step 3: Situation Analysis for Identifying Issues Related to DRR

Situation analysis is done in order to realize the opportunities and challenges in a given community, which may be a hamlet or a village. Situation analysis is done using participatory techniques. The participatory techniques include a variety of approaches, tools and methods that are used in collaboration with local people to gather information about local conditions and situations. Few participatory techniques are suitable for gathering information, while others are designed to promote people's participation and involvement in implementing a DRR strategy. The two main participatory techniques are Rapid Rural Appraisal (RRA) and Participatory Rural Appraisal (PRA). Both RRA and PRA are designed to facilitate interaction between field workers and local people.

RRA emphasizes the importance of learning rapidly and directly from local people. RRA involves tapping local knowledge and gaining information and insight from local people using a range of interactive tools and methods (Beebe 2005). PRA involves field workers learning with local people with the aim of facilitating local capacity to analyze, plan, resolve conflicts, take action and monitor and evaluate according to a local agenda (Chambers 1992). Conventional approaches to collecting information generally involve field workers gathering data through questionnaire surveys and formal forest inventories.

Experience has revealed that conventional methods of gathering information in rural settings commonly fail to provide timely, reliable, cost-effective and useful information. RRA and PRA approaches differ from conventional approaches in that field workers learn directly from local people. They tap local knowledge and gain information and insight from local people using a range of interactive tools and methods. Properly applied, RRA and PRA will yield locally relevant, timely, reliable, cost effective and useful information.

PRA tools can also be used in RRA, but practitioners should be clear and knowledgeable about when and how to use the tools independently and collectively. The difference between RRA and PRA is that the PRA aims to catalyze local capacity to deal with local problems through field workers learning with local people. In contrast, RRA involves field workers learning from local people according to the field worker's agenda. RRA does not necessarily involve facilitating local capacity or empowering local people to act.

RRA tools include building rapport, cross checking, semi structured interviews, group meetings and workshops, sketch mapping and direct observation. In addition, forest biodiversity profile will be developed by rapid assessment, specifically for identifying vulnerability of villages to natural hazards and options available for implementing a DRR strategy, including investing in soft structures to act as coastal bio-shields using mangroves and other viable species.

The common PRA tools include building rapport, ranking, time charts, semi-structured walks along the coast to identify the proximity of villages to the coast, density of mangroves and coastal forests, participatory mapping, participatory use of photographs and satellite imageries, group meetings and workshops and direct observation.

The choice of the appropriate participatory technique depends on the circumstance. As a general rule, RRA is used to learn rapidly from local people and PRA is used when field workers need to empower local people to apply a DRR strategy and implement related activities. RRA is a very useful tool when field workers are planning for the efficient use of their own resources (e.g. finances, human resources and time), and when exploring the prospects for working in a local area. RRA allows field workers to collect reliable and useful information from local people quickly. PRA should not be used during the earliest stages unless field workers are prepared to continue to support the user group after local interest has been raised. PRA should generally not be used when the intention is not to empower local people but to plan for the efficient use of externally funded resources.

Both RRA and PRA approaches are used in facilitating the user group planning. Before field workers can empower stakeholders and coastal communities with the roles, rights and responsibilities, they need to collect social and physical resource information about the local area and the people who live there. RRA can be used to accomplish this without unduly raising the expectations of local people. When the community is identified and field workers have a sense of the local situation, PRA approaches can be used.

These participatory methods are ideal for encouraging collaboration between field workers and resource-user groups in implementing the DRR strategy and on the initiative of investing in soft structures as a means of DRR strategy. Community members and stakeholders themselves can use many PRA tools to monitor the usefulness and the need for sustainability of DRR projects. RRA techniques are useful for field workers to monitor activities in which the projects may have an interest (even if the resource user group does not); for example, the use of government funds for plantation establishment as part of DRR strategy.

Step 4: Training and Capacity Building in DRR

Awareness on concept, benefits and opportunities of investing in soft structure as a DRR strategy is lacking in many of the sectors that operate in the coastal regions of many countries. Training and capacity building of stakeholders, particularly in the public works, revenue and local governance units is needed. Training in how to use planning tools such as Strategic Environmental Assessment (SEA), Social Impact Assessment (SIA) and Environmental Impact Assessment (EIA) is essential. Also, updated information and data on the impacts of natural hazards on the coastal areas and the coastal communities should be generated. Trainings should be conducted for building capacity on implementing DRR action plans for each sector to address the impacts of natural hazards. Guidelines for implementing these action plans should be developed and used. "Train the trainers" methods can be used to spread the capacity more widely to ensure spatial adaptation and implementation of a DRR strategy along the coasts of vulnerable communities.

Step 5: Preparation of Micro-plan for Implementing DRR Plan and Programme

Situation analysis helps in identifying the vulnerability status of a village or hamlet and effects and impacts of natural hazards on the biodiversity and ecosystem services and the livelihoods of local community. Micro-plans for a village or a group of vulnerable villages should be prepared. An action plan for a DRR strategy should be included in the micro-plan for implementation by the stakeholders. A perspective action plan for implementing a DRR strategy including key stakeholders the community and relevant sectors namely forestry, public works, fisheries, agriculture and aquaculture should be developed. These action plans should be implemented through the local institutions and local communities.

Step 6: Joint Implementation of Action Plans

Village level institutions should be made responsible for implementing the DRR activities pertaining to each village or a cluster of villages. Budgeted amount for activities will be deposited in a bank account of a community-level institution.

The bank account will be jointly held by the village head, representative of public works department or a forestry department. Required budget amounts can be withdrawn after passing a resolution in the committee meetings and with two signatures, which are mandatory to maintain transparency. In order to ensure effective implementation of actions plans, it should be ensured that: (1) an economic stake is created for the local community, (2) there is coordination among different departments, academic, research institutions and NGOs involved in the DRR activities, (3) there is strengthening of existing traditional management institutions, and (4) there is sharing of costs and responsibilities by revenue, public works and welfare departments. Additionally, adequate time and effort should be allowed for the field staff of implementing institutions to prepare and train themselves and other stakeholders. Golden rules and norms on do's and don'ts of implementing action plans should be developed and discussed with the community and other stakeholders.

Step 7: Monitoring and Evaluation

Monitoring and Evaluation (M&E) is an important stage in any DRR project. Not only does M&E ensure the effectiveness of said strategies, but also the longevity of the program. Periodical reviews of projects help determine if their outcomes are producing tangible benefits to the community, or if there are problems with projects that need to be addressed, or if the scale of the projects need to be changed. Like with other aspect of DRR strategies (i.e., planning and implementation), M&E should include community members and local stakeholders. To ensure community involvement, effective communication of scientific knowledge is required.

Step 8: Process Documentation and Replication of Best Practices of DRR Strategy

Many of the field level projects have reports done by the staff members who worked in the project and has limited capacity in writing reports and also are handicapped due to lack of time to do effective reporting. Also whatever has been reported in the Mid Term Evaluations and Final Evaluations remain as grey literature are not available for replication of the lessons learned, both positive and otherwise, including the best practices.

Process documentation "on investment in soft structures as a DRR strategy" will reveal the challenges in implementation and also identify the best practices that helped in realizing the outcomes of the project. This will help in the replicating the approach at national levels and upscaling the project benefits.

A Case Study of Community-Based Joint Mangrove Management

The Joint Management Model

In 1977, a devastating cyclone damaged the Machilipatnam coast in India, which affected the livelihoods of fishing community living along the mangroves of Krishna district of Andhra Pradesh. Another cyclone in 1996 marred the Godavari mangroves of the Kakinada coast, and this cyclone made local stakeholders realize that mangroves are nature's green wall that protects communities from cyclones and storms. After the Asian tsunami of December, 26, 2004, it was once again realized in the southern parts of Andhra Pradesh, that mangroves play a critical role in mitigating the effects of not only cyclones and storms, but also tsunamis. It was also understood that, wherever the mangroves were dense and high in spatial distribution, the effect of natural hazards was less intense, while impacts were more severe in places of no or sparse mangrove vegetation.

Restoration, conservation and management of the mangroves was difficult in the past because of the dependency local communities had on mangroves as a source of income, for which there were no alternatives. In addition, a lack of a sound understanding of the hydrological, geomorphologic and ecological aspects of mangrove ecosystems was also a reason for the degradation of these fragile forests.

Under these conditions, there was a need for integrated conservation and development of the mangrove forests, for which a multi-stakeholder participatory Integrated Conservation and Development (ICD) approach was developed and implemented in fishermen villages in Andhra Pradesh. The ICD approach was defined as a Joint Mangrove Management (JMM) model. The JMM envisages activities such as formation of mangrove conservation councils, gender mainstreaming in planning and implementation, joint preparation and implementation of micro plan for socio-economic development of community and restoration, conservation and management of mangroves by the local community, NGOs and Forest Department.

The community participated mangrove restoration and management in Godavari and Krishna mangroves resulted in the restoration of 520 ha of degraded mangroves by undertaking a contour survey, a geomorphological and hydrological survey, canal construction and planting of mangrove saplings. An area of 9442 ha of mangrove forest, including the restored area was brought under Joint Mangrove Management by eight village level institutions, namely Eco Development Committees (EDC) or Forest Conservation Committees (FCC/*VSS*). Village developmental activities, socio-economic and women in development activities were undertaken to meet the objectives of the villagers and the JMM model. To ensure sustainable participation of villager's dependent on mangroves, awareness generation exercises were concentrated on the theme of the ecological benefits accruing from the well-stocked mangroves in the form of protection from cyclonic storms and tidal waves and through enrichment of economically valuable fishery resources.

Joint Mangrove Management, Andhra Pradesh India

This project developed an integrated participatory natural resources management model for mangroves called Joint Mangrove Management (JMM) involved different stake holders, such as the Forest Department, NGO's and particularly the key stakeholders from the local community and micro-level institutions. The model involved a twin approach consisting of a social aspect for community mobilization, including awareness generation, addressing of poverty issues, capacity building for community empowerment, and land based alternatives for mangrove dependency. The other approach targeted the technical aspects, dealing with the causes of degradation, techniques for mangrove restoration involving geomorphological and hydrological aspects, and sustainable management of mangroves through mangrove management units. In Andhra Pradesh, this intervention affected nearly 1835 families of 5–7 members per family and 48 women Self Help Groups (SHG) and two youth groups that are dependent on mangroves in their day-to-day life.

The JMM model was implemented in eight villages, namely Matlapalem, Corgani—Dindu, Kobbarichettupeta, Gadimogga and Bhairavalanka in Godavari, and Dheenadayalapuram, Zinkapalem and Nali in Andhra Pradesh. An area of 520 ha of degraded mangroves were restored in Godavari and Krishna. Through the JMM model, an area of 9442 ha of pristine mangroves were brought under protection and management by these eight villages. Restoration of mangroves stopped further degradation of adjoining mangroves and also increase the fishery resources. The bio-diversity of the area improved, as crab populations in the restored areas increased due to the recharged water regime. Since, the work involved intensive labour, it provided employment opportunities to the members of the village level institutions. Based on the trainings provided during the JMM model project other NGOs, namely Sravanthi and Action in Godavari area and Sangamithra Service Society and Coastal Community Development Program in Krishna, have restored an additional nearly 215 ha of degraded mangroves.

In terms of the cost-benefit ratio of the restoration activity, mangroves restoration can be cost intensive, requiring funds for community mobilization and organization, and planning and implementation, surveying, nursery raising, canal construction, planting, and long term monitoring. The restoration of 10 ha of mangroves can cost 350,000 Indian rupees (i.e. $54,000 USD). Although the initial investment the cost might be high, the economic and ecological benefits to the local community are extensive over the long run. For example, in Andhra Pradesh, the community has benefited from re-establishment of the crab population and the growth of fodder grass. As the biodiversity returns and the denuded patches have been covered, populations of larger animals like otters and bird populations have substantially increased. Establishing the water regime was useful in tackling further degradation of mangroves and promotion of ecological heath and community resilience to disaster events.

Concluding Notes

This chapter illustrates how Nature-based Solutions (NbS), with specific reference to mangrove forests, can be used as a Disaster Risk Reduction (DRR) initiative. Information was presented on the theoretical knowledge associated with NbS, particularly planting guidelines and different vegetation species, as well as on the conceptual background of resilience and the Sendai Framework. Empirical evidence proving the effective nature of mangroves forests against natural hazards was explained using context specific examples of villages in India following natural disasters. In addition, notes on ecological restoration of degraded mangrove ecosystem and a framework for community involvement to realise a DDR strategy was also presented. And finally, an example of a successfully executed community based mangrove restoration project was discussed.

Nature-based solutions, which are embedded in concepts of socio-ecological systems theory and underpinned with concepts of ecosystem based services (i.e., ecological and social co-benefits), are effective approaches for protecting communities against imminent hazards. Since coastal communities are among the most vulnerable to natural hazards it is pertinent for them to build their capacity, evaluate their vulnerability, and to participate in designing and implementing resiliency approaches to cope with the risks posed by environmental and climate variability. Therefore, it is the hope that coastal community leaders, policy makers and decision takers associated with coastal protection strategies will invest in NbS, with particular attention placed on protection and augmentation of mangrove forest ecosystems.

Acknowledgements This synthesis is based on the experiences of work in multiple organizations and with heterogeneous community in the Asia-Pacific region. Largely influenced from a project on Coastal Wetlands: Mangrove Conservation and Management implemented in the Indian east coast by MSSRF supported by India-Canada Environment Facility. The senior author would like to acknowledge the working experience with colleagues from M.S. Swaminathan Research Foundation, FAO, UNDP and ADB and for the lessons learned from the fishermen community.

References

Anim DO, Nkrumah PN, David NM (2013) A rapid overview of coastal erosion in Ghana. Int J Sci Eng Res 4(2):1–7

Arkema KK, Abramson SC, Dewsbury BM (2006) Marine ecosystem-based management: from characterization to implementation. Front Ecol Environ 4(10):525–532

Baig SP, Rizvi A, Josella M, Palanca-Tan R (2016) Cost and benefits of ecosystem based adaptation: the case of the Philippines. IUCN, Gland, Switzerland, viii + 32 p

Banerjee K, Chakraborty S, Paul R, Mitra A (2018) Accumulation of metals by mangrove species and the potential for bioremediation. In: Nagabhatla N, Metcalfe CD (eds) Multifunctional wetlands: pollution abatement by natural and constructed wetlands. Springer, New York, p 388

Beebe J (2005) Rapid assessment process. In: Kempf-Leonard K (ed) Encyclopedia of social measurement. Elsevier, New York, NY, pp 285–291, ISBN: 9780123693983

Boonsong K, Piyatiratitivorakul S, Patanaponpaiboon P (2003) Potential use of mangrove plantation as constructed wetland for municipal wastewater treatment. Water Sci Technol 48(5):257–266

CEB (2013) United nations plan of action on disaster risk reduction and for resilience. http://www.preventionweb.net/files/33703_actionplanweb14.06cs1.pdf

Chambers R (1992) Rural appraisal: rapid, relaxed and participatory. IDS Discussion Paper 311

Cohen-Shacham E, Walters G, Janzen C, Maginnis S (eds) (2016) Nature-based solutions to address global societal challenges. IUCN, Gland, Switzerland, xiii + 97 p

Danielsen F, Sørensen MK, Olwig MF, Selvam V, Parish F, Burgess ND, Hiraishi T, Karunagaran VM, Rasmussen MS, Hensen LB, Quarto A, Suryadiputra N (2005) The Asian tsunami: a protective role for coastal vegetation. Science 310:643

Das S, Vincent JR (2009) Mangroves protected villages and reduced death toll during Indian super cyclone. Proc Natl Acad Sci 106(18):7357–7360

Donato DC, Kauffman JB, Murdiyarso D, Kurnianto S, Stidham M, Kanninen M (2011) Mangroves among the most carbon-rich forests in the tropics. Nat Geosci 4(5):293–297

Doswald N, Osti M (2011) Ecosystem-based approaches to adaptation and mitigation—good practice examples and lessons learned in Europe. https://www.bfn.de/fileadmin/MDB/documents/service/Skript_306.pdf

Duke NC, Meynecke JO, Dittmann S, Ellison AM, Anger K, Berger U, Cannicci S, Diele K, Ewel KC, Field CD, Koedam N, Lee SY, Marchand C, Nordhaus I (2007) A world without mangroves? Science 317:41–42

EJF (2006) Mangroves: Nature's defence against tsunamis—a report on the impact of mangrove loss and shrimp farm development on coastal defences. Environmental Justice Foundation, London, UK

Gallopín GC (2006) Linkages between vulnerability, resilience, and adaptive capacity. Glob Environ Chang 16(3):293–303

Giri C, Ochieng E, Tieszen LL, Zhu Z, Singh A, Loveland T, Masek J, Duke N (2011) Status and distribution of mangrove forests of the world using earth observation satellite data. Glob Ecol Biogeogr 20(1):154–159

Holling CS (1973) Resilience and stability of ecological systems. Annu Rev Ecol Syst 4(1):1–23

IPBES (2013) Decision IPBES-2/4: conceptual framework for the Intergovernmental Science-Policy Platform on Biodiversity and Ecosystem Services. http://www.ipbes.net/sites/default/files/downloads/Decision%20IPBES_2_4.pdf

Issaka S, Ashraf MA (2017) Impact of soil erosion and degradation on water quality: a review. Geol Ecol Landsc 1(1):1–11

IUCN France (2016) Nature-based solutions to address climate change. France, Paris

Kathiresan K, Rajendran N (2005) Coastal mangrove forests mitigated tsunami. Estuar Coast Shelf Sci 65(3):601–606

Leslie HM, McLeod KL (2007) Confronting the challenges of implementing marine ecosystem-based management. Front Ecol Environ 5(10):540–548

Lewis RR (2005) Ecological engineering for successful management and restoration of mangrove forests. Ecol Eng 24(4):403–418

Lewis RR, Brown B, Quarto A, Enright J, Corets E, Primavera J, Ravishankar T, Stanley O, Djamaluddin R (2006) Five steps to successful ecological restoration of mangroves. YARL and the Mangrove Action Project, Yogyakarta, Indonesia

MEA (2005) Ecosystems and human well-being: synthesis. Island Press, Washington, DC

Mitsch WJ (2012) What is ecological engineering? Ecol Eng 45:5–12

MOCZM (2014) Planting vegetation to reduce erosion and storm damage. http://www.mass.gov/eea/docs/czm/stormsmart/properties/ssp-factsheet-3-vegetation.pdf

Olivier J, Probst K, Renner I, Riha K (2012) Ecosystem-based Adaptation (EbA). https://www.giz.de/expertise/downloads/giz2013-en-ecosystem-based-adaptation.pdf

Ozment S, DiFrancesco K, Gartner T (2015) The role of natural infrastructure in the water, energy and food nexus, Nexus Dialogue Synthesis Papers. IUCN, Gland, Switzerland

Pendleton L, Donato DC, Murray BC, Crooks S, Jenkins WA, Sifleet S, Craft C, Fourqurean JW, Kauffman JB, Marbà N, Megonigal P, Pidgeon E, Herr D, Gordon D, Baldera A (2012) Estimating global "blue carbon" emissions from conversion and degradation of vegetated coastal ecosystems. PLoS One 7(9):e43542

Peng Y, Li X, Wu K, Peng Y, Chen G (2009) Effect of an integrated mangrove-aquaculture system on aquacultural health. Front Biol China 4(4):579–584

Perrings C, Folke C, Mäler KG (1992) The ecology and economics of biodiversity loss: the research agenda. Ambio 134:201–211

Polidoro BA, Carpenter KE, Collins L, Duke NC, Ellison AM, Ellison JC, Farnsworth EJ, Fernando ES, Kathiresan K, Koedam NE, Livingstone SR, Miyagi T, Moore GE, Nam VN, Ong JE, Primavera JH, Salmo SG III, Sanciangco JC, Sukardjo S, Wang Y, Yong JWH (2010) The loss of species: mangrove extinction risk and geographic areas of global concern. PLoS One 5(4):e10095

Ramasubramanian R, Ravishankar T (2004) Mangrove restoration in Andhra Pradesh in India. MSSRF, India, 26 pp

Ravishankar T, Gnanappazham L, Ramasubramanian R, Navamuniammal N, Sridhar D (2003) Atlas of mangrove wetlands of India: part 2—Andhra Pradesh, India. MSSRF, India. MSSRF/MG/03/15, 136 pp

Ravishankar T, Ramasubramanian R (2004) Community-based reforestation and management of mangroves for poverty reduction in the east coast of India. In: Sim HC, Appanah S, Lu WM (eds) Proceedings of the workshop forests for poverty reduction: can community forestry make money? FAO of the UN. 137–142. FAO RAP 2004/04, ISBN: 974-7946-51-3

Ravishankar T, Ramasubramanian R (2004) Manual on mangrove nursery raising techniques. MSSRF, India. MSSRF/MA/04/15, 48 pp

Ravishankar T, Navamumiammal M, Gnanappazham L, Nayak SS, Mahopatra G, Selvam V (2004) Atlas of mangrove wetlands of India: part 3—Orissa, India. MSSRF, India, 102 pp

Roth R (2013) Natural infrastructure: a climate-smart solution. https://www.climatesolutions.org/sites/default/files/uploads/natural_infrastructure_web.pdf

Selvam, V, Ravishankar T, Karunagaran VM, Ramasubramanian R, Eganathan P, Parida AK (2006) Toolkit for establishing coastal bioshields. MSSRF, INdia. MSSRF/MA/05/26, 117 pp

Spalding M, McIvor A, Tonneijck FH, Tol S., van Eijk P (2014) Mangroves for coastal defence. Guidelines for coastal managers & policy makers. Wetlands International and the Nature Conservancy, 42 p

Sthiannopkao S, Takizawa S, Wirojanagud W (2006) Effects of soil erosion on water quality and water uses in the upper Phong watershed. Water Sci Technol 53(2):45–52

Tam NF, Wong YS (1997) Accumulation and distribution of heavy metals in a simulated mangrove system treated with sewage. Hydrobiologia 352(1):67–75

Tanaka N, Nandasena NAK, Jinadasa KBSN, Sasaki Y, Tanimoto K, Mowjood MIM (2009) Developing effective vegetation bioshield for tsunami protection. Civ Eng Environ Syst 26(2):163–180

UNEP (2007) Mangroves of Western and Central Africa. UNEP-Regional Seas Programme/UNEP-WCMC

UNISDR (2009) UNISDR Terminology on Disaster Risk Reduction. https://www.unisdr.org/files/7817_UNISDRTerminologyEnglish.pdf

UNISDR (2015) Sendai framework for disaster risk reduction 2015–2030. http://www.preventionweb.net/files/43291_sendaiframeworkfordrren.pdf

Vayda AP, McCay BJ (1975) New directions in ecology and ecological anthropology. Annu Rev Anthropol 4(1):293–306

Wu Y, Chung A, Tam NFY, Pi N, Wong MH (2008) Constructed mangrove wetland as secondary treatment system for municipal wastewater. Ecol Eng 34(2):137–146

Yang Q, Tam NF, Wong YS, Luan TG, Su WS, Lan CY, Shin PKS, Cheung SG (2008) Potential use of mangroves as constructed wetland for municipal sewage treatment in Futian, Shenzhen, China. Mar Pollut Bull 57(6):735–743

Chapter 6
The Role of Constructed Wetlands in Creating Water Sensitive Cities

Shona K. Fitzgerald

Introduction

This chapter explores the use of constructed wetlands for improving the sustainability, resilience and liveability of cities. The first section highlights the current challenges facing cities and the proposed characteristics of the cities of the future that are needed to overcome these challenges. Three sets of principles to create these future cities are presented from UN Habitat, the International Water Association and the Australian Cooperative Research Centre for Water Sensitive Cities. The second section presents a case study of a small urban retrofit project located near Sydney, Australia. In addition to water quality improvement, the wetland provided many other benefits to the environment, the local community and the organisations coordinating the delivery of the wetland. For instance, the project improvements to vegetation and riparian zone condition were quantified through hedonic valuation to increase the value of the local houses by a total of $16 million (AUD). The third section then examines these benefits in terms of the progress of a city towards becoming sustainable, resilient and liveable. It proposes that benchmarking the state of urban water management and using indicators to measure the success of interventions such as constructed wetlands are important components of achieving this shift to a future city state.

S.K. Fitzgerald (✉)
United Nations University—Institute for Water, Environment and Health, Hamilton, ON, Canada

Sydney Water, Sydney, NSW, Australia
e-mail: shona.fitzgerald@sydneywater.com.au

© Springer International Publishing AG 2018
N. Nagabhatla, C.D. Metcalfe (eds.), *Multifunctional Wetlands*, Environmental Contamination Remediation and Management, DOI 10.1007/978-3-319-67416-2_6

The Changing Face of Cities

Population growth and climate change are transforming our cities. In 2015, the world's population reached 7.3 billion and is currently growing by 1.18% per year (United Nations DESA, Population Division 2015). At this growth rate, it is projected that by 2030, globally there will be 8.5 billion people. The trend in urbanisation is similarly increasing. In 2014, the global urban population exceeded 50% and continues to rise. It is expected that 66% of the world population will be urban by 2050 (United Nations DESA, Population Division 2014), and this will put increasing strain on cities and the resources that sustain them.

Population growth and urbanisation present significant challenges to sustainable urban development, particularly in terms of environmental sustainability, social equity and governance. Many of the issues associated with these challenges have been summarised in the issue paper on 'urban infrastructure and basic services, including energy' that was prepared in advance of discussions at the Habitat III conference in October 2016 (United Nations 2015). As highlighted in the issue paper, environmental sustainability is challenged by the dense population in urban areas resulting in a high level of resource consumption. Moreover, owing to generally high incomes in urban areas, cities have higher consumption patterns. In urban planning, low socio-economic areas often experience disparity in the accessibility and affordability of services such as transport, energy and water. This inequity in infrastructure services is often due to weaknesses in policy, the planning approach and institutional capacity. It often means that those living in informal settlements and slums are more exposed to the impacts of disasters and pollution, leading to poorer public health outcomes. It is suggested that extensive policy reform can enable better efficiencies in resource use and promote more effective methods for infrastructure planning and service delivery. In particular, the issue paper highlights the need for changes to regulation and attracting and enabling private sector investment.

Climate change adds to the significant challenges of sustainability and resilience of cities, and in particular, the challenge of sustainable urban water management. With climate change the world has experienced an increase in the frequency and intensity of drought, heat waves and rainfall events (Pachauri et al. 2014). These extreme weather events result in large impacts to infrastructure, and to the livelihoods and wellbeing of people. The impact of drought is largely felt in the change to water and food supplies. Drought reduces supply, but also alters the geology of catchments, which changes runoff patterns and promotes wash off of contaminants (e.g. nutrients, organic matter) when rainfall does occur. In this way and through flooding and changing rainfall patterns the quality of source water is threatened (Khan et al. 2015). Heat waves are particularly impactful in cities, resulting in increased mortality and morbidity in urban areas (Bi et al. 2011).

To meet these challenges of population growth, urbanisation and climate change a shift is required in the function of cities and therefore in the execution of urban planning and design. The following sections examine three theoretical frameworks that describe the role that water plays in realising this required shift, and provide guiding principles for urban water management. These include, the new vision of a city set out by UN Habitat in the New Urban Agenda (United Nations 2016), the principles of 'Cities of the Future' proposed by the International Water Association (Binney et al. 2010) and 'Water Sensitive Cities' described by the Australian Cooperative Research Centre for Water Sensitive Cities (Brown et al. 2009).

Future Cities

The 17 Sustainable Development Goals (SDGs) were ratified by the United Nations (UN) in September, 2015. Across the 17 goals, there are 169 targets, which create an action plan for the universal ambition to eradicate poverty and hunger, to protect the planet and to create a world where all people can lead prosperous and peaceful lives (United Nations General Assembly 2015). The SDGs are ambitious, but they advance our dialogue around sustainable development as a universal objective that is integrated between social, economic and environmental outcomes. This agenda also sets out, for the first time, a stand-alone urban development goal in SDG 11. This goal aims to make cities and human settlements inclusive, safe, resilient and sustainable. The inclusion of an urban development goal recognises the importance of cities in facilitating and achieving sustainable development.

Sustainable urban development has long been a focus of the UN, formalised at the first Habitat Conference on Human Settlements in 1976. Habitat I declared that improving the quality of life of those in urban settlements is the most important objective of urban development. This included ensuring availability of food, shelter, clean water, employment, health and education, and that these be achieved in a fair, just, equitable, participatory and peaceful way that respects the country's sovereignty and protects the environment (United Nations 1976). Forty years on, UN Habitat has developed the New Urban Agenda, which was discussed and adopted at the Habitat III conference in Quito in October 2016. The vision outlined in this Agenda not only recognises the need to develop cities in a way that will achieve the quality of life articulated at Habitat I, but shifts the thinking of urbanisation from an outcome of development to urbanisation as a transformative power and a strategy for development. While Habitat I focused on the quality of life afforded by establishing basic infrastructure and services, the New Urban Agenda seeks to achieve a quality of life above and beyond this, which is described as cities being compact, inclusive, equitable, cohesive, participatory, resilient, sustainable and productive (United Nations 2016).

There are three interlinked principles articulated in the New Urban Agenda in order to achieve this vision:

1. Leave no-one behind, urban equity and poverty eradication
2. Sustainable and inclusive urban prosperity and opportunities for all
3. Foster ecological and resilient cities and human settlements

The first of these principles will include ensuring equitable access to physical and social infrastructure and basic services, such as adequate housing, drinking water and sanitation. The second will focus on urban development to enable business, jobs and livelihoods, thereby developing prosperous urban economies. The third principle will result in protection of ecosystems and biodiversity as well as cities achieving sustainable consumption patterns and resilience to shocks, such as natural disasters. Urban water management plays a large role in achieving the New Urban Agenda's vision, for example, through ensuring drinking water and sanitation, providing public spaces, improving environmental sustainability and resilience. Each of the above principles also recognises the importance of enabling and strengthening participation from the community to achieve inclusivity and equity. This presents a challenge for urban water management to lift the focus from service provision and the associated physical infrastructure, and instead to think more broadly about water's diverse contribution to people's quality of life and to encourage community engagement in water management. The community's connection with water is important for an understanding of sustainable management of the water cycle and also for health and wellbeing.

This vision of the city in the New Urban Agenda is reflected in the water industry's work in recent years attempting to define and set out the transition to 'Cities of the Future'. The industry has sought to define the role of urban water management in creating future cities which are liveable, sustainable and productive. The International Water Association's (IWA) Cities of the Future Program has the objective of recognising that water, and its interactions with other urban sectors, is central to urban development. In light of this, it encourages urban water management to be redesigned to not only deliver sustainable water services, but to also enhance life within and beyond the urban environment (Binney et al. 2010). It suggests this redesign focuses on four themes: liveability and sustainability; recognising and capitalising on the many values of water; community choice and knowledge sharing; and an adaptive and collaborative water sector. To define how this can be achieved, the IWA facilitated a workshop in 2010 to develop a set of principles for each of the themes that are central to a City of the Future; these are shown in Fig. 6.1. The principles of Cities of the Future have since been expanded to the principles for Water-Wise Cities (International Water Association 2016). The Water-Wise City vision includes 17 principles under the four key themes of regenerative water services, water sensitive urban design, basin connected cities and water-wise communities. There are many similarities between the principles of Water-Wise Cities and those of Cities of the Future shown below.

The Cities of the Future principles help to articulate the role that water can play in meeting the higher level principles of the New Urban Agenda. The principles shown in Fig. 6.1 paint a vision of a city that ensures equitable access to water and wastewater services while at the same time recognising the many values of water, such as its role in providing healthy waterways, green spaces, social connection and reducing urban heat. This aligns with the New Urban Agenda's commitment to end poverty and ensure urban equity. This happens in part through provision of physical and social infrastructure which protects ecosystems and biodiversity and enhances

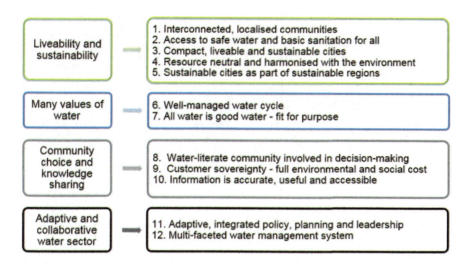

Fig. 6.1 Four themes and 12 principles for a City of the Future, as identified by Binney et al. (2010)

liveability, health and wellbeing. Principles 4 and 6, in particular, capture the need to focus on preserving and improving ecosystems and integrating urban water management, which are essential to realise many of the other principles. A City of the Future consumes resources at a rate that does not deplete the environment and recognises the environmental and social benefits of compact communities. The principles state that the decisions around water management should be collaborative with the community and across the water sector, and should be based on a holistic understanding of the costs and benefits. This again aligns with the New Urban Agenda principle of environmental sustainability, as well as with the underlying commitment of the Agenda being people-centred and promoting participatory, cohesive, inclusive and safe societies. The principles also recognise the need for policy that supports an integrated approach. The lack of appropriate policy and regulation was stated as a key challenge in the Habitat III issue paper (United Nations 2015). The Cities of the Future principles can therefore help to frame how urban water management can contribute to the New Urban Agenda.

While participating in the conversation to develop The Cities of the Future principles, the Australian water industry has also spent the past decade developing the concept of a Water Sensitive City. The pursuit of water sensitive cities is now an increasingly common goal for urban water managers in Australia. At the core, water sensitive cities are sustainable, resilient and liveable and consist of informed citizens engaged in water management decisions. Researchers in Australia have clearly defined the elements of a water sensitive city and the key pillars for practice for creating water sensitive cities. These pillars are also a means of grounding the high level principles of the New Urban Agenda. Moreover, they describe the required knowledge and social values and define actions to achieve the principles of the Cities of the Future.

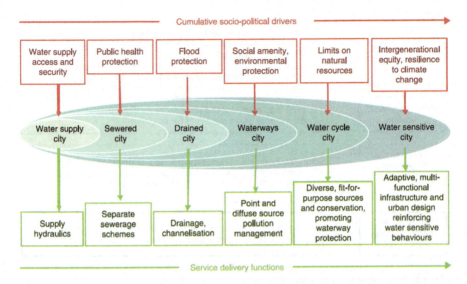

Fig. 6.2 Urban water management transitions framework taken from Brown et al. (2009)

The conceptual goal of building a water sensitive city has been well established, though, there is currently no example of a water sensitive city in the world (Brown et al. 2016). The difficulty in establishing water sensitive cities is owing to challenges such as a clear, shared vision, regulation and institutional reform and funding (Ison et al. 2009). To help overcome some of these difficulties, Brown et al. (2009) undertook extensive research from 2002 to 2008 to better understand the historical, current and future stages of Australian urban water management in cities. Through this research it was possible to identify the values underpinning a water sensitive city and the knowledge and institutional structures that are required to transition to a water sensitive state. The research identified six urban water management phases, which are expressed in the 'urban water transitions framework' (Fig. 6.2).

Each phase has distinct cognitive, normative and regulative aspects, that is, each is influenced or defined by the current knowledge and technical skills, the values of water and the rules and systems. The initial three phases, water supply, sewered and drained cities were similar in their cognitive approach of using centralised infrastructure and in their normative perspective that the government would supply water, protect public health and minimise urban flooding. In each of these three phases the environment was perceived as benign and therefore water withdrawal and waste disposal to waterways was effectively limitless. The following three phases show a shift in the normative perception of the environment and of the role of water in a city. There is also a change in the cognitive tools, which reflects the water industry adapting to the changing values of water and the environment. For example, the waterways city was the first phase to recognise the pollution of urban waterways, particularly from stormwater, and sought alternative means of managing stormwater to repair and protect the environment. The next phase was moving to the water cycle city, which recognised the limits around water supply. So, in addition to protecting waterway health, includes the implementation of water conservation, fit-for-purpose water supply, and sensitivity to the relationship between water, wastewater, energy and nutrients. The end state depicted on the continuum is the water sensitive city.

The water sensitive city, while encompassing the normative values of the various city states around water and the environment, goes further to incorporate societal values around sustainability, intergenerational equity and resilience to climate change (Brown et al. 2009). These values form a vision of a water sensitive city which can be described as sustainable, liveable and resilient. Sustainability has been described in many ways and generally captures the need to use non-renewable resources at a rate that they can be replenished, to maintain pollution within the environmental assimilation capacity and implement fair and just social and economic development (Ruth and Franklin 2014). To build liveable cities, the tangible elements that create liveability are contextual and subjective, however, the guiding principles are consistent. That is, liveability is having built infrastructure and ecosystems that provide access to goods and services as required and desired in the city context (Ruth and Franklin 2014). Resilience is often described as the ability to absorb shocks. However, as described by Folke (2006), a city's resilience must also encompass an adaptability to ensure capacity for renewal, re-organisation and development. For infrastructure and ecosystems to provide sustainability, liveability and resilience we need to plan our cities in a way that meets essential needs, protects the environment, ensures climate resilience and connects people. As seen in Fig. 6.2, it is proposed that the cognitive approach that is required is use of adaptive, multi-functional infrastructure and urban design, which is also reflected in the Cities of the Future principles.

To translate these aspects of a water sensitive city into the practice of water management Wong and Brown (2009) developed and described three pillars of practice:

1. Cities as water supply catchments
2. Cities providing ecosystem services
3. Cities comprising water sensitive communities

Embracing cities as supply catchments means diversifying water sources within a city. This improves the security of water supply and therefore the city's resilience to drought. Moreover, having multiple alternative sources within a city allows for dynamic decisions about supply options in terms of cost, source water quality and environmental sustainability. It can also promote fit-for-purpose water treatment and use. The alternative water sources within a city include groundwater, desalinated water, stormwater from urban runoff, rainwater from roof runoff and recycled wastewater.

The role that cities play in protecting and enhancing ecosystem services will be increasingly important with a changing climate. City spaces have the capacity to be used to contain pollution from diffuse sources, regulate microclimate, produce food, and act as carbon sinks.

The third pillar recognises the importance of community knowledge and engagement to enable change. As has been seen in the transition framework the value the community places on water and their engagement in managing urban water supply are essential socio-political drivers for change. One way that these pillars are applied is through the practice of water sensitive urban design.

Water Sensitive Urban Design

Water sensitive urban design (WSUD) aims to build sensitivity to water into urban design (Wong 2006a). It is well integrated with the concept of Ecologically Sustainable Development, as shown in Fig. 6.3.

Ecologically Sustainable Development considers physical, social and economic contexts in sustainably designing the various services of a city, such as waste management, transport infrastructure, etc. (Wong 2006b). WSUD, however, is specifically focused on the interaction between the urban water cycle and the built and natural urban landscapes. The urban water cycle includes the drinking water, sewerage and stormwater systems and WSUD aims for the integrated management of

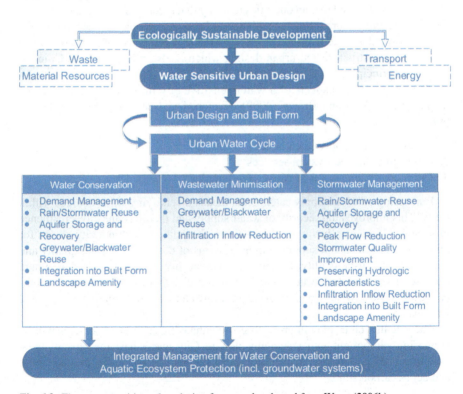

Fig. 6.3 The water sensitive urban design framework, adapted from Wong (2006b)

these water streams. Figure 6.3 shows some of the water cycle management strategies aimed at environmental protection and the integrated and sustainable use of water sources. WSUD also seeks to connect this urban water management with urban design. This means that urban planning and development would consider the management, protection and conservation of the urban water cycle, and integrate this with the community values and aspirations that govern urban design decisions. The results of designing cities in this way could include improved stormwater management, water supply security, urban biodiversity, amenity and recreation, microclimate benefits and reducing the carbon footprint of cities (Wong et al. 2011).

The natural water cycle has been significantly changed by urbanisation, in particular, changing the local water balance and downstream water-related ecosystems. WSUD technologies have been developed to mitigate the impacts of this change. For example, large volumes of potable water imported for drinking and wastewater that is exported change the local water balance. WSUD technologies can be used to reduce imports through water efficiency and reduce exports through reuse. Stormwater also plays a central role in the local water balance and in waterway health. Compared to the natural state, the urban catchment has been affected by the increase in impervious surfaces and the diversity of land uses. The imperviousness of cities results in a large increase in the volume and velocity of stormwater runoff and a decrease in evapotranspiration and infiltration. For example, a study in Australia from 1988 showed the average runoff in the Gungahlin urban catchment, with 25% impervious surface, was found to be six times the amount from the Gungahlin rural catchment, and the peak flow was ten times as large (Fletcher et al. 2004). This alters the natural hydrology of the catchment. Furthermore, as championed in the 1970s, the state of a waterway is directly related to the condition of its catchment (Hynes HBN 1975). The activity within the catchment, for example residential, parkland or industrial, will influence the quality of the stormwater runoff. Management of stormwater runoff is therefore a key element to managing and protecting the urban water cycle.

Urban stormwater management must include strategies to re-establish the natural hydrology of waterways and address waterway pollution (Wong et al. 2011). Traditional stormwater management strategies were based on the principle of moving stormwater away from cities as quickly as possible to prevent urban flooding and remove pollution. While the constructed infrastructure, such as piped stormwater and channelising creeks and rivers can reduce flooding risk, it still maintains less infiltration and evapotranspiration in the local catchment and greater volumes of polluted water entering downstream waterways. Instead of these traditional approaches, it is possible to use WSUD technologies, which can reduce runoff rates to downstream waterways and reduce pollution, but can also have other benefits in urban water management, such as reducing potable water demand through capture and use of stormwater, and by creating green space and amenities in the cityscape. WSUD technologies seek to reduce runoff rates and volume with solutions that slow, retain and reuse stormwater. These technologies also promote pollution reduction through treatment options that reduce the amount of sediment, nutrients, metals, and biological and organic components exported downstream.

In Australia, 89% of the 24 million people live in urban areas where there is a high percentage of impervious surfaces (United Nations DESA, Population Division 2014). Stormwater runoff in Australia has historically been managed through drainage infrastructure aimed at removing the water from urban areas as quickly as possible to minimise flooding. WSUD technologies for managing stormwater in Australia are increasingly adopted, though the majority of stormwater is still untreated and discharged directly to local waterways. This has resulted in many urban creeks and rivers suffering from erosion and high levels of pollution. Moving away from the traditional drainage approach is challenging as it often requires large capital investment to retrofit existing infrastructure. However, there are many examples of employing WSUD technology with the objective to improve waterway health and riparian zone condition. Implementing these technologies requires a good understanding of the stormwater pollutants and pollutant loads from the catchment.

Urban stormwater contains a range of pollutants, from gross pollutants (e.g. litter and debris), to trace metals and nutrients associated with fine sediment, to dissolved pollutants (Fletcher et al. 2004). Major Australian cities have separate stormwater and sewerage systems and so, studies have been undertaken to understand the quality of urban stormwater in order to design appropriate treatment measures for these systems. Table 6.1 shows the typical water quality parameters of stormwater compared to urban stream water quality (Wong 2006b). It can be seen that urban runoff water quality can contain high concentrations of all listed pollutants compared to the urban stream. Large volumes of stormwater can therefore add significant pollutant loads to waterways. These pollutants can have detrimental impacts on waterway health through deposit of sediment, algae growth due to increased nutrients, reduced dissolved oxygen and toxicity of metals and pesticides to fish and aquatic insects.

Table 6.1 Typical water quality parameters for urban stormwater runoff and urban streams (from Wong 2006b)

Variable (mg/L unless otherwise indicated)	Urban runoff	Typical urban stream water quality[a]
Suspended solids	250 (13–1620)	2.5–23
Biological oxygen demand	15 (7–40)	1.0–4.0
Lead	0.01–2.0	<0.002–0.024
Zinc	0.01–5.0	0.009–0.14
Copper	0.4	0.001–0.017
Chromium	0.02	–
Cadmium	0.002–0.005	<0.0005
Faecal coliforms (organisms/100 mL)	10^4 (10^3–10^5)	$0.4–7.4 \times 10^3$
Total phosphorus	0.6 (0.1–3)	0.02–1.2
Ammonium	0.7 (0.1–2.5)	0.002–0.16
Oxidised nitrogen	1.5 (0.4–5)	0.34–3.2
Total nitrogen	3.5 (0.5–13)	0.39–4.9

[a]From Melbourne Urban Streams—Melbourne Water data

Characterising the pollutants of a particular catchment can help inform the most appropriate WSUD solution.

Where stormwater is treated, the treatment measures take into account both the untreated stormwater quality and the target outcome, whether it be flood mitigation, waterway health or stormwater harvesting. There are various relevant water quality guidelines in Australia, which help to set the target treated water quality. For example, the national guidelines for fresh and marine water quality (ANZECC, ARMCANZ 2000) provide general indicators of the health of a waterway and can be used to trigger management actions. The actions should be aligned with the determined local guidelines and targets, which are then used to manage water quality. The principles of WSUD encourage options for reducing the use of mains drinking water and instead finding fit-for-purpose water solutions, which can sometimes be met through stormwater harvesting. In the case that the stormwater is reused then the required water quality will be dictated by the end use. The Australian Guidelines for Water Recycling specify the treatment requirements for various recycled water uses (NRMMC, EPHC, NHMRC 2009). Similar to these water quality targets, there are also state- and local-based targets for stormwater harvesting, such as the NSW Guidelines for Management of Private Recycled Water Schemes (NSW Government Department of Water and Energy 2008).

Typical WSUD infrastructure for stormwater management includes roof gardens, sediment basins, bioretention swales, bioretention basins, sand filters, swales/buffer systems, wetlands, ponds, infiltration measures, rainwater tanks and aquifer storage and recovery (Francey 2005). Table 6.2 gives a general outline of the appropriate end uses for different WSUD technologies. This understanding of the treatment capability of different WSUD technologies coupled with an understanding of the specific influent stormwater quality can help prioritise WSUD technologies for fit-for-purpose treatment.

Constructed Wetlands

Constructed wetlands are commonly used around the world for treating wastewaters, such as sewage, industrial wastewater and stormwater. In Australia, constructed wetlands have also developed as a key technology for retaining and treating urban runoff. While the literature defines wetlands as anything from intertidal rocky shores to rivers, the term constructed wetlands in reference to stormwater treatment generally describes an aquatic environment consisting of marsh, swamp and pond elements (Fletcher et al. 2004). Constructed wetlands are engineered systems built to mimic the natural environment. They are designed to speed up the natural processes, thereby downsizing land requirements. Constructed wetlands use natural physical, chemical and biological processes to alter the hydraulics and improve the water quality of urban runoff. In this way, constructed wetlands serve multiple purposes, including protection of downstream waterways from pollutants and peak flows, creation of new habitat for flora and fauna, and the opportunity for stormwater

Table 6.2 Stormwater treatment technologies achieving specific end-use water quality requirements, taken from Wong et al. (2011)

End use		Municipal use with restricted access and drip irrigation	Municipal use with unrestricted access	Dual reticulation with indoor & outdoor use	Drinking water	
Pre-treatment	Screens					Before storage
	GPTs					
Preliminary	Oil & sediment separators					
	Swales					
	Tanks					
	Sediment basins					
	Ponds and lakes					
Secondary	Infiltration systems					
	Wetland					
	Biofilters					
	Stormwater filters					
Advanced	Sand filters					After storage
	Aquifers					
	Suitable drinking water technologies (e.g. reverse osmosis and advanced oxidation)					
		Water quality level achieved when disinfection is employed (e.g. chlorination)				
		Currently requires disinfection but this requirement may be removed in a near future with the advancement of WSUD technologies				

harvesting. Constructed wetlands are currently used extensively in Australia as a WSUD strategy for treatment of urban stormwater (Deletic et al. 2014). It is recognised that constructed wetlands play an important role in creating and protecting ecosystems and in diversifying urban water supply. Moreover, constructed wetlands can also play a role in enhancing the liveability of a city.

The following section describes a case study for a constructed wetland near Sydney, Australia named the Cup and Saucer Wetland. The wetland was constructed as part of a suite of projects aimed at improving the health of an urban river. The qualitative and quantitative outcomes are examined in terms of the environmental impacts and the other benefits to liveability. Quantitative outcomes include water quality data, liveability indicator metrics and economic analysis. Qualitative outcomes are largely anecdotal. The overall wetland outcomes are then considered in context of how this case study has helped to progress to a water sensitive city.

Cup and Saucer Wetland

The Cup and Saucer Wetland was built at Heynes Reserve, approximately 11 km south-west of Sydney. Heynes Reserve is at the confluence of Cup and Saucer Creek and the Cooks River, as shown in Fig. 6.4. Also shown in the figure is the water balance for the Cooks River catchment, indicating that 68% of the annual rainfall (72,252 mL/year) runs off to the Cooks River. The Cup and Saucer Creek is one of the tributaries delivering stormwater to the Cooks River. Cup and Saucer Creek drains water from an urban catchment covering approximately 503 ha, of which 49% is impervious. The Creek was converted into a 3.6 km concrete stormwater channel in the 1930s, which was in line with the urban drainage infrastructure of the time; a trend which has resulted in 89% of the Cooks River tributaries being converted from natural waterways to concrete and brick drainage structures (Cooks River Catchment Association of Councils 1999). The Cup and Saucer Wetland was a small retrofit project built in 2010. The primary aim was to treat a portion of stormwater from the Cup and Saucer Creek before discharge to the Cooks River, which then flows into the marine environment of Botany Bay.

The wetland project was one in a suite of water quality improvement projects under the Cooks River Urban Water Initiative. Environmental protection and improvement has long been a priority for the Botany Bay catchment as it is home to numerous endangered species and communities, a RAMSAR listed wetland, migratory bird species and both Aboriginal and European heritage sites. The Cooks River, though only 9% of the Botany Bay catchment area, contributes 18% of the Total Suspended Solids (TSS) load to Botany Bay and 16% and 18% of the Total Nitrogen (TN) and Total Phosphorus (TP) loads respectively. The Cooks River catchment is highly urbanised with 72% residential use, 20% commercial use and 8% parkland. Seventy one percent of the catchment has no vegetation and 23% of the catchment is in degraded ecological condition, making it an important target for Botany Bay environmental improvement measures (Kelly and Dahlenburg 2011).

The Cooks River Urban Water Initiative was a group of projects that sought to improve the health of the Cooks River (SMCMA 2011). These projects were completed between 2008 and 2011, and aligned with the broader objectives of improving Botany Bay water quality. The objectives are presented in the Botany Bay and Catchment Water Quality Improvement Plan (Kelly and Dahlenburg 2011), which outlines the water quality improvements required to meet the environmental values set out by the local community. The Cooks River Urban Water Initiative supported these water quality improvement goals through two key objectives: to reduce the impact of stormwater runoff; and wetland remediation. WSUD was recognised as a

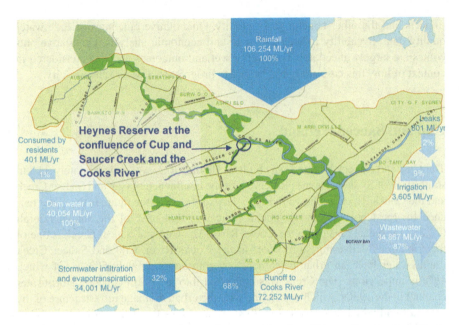

Fig. 6.4 Water cycle in the Cooks River catchment and location of Heynes Reserve (from Cooks River Alliance 2014)

key vehicle for achieving these goals and a broad range of options were explored. Projects were scoped with input from the Cooks River community and considered previous studies and plans, such as the Cooks River Stormwater Management Plan. Fourteen major projects were prioritised, including, six gross pollutant traps (GPTs), four biofiltration systems, two constructed wetlands, one wetland refurbished, and bank edge modification to create saltmarsh at two locations (SMCMA 2011).

The Cooks River Stormwater Management Plan (Cooks River Catchment Association of Councils 1999) identified the Cup and Saucer Creek as a hotspot for pollution and reported water quality results from 1990 to 1991 with elevated levels of nutrients, zinc, iron, copper, chromium, lead and nickel. More recent data (Table 6.3) shows the Cup and Saucer Creek accounts for 2.4%, 21% and 18% of the annual TSS, TN and TP load to the river. As stormwater runoff is a major pollution source in this catchment, it highlighted that there was an opportunity for stormwater management of the Cup and Saucer Creek catchment to improve the downstream water quality. The Cooks River Stormwater Management plan (Cooks River Catchment Association of Councils 1999) recommended an investigation into constructing a small wetland at Heynes Reserve. However, it wasn't until the Cooks River Urban Water Initiative that the Cup and Saucer Wetland was undertaken.

Heynes Reserve was a low-use open space and was identified, with considerable support from the community, as a good location for a small wetland (Thompson Berrill Landscape Design 2009). The Cup and Saucer Wetland was designed to treat only a portion of the flow from Cup and Saucer Creek as the wetland size and there-

Table 6.3 Cup and Saucer Creek contribution to Cooks River and Botany Bay annual pollutant loads

Waterway	Area (ha, %)	Total suspended solids (T/year, %)	Total nitrogen (kg/year, %)	Total phosphorus (kg/year, %)	Reference
Cup and Saucer Creek	503	100	13,530	1420	MUSIC modelling report
Cup and Saucer contribution to Cooks River catchment	5.2%	2.4%	21%	18%	Botany Bay Water Quality Improvement Plan
Cup and Saucer contribution to total Botany Bay catchment	0.5%	0.4%	3.3%	3.2%	Botany Bay Water Quality Improvement Plan

fore inflow volume was limited by the size of the reserve. The wetland size is 0.225 ha and a model using the Model for Urban Stormwater Improvement Conceptualisation (MUSIC) software was developed to determine the optimal flow rate for this size wetland (Martens & Associates Pty Ltd 2009). It was estimated that the wetland would need to have dimensions of between 7.5–30 ha to achieve the load reduction targets of the Botany Bay water quality improvement plan (Cunningham and Birtles 2013). This highlights two challenges, the availability of space for urban retrofit projects and the need for many integrated interventions across a catchment to achieve regional waterway health goals. Despite its small size, the Cup and Saucer wetland was pursued as part of an integrated approach and due to the water quality treatment (albeit limited), ecological and aesthetic values. In this way, it functioned as a demonstration of how WSUD could be retrofitted into the urban landscape.

The MUSIC modelling results of the 0.225 ha wetland showed that the best treatment results would be achieved with a maximum inflow of 0.2 m^3/s. At this flow the wetland would achieve an annual load reduction of 5000 kg/year (5.1%), 130 kg/year (0.9%) and 40 kg/year (3.7%) for TSS, TN and TP respectively (Martens & Associates Pty Ltd 2009). A 2-year average recurrence interval (ARI) event for Cup and Saucer Creek has a peak flow of 43.9 m^3/s, the wetland is therefore designed to be most effective in treating smaller rainfall events with intensity <2-year ARI (PB MWH Joint Venture 2009, Sydney Water 1999).

A schematic of Cup and Saucer Wetland is shown in Fig. 6.5, with numbering of the key features as follows:

1. Inlet to the wetland for flow diverted from Cup and Saucer Creek
2. Sedimentation forebay
3. Cell 1 with various marsh and ephemeral zones
4. Pipe between cells
5. Cell 2 with various marsh and ephemeral zones

Fig. 6.5 Schematic of the Cup and Saucer Wetland near Sydney, Australia

6. Outlet pipe to Cooks River
7. Overflow spillway to Cooks River

Cup and Saucer Wetland was designed with two main cells. Water is gravity-fed from Cup and Saucer Creek, with flow controlled by an offtake weir. There is a rocked energy dissipater at the outlet pipe to the sediment forebay. The forebay has a 100 m^2 base and has been designed to capture 5000 kg/year of sediment, which is expected to require dredging every 10 years (Martens & Associates Pty Ltd 2009).

The flow then passes over a porous rock weir to the subsequent macrophyte zone of Cell 1. The flow travels through Cell 1 and is then piped to Cell 2 with a rocked headwall at the pipe outlet. The bathymetry of the Cell 1 and Cell 2 varies in depth creating shallow (depth 0.2 m), deep (depth 0.2–0.4 m) and submerged (depth 0.4–0.9 m) marsh zones, an ephemeral zone up to 0.3 m above the normal top water level (NTWL) and the wetland fringe at 0.3–0.6 m above NTWL. The density of vegetation in these zones is shown in Table 6.4. The key for identifying the layout of these zones in each cell is shown on the bottom right of Fig. 6.5.

The outflow pipe from Cell 2 has a grated inlet, which controls the water level. For overflow during storm events, the design includes rocked spillways 400 mm above the normal top water level between Cell 1 and Cell 2, and also between Cell 2 and a constructed passage to the Cooks River (Sydney Water 2010).

The wetland was planted with 27,650 aquatic and terrestrial local native plants, which have thrived and are now self-seeding. The terrestrial areas around the wetland were planted with a mixture of 27 different species consisting of trees,

Table 6.4 Number of plants and species for wetland zones

Planting zone	No. plants per m²	No. of species	Total no. plants
Rocked overflow	4	3	512
Wetland margin (0.3–0.6 m above NTWL)	6	5	2016
Ephemeral (up to 0.3 m above NWTL)	6	6	1974
Shallow marsh (depth 0.2 m)	6	8	7530
Deep marsh (depth 0.4 m)	6	6	2766
Submerged marsh (depth 0.4–0.9 m)	2	6	440

large and small shrubs, ground layer and tuft plants. The mixture of species and density per square metre varied across the wetland zones, as shown in Table 6.4.

Other design features of the wetland include a viewing area, seating, outdoor classroom, interpretive signs and a vegetated fauna passage to facilitate the movements of turtles, eels and other animals between Cooks River and the wetland. Figure 6.6 shows the Heynes Reserve before construction and the wetland after a 1 year establishment period. The benefits of the wetland were assessed in terms of water quality improvements, anecdotal evidence, performance against liveability indicators and economic analysis.

Water Quality

The low frequency of data points and lack of flow data presents difficulties in assessing the pollutant load and performance of the wetland. Standard practice indicates that at least 20 events should be monitored to understand the performance of a WSUD intervention (Landcom 2009). In this case, water quality of the wetland is generally tested twice per year, with some periods of more frequent influent sampling. The inflow to the wetland is not monitored. The lack of water quality data limits the quantitative assessment of the wetland performance against the design for water quality improvement. The available water quality data has been presented below to give an indication of the wetland performance.

Influent Water Quality

Water sampling was undertaken by the Georges River Combined Councils' Committee and the Cooks River Alliance. The wetland influent water quality was measured upstream of the offtake; TN, TP and turbidity sampling results are shown in Figs. 6.7 and 6.8, and 6.9. The figures also show the influent concentrations, the concentrations that were assumed as MUSIC model variables for base flow and for storm flow conditions and the daily rainfall on the sampling days. As seen in

Fig. 6.6 Heynes Reserve before construction and the wetland 1 year after construction

comparing the sampling results with the daily rainfall, the concentration across all three parameters does not show a consistent pattern with rainfall. In fact, the maximum values for TN, TP and turbidity were all recorded on days without rainfall. This indicates that there could be a high pollutant load for both the base and storm flows. There was no available flow data to understand the load to the wetland.

Effluent Water Quality

MUSIC modelling indicated the wetland as designed would achieve an annual load reduction of 5.1%, 0.9% and 3.7% for TSS, TN and TP, based on an inflow of 0.2 m^3/s. As the load to the wetland is unknown, the effluent water quality results are instead examined in terms of concentration. There are two sampling events per year. Figures 6.10, 6.11 and 6.12 show water quality results from six effluent grab samples between November, 2011 and March, 2014.

All results are highly variable. However, there is insufficient data to correlate this variability to wetland performance. In each data set there is a high concentration for the sample collected in April, 2012. There was no rainfall in the catchment on this day, nor in the week preceding, and so it could be evidence of an upstream pollution event or an issue with the wetland performance. It should be noted that these effluent sampling results were shortly after the wetland establishment period in 2010.

As well as the sampling described above, monthly *E. coli* monitoring was carried out between January to June, 2011, and the minimum, maximum and median values reported on the Cooks River Valley Association annual water quality report

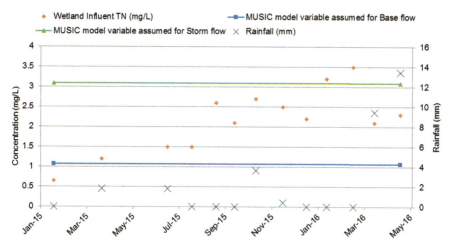

Fig. 6.7 Influent measured and modelled total nitrogen concentration with daily rainfall

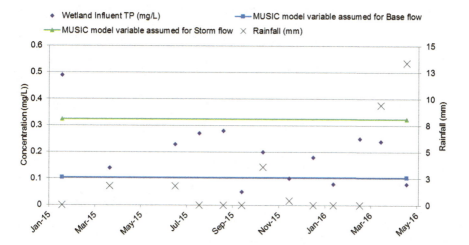

Fig. 6.8 Influent measured and modelled total phosphorus concentration with daily rainfall

(Cooks River Valley Association 2011). These results showed that the median removal of *E. coli* for the five dry weather samples was 91%. However, these do not give an indication of wet weather *E. coli* removal.

Without flow data or regular influent and effluent water quality testing, it is difficult to ascertain the wetland performance compared to the design intent. Instead, the benefits have been understood through identifying a number of other benefits offered by the wetland, which are outlined below (Cunningham and Birtles 2013).

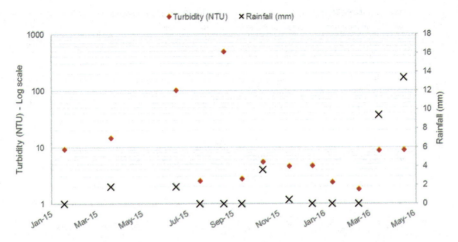

Fig. 6.9 Influent measured turbidity with daily rainfall. Note log-scale on y-axis

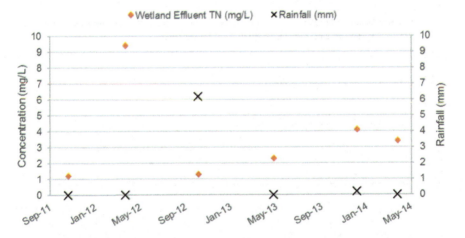

Fig. 6.10 Wetland effluent Total Nitrogen concentration with daily rainfall

Reported Benefits

Community Involvement

The community has been actively involved throughout this project. Community support for the project was evident from the outset when 86 resident surveys on the concept design were returned indicating 90% agreement with the option to construct a wetland to treat stormwater (Thompson Berrill Landscape Design 2009). This support was consistent throughout the project to the opening of the wetland, which 250 people attended (SMCMA 2011). Engagement with the community included: consultation with local community members and groups during the design

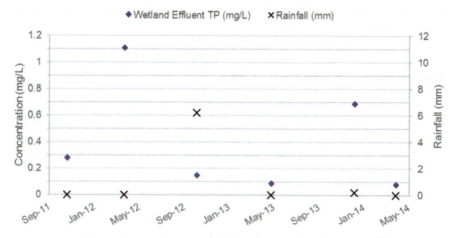

Fig. 6.11 Wetland effluent Total Phosphorus concentration with daily rainfall

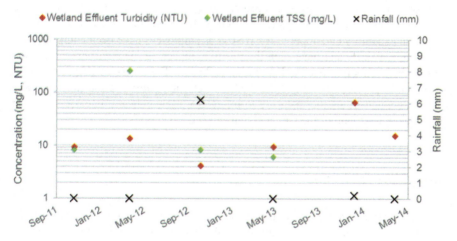

Fig. 6.12 Wetland effluent turbidity and Total Suspended Solids (TSS) concentration with daily rainfall. Note log-scale on y-axis

stage through surveys and meetings; groups and schools touring the site during construction and assisting with terrestrial planting; volunteers taking ownership in maintaining the wetland through communal volunteering days and conducting water quality sampling. This community engagement was facilitated by the local councils and community groups, such as the Cooks River Alliance.

In addition to the groups who have been actively involved in the wetland design, construction and maintenance, there have also been benefits to the broader community. The reserve is used more frequently than previously and direct feedback and monitoring of print and social media has shown positive reflections on the value of the public space (Cunningham and Birtles 2013). Moreover, interpretive signage and a video produced about the wetland educate the community on the wetland function, design and construction.

Habitat, Biodiversity and Local Climate

The freshwater wetland has created a unique habitat that has improved biodiversity. The planted vegetation has thrived since construction and is now self-seeding. Anecdotal and photographic evidence suggests the wetland is well populated with native birds, macroinvertebrates, frogs and turtles. It is expected that the vegetation has helped to reduce urban heat in the area. Heat mapping has been planned to measure this.

Interagency Rapport and Organisational Capability and Appetite

The organisations involved in this project have reported strengthened rapport with partner organisations and with the community. The project has increased the organisational capability and appetite for WSUD as employees have been equipped with the required skills and have seen the positive impact on the environment and the community. This improved collaboration and capability has since helped to facilitate similar projects. For example, further river bank naturalisation in the Botany Bay catchment as another measure to improve waterway health is currently being undertaken upstream of the Cup and Saucer Wetland site.

Liveability Indicators

The qualitative benefits listed above have anecdotally shown the broad range of outcomes. However, it is difficult to use this anecdotal evidence to demonstrate the business case for investing in similar future solutions. One way that the Australian water industry is trying to better measure the outcomes of these sorts of projects is through liveability indicators to capture and quantify, in a simplified way, the environmental and social benefits in addition to a financial evaluation. There are various indicators available; this paper focuses on the liveability indicators published in 2016 by the Water Services Association of Australia (WSAA 2016). The WSAA indicators aim to help water utilities to contribute to amenity and community wellbeing, productivity, sustainability and future focus. While water quality outcomes are measurable, these indicators help to quantify the other benefits that have been outlined. The relevant indicators with their metric have been included in Table 6.5 and are compared against the Cup and Saucer Wetland outcomes.

6 The Role of Constructed Wetlands in Creating Water Sensitive Cities

Table 6.5 Performance of Cup and Saucer Wetland measured against liveability indicators

Indicator	Metric	Cup and Saucer Wetland outcome
Days exceeding critical heat threshold	Number of days per annum exceeding critical heat threshold	It is difficult to measure this metric on a local scale. There are plans for heat mapping of the area around the wetland to understand the microclimate impacts
Infrastructure land available for community purposes	Number of hectares of land available for community access or Number of licences for recreational use of land	0.6 ha of land available
Length of paths and cycleways providing connectivity	Kilometres of path on utility land	0.28 km walking and bike paths
Contribution to water sensitive urban design/stormwater harvesting	Record of WSUD features directly contributing or Volume of water treated through WSUD feature or Money contributed to schemes	Up to a maximum of 17.28 mL/day of water treated through a WSUD feature
Value-add of projects delivered through collaboration	Value added ratio of dollars spent to dollars of value delivered (not currently measured by utilities)	Collaboration between utility and local councils, but value add not measured
Native vegetation gain	Total area vegetation gain	5000 m^2 of vegetation gain
Community water literacy	*Not currently measured by utilities but potential to be a useful metric*	Interpretive signage and community involvement through wetland maintenance
River health	*Water utilities have limited influence over this indicator but it is a valuable context or outcome metric to track*	Treatment performance outcomes not sufficiently measured to track project benefits to river health

These indicators provide a simple means for post-implementation evaluation of the broader benefits of the Cup and Saucer Wetland. They can also provide incentives for measuring outcomes, such as water quality, which was lacking in the Cup and Saucer Wetland case study. Not only do these indicators provide a means of understanding value, but they are also important in setting direction and purpose at the beginning of a project.

Economic Analysis

A more sophisticated post-implementation evaluation was undertaken to understand the economic benefits of the improvements at Heynes Reserve (Thomy et al. 2016, Morrison et al. 2016). These improvements include the wetland construction and improvements to the Cooks River riparian zone near the confluence with Cup and Saucer Creek. The economic benefits study used the housing market data to indicate amenity values. The study methodology used hedonic valuation, based on the assumptions that people pay extra for houses that have views, that are close to parks and that are close to rivers that are in good health. The increase in housing value for 4711 properties surrounding Heynes Reserve was modelled, based on the improvement to vegetation cover and riparian condition and the proximity of the house to the improvement. The calculation showed that the total increase in housing value was in the order of $16 million (AUD).

Cup and Saucer Wetland Outcomes

Through anecdotal evidence, indicators and economic analysis the various benefits to liveability are clear. However, the environmental outcomes for water quality and biodiversity were not sufficiently quantified to assess if there has been any improvement. The case study has shown some challenges in constructing wetlands in urban areas. The water quality outcomes were largely limited by the available space for the retrofit project. There was also little post-implementation evaluation of the environmental outcomes. Extensive environmental monitoring is not required through regulation and the funding is often limited. Water quality monitoring is more often conducted for research purposes rather than for wetland performance verification. Through the Cooks River Urban Water Initiative, Cup and Saucer Wetland was one of a suite of projects aimed at taking an integrated approach to improving the health of the waterway. This was a good approach for overcoming the challenge of space availability for large-scale WSUD interventions. The liveability metrics and economic analysis built a strong case for the value of this project, however, the environmental outcomes should be better understood for similar WSUD projects to move from demonstrative to standard practice in the future.

Global Future Cities

The issue paper on 'urban infrastructure and basic services, including energy' from UN Habitat conference III (United Nations 2015) outlined the key actions for meeting the challenges facing cities:

- Policy reform
- Building viable and well-managed institutions

- Developing effective and integrated infrastructure planning
- Developing new business models
- Creating strategic partnerships to foster and apply technological innovation

The Cup and Saucer Wetland has demonstrated how integrated infrastructure planning and strategic partnerships are occurring at local scales. The outcomes of the wetland also inspired institutional commitment and improved the capacity of the involved utility and councils to engage in further WSUD projects. However, there are evidently still some gaps in policy, regulation and funding needed to increase the number and effectiveness of these integrated projects. This section shows the use of benchmarking to better define the specific gaps and challenges for urban water management. It looks at the principles of water sensitive cities and examines the key challenges of transitioning to a water sensitive city state.

Waterways City

As described in Brown et al. (2009) the waterways city state emerged in Australia in the late 1960s and 1970s. This occurred when there was a shift away from seeing the environment as benign and instead, normative views focused on environmental protection and valuation of the visual and recreational benefits of water. This shift was influenced by visible pollution of waterways and beaches and also by the global environmentalism movement. It resulted in measures such as environmental regulation for wastewater discharges and systems to protect urban waterways from stormwater pollution. This included an increase in WSUD solutions being applied to manage both stormwater quantity and quality, but, they are still not yet standard practice.

Morison (2008) identified the key barrier to broader implementation of WSUD solutions is the lack of capacity and commitment of local authorities. Morison's study looked at local municipalities' stormwater management capacity and commitment in terms of funding, training, expertise and regulation, as well as their WSUD capacity and commitment. The results showed a great diversity according to geography, density, planning regulation and economic status. This is especially problematic for the consistent management of a shared catchment in the context of transition to the waterways city paradigm and highlights the need for integrated planning.

The Cup and Saucer Wetland is a good example of progress towards a waterways city. It was driven by a need to improve environmental health and creation of a functional green space for the public. Perhaps one of the greatest successes of this project is the commitment and collaboration between multiple local municipalities. Cup and Saucer Wetland was a partnership between Sydney Water, Canterbury City Council and the Sydney Metropolitan Catchment Management Authority, with part funding from the Australian Federal Government. The project also involved significant input and support from local community groups such as the Cooks River Alliance. The successful collaboration has resulted in increased capacity and organisational appetite for WSUD projects. This appetite was largely encouraged by the strong community engagement and economic valuation. These two outcomes have clearly demonstrated the potential liveability benefits of small urban retrofit projects. The increased organisational commitment to WSUD projects has been seen in further stormwater interventions at higher points in the catchment to ensure a whole-of-catchment approach to stormwater management. This project has successfully promoted the normative values of a waterways city, and is starting to shift the approach to local water management thereby further encouraging WSUD solutions as standard practice.

Water Cycle City

The subsequent city state described on the continuum is the water cycle city (Brown et al. 2009). The water cycle city builds on the environmental values articulated in the waterways city vision to recognise that there are limits to water sources and to a waterway's ability to absorb pollutants. Moreover, this city state incorporates broader values of social, economic and environmental sustainability. The features of a water cycle city may include water conservation, wastewater recycling, stormwater harvesting and reuse. These features could be delivered through both centralised and decentralised technologies with sensitivity to other variables such as energy and nutrient flux within the urban setting. The delivery of such a city requires an interdisciplinary approach with input from business, communities and government. While regulation and water scarcity have driven this approach in some areas, there still lacks systematic, institutional effort to achieve this vision (Mukheibir 2015). In Australia, this leads to difficulties implementing green infrastructure and decentralised systems. These difficulties are compounded by the traditional water supply, sewered and drained phases of urban water management that favoured centralised, government funded infrastructure for discrete water, wastewater or stormwater services.

There are some examples of the use of constructed wetlands in Sydney to move towards a Water Cycle City. For example, the Sydney Park Stormwater Harvesting Scheme uses constructed wetlands to realise the targets laid out in the City of Sydney's Decentralised Water Master Plan, in particular to: reduce the sediment by 50% and nutrients by 15% entering waterways from stormwater; reduce council water consumption by 10% of 2006 levels; and provide 30% of the local government area water demand from recycled sources (City of Sydney 2012). The City of Sydney Council has shown leadership in taking a long-term approach and using decentralised solutions to diversify the water supply options, while at the same time improving waterway health, increasing biodiversity and enhancing the amenity of the area. The scheme was driven by goals set in the Sustainable Sydney 2030 Community Strategic Plan and in the City of Sydney's Decentralised Water Master Plan and has been designed to provide drought-resilient water supplies for the council's parklands, street cleaning and other non-essential uses. The development of these plans was led by the council with broad community involvement. The plans take account of and build on the regional and state planning and require collaborative implementation (City of Sydney 2014). The scheme has been recognised as an example of the required political and institutional commitment of the City to build capacity, and the collaborative and consultative approach that is needed to overcome institutional challenges and find alternative funding sources to implement water cycle city projects (Mukheibir and Howe 2015).

Transitioning to a Water Sensitive City

The Cup and Saucer Wetland and Sydney Park Stormwater Harvesting Scheme examples show positive change in terms of effort and commitment to sustainable management of stormwater and diversification of urban water sources. However, the creation of water sensitive cities requires a major, city-wide transformation in the perceptions and value of water and away from the conventional technical approach to urban water management. Some of the challenges to such a transformation include identifying a common vision for the future, community participation, institutional capacity and governance and understanding the true costs and benefits (Ison et al. 2009). Understanding the true costs and benefits provides evidence and incentive to move away from traditional water infrastructure or investment decisions based solely on capital and operational costs. Meeting these challenges in an integrated way across a city needs widespread and unified effort and so it is important to benchmark the state of the city's urban water management, prioritise actions and track progress.

Brown et al. (2016) have developed a transitions framework to assist with prioritising actions for this transformation to occur. According to this framework, the first step is to set out the different areas where change and capacity building must occur. These areas have been defined as 'domains of change' and encompass:

- Actors: Key networks of individuals
- Bridges: (Semi) Formalised organisations, structures and processes for coordination and alignment
- Knowledge: Research, science and contextualised knowledge
- Projects: Experiments, demonstrations and focus projects
- Tools: Legislative, policy, regulative and practice tools

As a city transitions to being water sensitive, each of these domains becomes more sophisticated and the city becomes more capable of progress for positive change. In each of these domains, there are six expected stages of transition from identifying and quantifying the problem, agreeing on and building confidence in the possible solutions, implementing the required policy and governance to embedding the new practices. In Table 6.6, these transition stages and the progress in each domain of change are seen on the left side. This table is coined the transition dynamics framework and has been developed with the intention of using these indicators to benchmark a city's progress towards being a water sensitive. Table 6.6 also

Table 6.6 Transition dynamics framework from Brown et al. (2016), with examples of case study contributions to city state transitions

Transition phase	Domains of change					Description of Cup and Saucer Wetland contribution to transitioning to a Waterways City
	Actors	Bridges	Knowledge	Projects	Tools	
1. Issue emergence	Issue activists	N/A	Issue discovery	High profile scientific studies	N/A	Issue of water pollution well established in 1990s. Water quality testing programs completed for Cooks River in 1990s
2. Issue definition	Science leaders	Science-industry	Cause-effect	Laboratory-based and scientific solution prototypes	N/A	Results of water quality testing informed stormwater management plans developed by 13 local councils, the catchment management committee, the metropolitan water utility and the state road and traffic authority
3. Shared understanding and agreement	Technical solution coalition	Science-industry-policy	Basic technological solutions	Minor scientific field demonstrations	Draft best-practice guidelines	Stormwater management plan (Cooks River Catchment Association of Councils 1999) options were to maintain natural waterways, source control and "end-of-pipe" solutions such as GPTs, detention basins, litter booms and wetlands. The plan set an action to build Cup and Saucer Wetland. Requirement to improvement of waterway reflected in Botany Bay water quality improvement plan (Kelly and Dahlenburg 2011)

4. Knowledge dissemination	Informal policy coalition	Science-industry-policy-capacity building	Advanced technological solutions	Major scientific field demonstrations	Best-practice guidelines, targets	WSUD technologies well developed and best-practice guidelines in place such as CSIRO WSUD Engineering Procedures (Melbourne Water 2005). Cooks River Urban Water Initiative (2008–2011) used a portfolio of best-practice solutions to address water quality issue
5. Policy and practice diffusion	Policy and decision coalition	Science-industry-policy-capacity building	Modelling solutions, capacity building	Numerous industry-led field experiments	Legislative amendment, market offsets, national best-practice guidelines, regulatory models	Cup and Saucer wetland built in 2010 with design based on MUSIC modelling results. Industry-led application of WSUD resulting in increased institutional and community capability and appetite for WSUD projects. Needed more demonstrated water quality outcomes to support move to standard practice. WSUD intervention not strongly supported by regulatory models
6. Embedding new practice	Multi-agency coalition	Formalised institution	Next research agenda	Standard practice	Political mandate, coordinating authority, comprehensive regulatory models and tools	Successful multi-agency coalition, though mostly coordinated through informal bridges. Still require further embedding of WSUD retrofit projects to become standard practice. Changes to policy and regulation could provide incentive for investment and post-implementation evaluation of WSUD projects in the future

Table 6.7 Cup and Saucer Wetland outcome for each domain of change

Domain of change	End goal	Current phase	Cup and Saucer Wetland outcome
Actors	Multi-agency coalition	Multi-agency coalition	Good multi-agency coalition between Sydney Water, councils and community groups
Bridges	Formalised institution	Science-industry-policy capacity building	No formalised bridging organisation. The collaboration was driven and coordinated by a few individuals
Knowledge	Next research agenda	Modelling solution, capacity building	Used established technology and modelling to develop the wetland. Still building capacity for this knowledge to be widespread in organisations. However, this project did increase the WSUD appetite and capacity
Projects	Standard Practice	Numerous industry-lead field experiments	This is one of numerous industry-led field experiments, however, the implementation of these projects still relies on opportunistic funding and is not yet standard practice
Tools	Political mandate, coordinating authority, comprehensive regulatory models and tools	Best-practice guidelines, targets	The relevant local council requires collection and onsite detention of stormwater in the Development Control Plan, but no specific requirements for WSUD

includes a brief description (in the right-hand column) that describes to what extent the case study example has contributed to the transition to a new state of water management. The transition dynamics framework has been developed to be used on a city-wide scale, however, the case study descriptions included are focused on the outcomes that Cup and Saucer Wetland has had on a local scale. Nevertheless, this is useful to visualise the use of this framework and to better understand the outcome of the case study.

For each domain of change, it is possible to benchmark how far through the transition phases a city is. For example, Table 6.7 summarises the outcomes of the Cup and Saucer Wetland project for each domain of change. This then highlights what domains of change are a priority for further action. The Cup and Saucer Wetland example indicates that the priority for action is to develop the necessary tools to transition from knowledge dissemination to policy and practice diffusion and embedding new practice. For example, in this case the relevant local council only required onsite detention of stormwater and did not have specific WSUD requirements in place. Prioritising these tools would include legislative amendment, regulatory models and political mandate to progress to water sensitive practices being the standard approach. The other highlighted gaps were that while there was good multi-agency coalition, there was no formal bridging institution for coordination and alignment of water sensitive initiatives. It also showed that this type of project is not yet standard practice, as there are still gaps in capacity building and lack of evidence from robustly evaluated industry-led field experiments.

Though not addressing the whole picture, this case study helps to demonstrate the use of the transition framework to understand the progress of a city towards transformation into a water sensitive city. The application of this framework for city-wide benchmarking of a city's water management state and its progress through the transition phases occurs through a thorough assessment of the state of urban water management by reviewing the literature and engaging with a diverse range of stakeholders, including water and related sector professionals, researchers, community and government. The assessment outcomes highlight which areas of the transition dynamics framework are well developed and which should be prioritised to further the city's efforts in becoming water sensitive.

After prioritising target areas for improvement, it is important to measure the success of the implemented projects. As articulated by Polyakov et al. (2015), the evaluation of the costs and benefits of WSUD technologies tends to occur before construction to ensure the chosen option gives the most benefit to the community, but there are few examples in the literature of comprehensive project evaluation after completion. Post-implementation evaluation is important to inform the decision-making process going forward. The use of indicators plays an important role in embedding water sensitive practices by providing a measure of values not commonly included in a business case. Currently, projects are often measured based on their financial viability, rather than their holistic value. In Australia, this is exacerbated by pricing regulation, which requires assessment of the least cost rather than the highest value options. Post-completion project evaluation is important as it demonstrates, and sometimes quantifies, the value the project economically, environmentally and socially. While project costs are easily measured, it is less straightforward to evaluate the environmental and social impact of a project. Indicators can be a useful tool in quantifying the costs and benefits and linking them back to the overall project goals. In this way indicators can be helpful for decision makers to understand the contribution of existing projects to creating water sensitive cities and to direct the focus of future projects. Indicators, such as WSAA's liveability indictors and the relevant SDG indicators can be used in conjunction with tools such as the Economics of Ecosystems and Biodiversity database developed by Sukhdev et al. (2010) to qualify and quantify the environmental and social benefits, which generally do not have a direct market value and are otherwise difficult to measure.

Benchmarking cities and tracking our progress in this way forms a powerful tool for comparative analysis and the opportunity to learn from the experience of other cities. In fact, it is envisaged that this benchmarking may facilitate prioritising actions for developing cities to "leapfrog" to a more advanced city state (Jefferies and Duffy 2011; Brown et al. 2016). While the Australian city state transitions that have been examined in this chapter have seen progressions between subsequent city states, the concept of "leapfrogging" proposes that it is possible to find alternative paths of development and therefore skip over one or more of the interim states.

This concept of leapfrogging to more advanced sustainable urban water management was explored by Poustie et al. (2016) with a pilot project in Port Villa on the South Pacific island of Vanuatu. Baseline benchmarking, including an understanding of the evolution of urban water management, showed that Port Villa is currently in a state between a water supply and sewered city. This city state is similar to many developing cities around the world, as identified through the UNESCO SWITCH benchmarking project (Brown et al. 2016). Poustie et al. (2016) noted the key elements that will potentially enable leapfrogging from this state to a more sustainable city state; these included the knowledge, engagement and motivation of the key stakeholders, appropriate policy, legislation and planning processes, accurate data and collaboration and data sharing between organisations.

The transition to water sensitive cities requires a major shift in current practice. It proposed that this shift can be facilitated by: benchmarking a city's state of urban water management; identifying the desired city state; using the transition framework to identify key gaps in the stages of transformation to the desired city state; and using relevant indicators to better understand the full economic as well as non-monetary costs and benefits of the applied water solutions. In Australia, urban water managers can prioritise future action in this way to continue progress to cities becoming water sensitive, while in developing countries this may help them to set a vision and actions to leapfrog to a more sustainable city state. For the shift in practice to occur strong champions of change are required. A key challenge for a champion of change in a developing city is to unite the vision of the urban water managers to embrace water sensitive practices, rather than traditional centralised approaches seen in developed cities. For these champions, case studies of waterways and water cycle city initiatives, such as those presented in this chapter, can be useful to inspire and can evidence the benefits for the city in terms of environmental, social and economic impacts. Moreover, constructed wetlands and other WSUD interventions are low-technology solutions and may be translatable to developing country and city contexts.

Conclusions

This chapter has outlined the future vision for cities to be liveable, sustainable and resilient. SDG 11 sets targets to achieve this across urban development sectors to ensure that cities are providing the infrastructure services as articulated in the Habitat I Vancouver Declaration, but are also enhancing the liveability of a city as described in the New Urban Agenda. The concept of Water Sensitive Cities sets goals specifically related to urban water management and its contribution to sustainable urban development. Central to achieving these goals is embedding WSUD practices. WSUD technologies help to provide ecosystem services to cities and diversify supply thereby improving a city's resilience to climate change. Cup and Saucer Wetland is an example of a technology that contributes to building a Waterways City. It was constructed as one of a suite of projects to improve the water quality of the Cooks River. The pollution reduction of the wetland was limited by the size of the space available. However, it was part of an integrated approach aimed at achieving the regional water quality improvements required in the Botany Bay Water Quality Improvement Plan. Although the wetland was not designed to achieve extensive pollution reduction, other benefits have been evidenced. These have included community involvement, improving the habitat, biodiversity and local climate. Moreover, the wetland project showed signficant increase to local housing prices, has demonstrated strong inter-agency collaboration and has built the appetite and capability of these agencies for future WSUD projects.

The wetland outcomes were assessed in terms of the contribution to the local area's transition towards a water sensitive city state. It was seen that more work needs to be done to embed WSUD as standard practice. The water sensitive cities transition dynamics framework helps understand the current state of transition and identify areas to prioritise action. This is complemented by indicators, such as WSAA's liveability indicators and relevant SDG indicators that can measure the progress and impact of water sensitive practices. The transition dynamics framework and indicators can be used in conjunction to set the direction for what actions a city should prioritise in terms of building inter-agency collaborations, furthering knowledge and technical capability, setting policy and regulation, and developing guidelines and tools. At the same time, the framework and indicators enhance understanding of the specific impact that WSUD solutions are having on a city and choosing technical solutions that will address the gaps in liveability, sustainability and resilience. The urban water management transition and transition dynamics frameworks and case study examples assist in building a case for how developing cities can leapfrog city states to transition to an advanced state of sustainable urban water management. It has been discussed that key elements to enabling this leapfrogging are the involvement, motivation, and commitment of the stakeholders, enabling policy and legislation, accurate and available data and sharing of knowledge and data between organisations. With these key elements, by using the theoretical framework of Water Sensitive Cities and adopting technologies such as constructed wetlands, urban water managers can assist cities in both developed and developing countries to transition to being liveable, sustainable and resilient.

Acknowledgments The author would like to acknowledge the contribution of Dr. Nidhi Nagabhatla and Dr Chris Metcalfe at the United Nations University Institute for Water, Environment and Health, the input and assistance from Daniel Cunningham, Phillip Birtles and Kaia Hodge at Sydney Water and the Cooks River Alliance for their provision of water quality data.

References

ANZECC, ARMCANZ (2000) Australian and New Zealand guidelines for fresh and marine water quality. Australian and New Zealand Environment and Conservation Council and Agriculture and Resource Management Council of Australia and New Zealand, Canberra

Bi P, Williams S, Loughnan M, Lloyd G, Hansen A, Kjellstrom T, Dear K, Saniotis A (2011) The effects of extreme heat on human mortality and morbidity in Australia: implications for public health. Asia Pac J Public Health 23(2):27S–36S

Binney P, Donald A, Elmer V, Ewert J, Phillis O, Skinner R, Young R (2010) IWA Cities of the Future Program—Spatial Planning and Institutional Reform. Discussion Paper for the World Water Congress, Montreal, QC, Canada

Brown RR, Keath N, Wong THF (2009) Urban water management in cities: historical, current and future regimes. Water Sci Technol 59(5):847–855

Brown RR, Rogers B, Werbeloff L (2016) Moving toward Water Sensitive Cities: a guidance manual for strategists and policy makers. Cooperative Research Centre for Water Sensitive Cities, Cooperative Research Centre for Water Sensitive Cities, Melbourne, Australia

City of Sydney (2012) Decentralised Water Master Plan 2012–2030. http://www.cityofsydney.nsw.gov.au/__data/assets/pdf_file/0005/122873/Final-Decentralised-Water-Master-Plan.pdf

City of Sydney (2014) Sustainable Sydney 2030: Community Strategic Plan 2014. http://www.cityofsydney.nsw.gov.au/__data/assets/pdf_file/0005/209876/Community-strategic-plan-2014.pdf

Cooks River Alliance (2014) Cooks River Alliance: Management Plan 2014. http://cooksriver.org.au/publications/cooks-river-alliance-management-plan-2014_final-draft_20-june-2014/

Cooks River Catchment Association of Councils (1999). Cooks River Stormwater Management Plan. Sydney, Australia

Cooks River Valley Association (2011) Cooks River Valley Association, Annual Water Quality Report 2010/11. http://www.crva.org.au/wp-content/uploads/2013/05/Jun2011-Report-Final.pdf

Cunningham D, Birtles P (2013) The benefits of constructing an undersized wetland. In: Proceedings of water sensitive urban design 2013: WSUD 2013, Canberra

Deletic A, McCarthy D, Chandrasena G, Li Y, Hatt B, Payne E, Zhang K, Henry R, Kolotelo P, Randjelovic A, Meng Z (2014) Biofilters and wetlands for stormwater treatment and harvesting. Cooperative Research Centre for Water Sensitive Cities, Monash University, Melbourne

Fletcher T, Duncan H, Poelsma P, Lloyd S (2004) Stormwater flow and quality and the effectiveness of non-proprietary stormwater treatment measures: a review and gap analysis. Cooperative Research Centre for Catchment Hydrology, Melbourne, Australia

Folke C (2006) Resilience: the emergence of a perspective for social–ecological systems analyses. Glob Environ Chang 16(3):253–267

Francey M (2005) WSUD engineering procedures: stormwater. Melbourne Water, CSIRO Publishing, Melbourne

Hynes HBN (1975) Edgardo Baldi Memorial Lecture: The stream and its valley. Verhandlungen der Internationalen Vereinigung fur theoretische und angewandte Limnologie 19:1–15

International Water Association (2016) The IWA Principles for Water Wise Cities. http://www.iwa-network.org/wp-content/uploads/2016/08/IWA_Principles_Water_Wise_Cities.pdf

Ison RL, Collins KB, Bos JJ, Iaquinto B (2009) Transitioning to Water Sensitive Cities in Australia: a summary of the key findings, issues and actions arising from five national capacity building and leadership workshops. NUWGP/IWC, Monash University, Clayton. http://www.watercentre.org/resources/publications/attachments/Creating%20Water%20Sensitive%20Cities.pdf

Jefferies C, Duffy A (2011) The SWITCH transition manual. University of Abertay, Dundee, UK

Kelly R, Dahlenburg J (2011) Botany bay water quality improvement plan. Sydney Metropolitan Catchment Management Authority, Sydney

Khan SJ, Deere D, Leusch FD, Humpage A, Jenkins M, Cunliffe D (2015) Extreme weather events: should drinking water quality management systems adapt to changing risk profiles? Water Res 85:124–136

Landcom (2009) Water sensitive urban design book 4: maintenance. Parramatta, Australia. http://www.landcom.com.au/publication/water-sensitive-urban-design-book-4-maintenance-draft/

Martens & Associates Pty Ltd (2009) MUSIC modelling: Cup and Saucer Creek stormwater treatment wetland project. Martens Consulting Engineers, Sydney

Melbourne Water (2005) WSUD Engineering Procedures: Stormwater. CSIRO Publishing, Australia

Morison PJ (2008) Creating a "waterways city" by addressing municipal commitment and capacity: the story of Melbourne continues. In: 11th international conference on urban drainage, Edinburgh

Morrison M, Duncan R, Boyle K, Thomy B, Weibin X, Parsons G, Bark R, Burton M (2016) River Health Project Report: the value of improving urban stream health in the Cooks River and Georges River Catchments. Appendix: Benefit analysis for improvements at Heynes Reserve, Earlwood, NSW. https://www.sydneywatertalk.com.au/21003/documents/48485

Mukheibir P (2015) Integrating 'one water' into urban liveability. Water: Journal of the Australian Water Association 42(6):40

Mukheibir P, Howe C (2015) Pathways to 'One Water': a guide for institutional innovation. Water Environment Research Foundation

NRMMC, EPHC, NHMRC (2009) Australian guidelines for water recycling: stormwater harvesting and reuse. Published for the Natural Resource Management Ministerial Council, the Environment Protection and Heritage Council, and the National Health and Medical Research Council. https://www.environment.gov.au/system/files/resources/4c13655f-eb04-4c24-ac6e-bd01fd4af74a/files/water-recycling-guidelines-stormwater-23.pdf

NSW Government Department of Water & Energy (2008) Interim NSW Guidelines for Management of Private Recycled Water Schemes. https://www.metrowater.nsw.gov.au/sites/default/files/publication-documents/Private_Recycled_Water_guideMay2008.pdf

Pachauri RK, Allen MR, Barros VR, Broome J, Cramer W, Christ R, Church JA, Clarke L, Dahe Q, Dasgupta P, Dubash NK (2014) Climate change 2014: synthesis report. Contribution of Working Groups I, II and III to the Fifth Assessment Report of the Intergovernmental Panel on Climate Change, IPCC, Geneva

PB MWH Joint Venture (2009) Cooks river flood study. Report prepared for Sydney Water, February 2009, Sydney, Australia

Polyakov M, Fogarty J, Zhang F, Pandit R, Pannell DJ (2015) The value of restoring urban drains to living streams. Water Res Econ 2015:211–222

Poustie MS, Frantzeskaki N, Brown RR (2016) A transition scenario for leapfrogging to a sustainable urban water future in Port Villa, Vanuatu. Technol Forecast Soc Chang 105:129–139

Ruth M, Franklin RS (2014) Livability for all? conceptual limits and practical implications. Appl Geogr 49:18–23

SMCMA (2011) Cooks river urban water initiative: final progress report. Sydney Metropolitan Catchment Management Authority, Sydney

Sydney Water (1999) Cup and Saucer Creek SWC 77: capacity assessment. Internal report, Sydney Water

Sydney Water (2010) Cup and Saucer Creek Wetland Technical Specification. Sydney Water

Sukhdev P, Wittmer H, Schröter-Schlaack C, Nesshöver C, Bishop J, ten Brink P, Gundimeda H, Kumar P, Simmons B (2010) The economics of ecosystems and biodiversity: mainstreaming the economics of nature: a synthesis of the approach, conclusions and recommendations of TEEB. European Communities

Thompson Berrill Landscape Design (2009) Cooks River Bank Naturalisation Project: final report, community consultation outcomes on the draft concept designs. Thompson Berrill Landscape Design Pty Ltd., Sydney

Thomy B, Morrison M, Boyle K, Bark R (2016) Hedonic price value of riparian vegetation in the georges and cooks river catchments of Sydney, Australia. Internal report presented by Charles Sturt University, Sydney, Australia

United Nations (1976) The Vancouver Declaration on Human Settlements. Habitat: United Nations Conference on Human Settlements, Vancouver, BC, Canada

United Nations (2015) Habitat III Issue Paper 18: urban infrastructure and basic services, including energy. Habitat III 2016, United Nations Conference on Housing and Sustainable Urban Development

United Nations (2016) New Urban Agenda. Habitat III, United Nations Conference on Housing and Sustainable Urban Development, Quito, Ecuador

United Nations DESA, Population Division (2014) World Urbanization Prospects: The 2014 Revision. (ST/ESA/SER.A/366)

United Nations DESA, Population Division (2015) World Population Prospects: The 2015 Revision, (ESA/P/WP.241)

United Nations General Assembly (2015) Transforming Our World: The 2030 agenda for sustainable development. United Nations, New York

Wong THF, Allen R, Brown RR, Deleti A, Gangadharan L, Gernjak W, Jakob C, Johnstone P, Reeder M, Tapper N, Vietz G (2011) Stormwater management in a water sensitive city: Blueprint 2011. Cooperative Research Centre for Water Sensitive Cities, Melbourne, Australia. ISBN: 9781921912009

Wong THF (2006a) Water sensitive urban design-the journey thus far. Australian Journal of Water Resources 10(3):213–222

Wong THF (2006b) Australian runoff quality: a guide to water sensitive urban design. Engineers Australia

Wong THF, Brown RR (2009) The water sensitive city: principles for practice. Water Sci Technol 60(3):673–682

Water Services Association of Australia (2016) Occasional Paper 31, Liveability Indicators: A report prepared for the water industry. https://www.wsaa.asn.au/sites/default/files/publication/download/Liveability%20Indicators%202016.pdf

Chapter 7
Methylmercury in Managed Wetlands

Rachel J. Strickman and Carl P.J. Mitchell

Introduction

This chapter focuses on mercury transformations in surface-flow artificial wetlands of a relatively small size that are managed for stormwater control or habitat creation. Referred to as stormwater or habitat wetlands, or collectively "artificial wetlands", these pondlike structures are an increasingly popular management strategy in urban and suburban watersheds of North America (Smith et al. 2002). There are many benefits to these low-impact, relatively low-cost wetlands, including reduction of multiple types of pollutant loads, protection from flooding and erosion, attenuation of pathogens, maintenance of a functional hydrologic cycle, habitat provision, and improved aesthetics (Brix 1997; Kivaisi 2001; Vymazal 2007; Malaviya and Singh 2012).

Artificial wetlands are a potentially important, but understudied component of the global mercury cycle. Like other aquatic environments, artificial wetlands are sites for the production of methylmercury (Strickman and Mitchell 2017), a neurotoxin that is synthesized from inorganic mercury (IHg) by a variety of microorganisms, primarily in waterlogged anoxic environments (Compeau and Bartha 1985; Kerin et al. 2006; Hamelin et al. 2011; Gilmour et al. 2013; Podar et al. 2015). Methylmercury (MeHg) is readily bioaccumulated, with both the total mercury burden, and the proportion of mercury in an organism present as MeHg, increasing with each trophic level in invertebrates, fish and mammals (Hall et al. 1997; de Wit et al. 2012; Edmonds et al. 2012), and reaching levels more than 10^6 times higher than those in the water of the aquatic habitat (Pickhardt et al. 2002). For individuals who regularly consume highly contaminated fish and shellfish, the neurotoxic effects of MeHg in humans can include tremors, visual and auditory problems, motor impairment, increased risk of epilepsy and attention-deficit hyperactivity

---

R.J. Strickman (✉) • C.P.J. Mitchell
University of Toronto, Scarborough Campus, Scarborough, ON, Canada
e-mail: rachel.strickman@googlemail.com

© Springer International Publishing AG 2018
N. Nagabhatla, C.D. Metcalfe (eds.), *Multifunctional Wetlands*,
Environmental Contamination Remediation and Management,
DOI 10.1007/978-3-319-67416-2_7

disorder, and mild to severe cognitive deficits depending on the dose, route, and age at exposure (Mergler et al. 2007; Yuan 2012; Boucher et al. 2012). Adverse effects can be significant even at very low doses (Karagas et al. 2012), particularly if exposure takes place during a vulnerable developmental stage *in utero* or as an infant (Mergler et al. 2007). For example, it is estimated that in the United States alone, as many as 300,000 children are born each year who have been exposed *in utero* to MeHg at levels exceeding those considered safe (Mahaffey et al. 2003). The resulting subclinical intellectual impairment is estimated to translate to lost earnings of up to $43.8 billion USD per year (Trasande et al. 2005). Currently, the WHO recommends that MeHg concentrations in fish consumed as food should not exceed 460 ng/g wet weight, with lower concentrations being desirable for more frequently consumed fish species or more sensitive human populations (World Health Organization 2017).

Methylmercury is also a health threat to wildlife, particularly to higher-trophic level organisms that depend on fish or aquatic insect prey, where exposure may lead to reproductive impairments, hormonal disturbances, behavioral changes, and possibly, acute toxicity. The concentrations of MeHg in a prey species that represents a health threat to its consumer will necessarily depend on the tropic level and the biology of the predator. The limited evidence available for insectivorous birds indicates that concentrations in insect prey above 100,000 ng/g are a risk for acutely toxic effects (Wiener et al. 2003), while total mercury concentrations in insects exceeding 970 ng/g have been associated with reduced reproductive success in birds (Brasso and Cristol 2007). The thresholds of reproductive effects or acute toxicity for piscivorous birds and mammals are lower, estimated at 400 ng/g and 1100 ng/g, respectively (Scheuhammer et al. 2007).

While there is considerable potential for stormwater and habitat wetlands to contribute to landscape-level MeHg contamination, this pathway has historically been under-investigated (Chumchal and Drenner 2015). This review focuses on the biogeochemistry of mercury in wetlands created for the primary goal of controlling stormwater or providing aquatic habitat. Mercury dynamics in other artificial wetlands, including hydroelectric reservoirs and rice paddies has been reviewed elsewhere (Mailman et al. 2006; Windham-Myers et al. 2014a; Rothenberg et al. 2014). Below, we will discuss the construction and management of stormwater and habitat wetlands, as well as the mercury cycle, with a particular focus on MeHg production. We will review the existing literature on the biogeochemical drivers and landscape-level impacts of MeHg production in stormwater and habitat wetlands, and will conclude with suggestions for the areas of highest priority for future research.

Management of Small Artificial Wetlands

The geographic focus of our previous research has been southern Ontario, Canada, where both stormwater and habitat wetlands are constructed by excavation of a basin, or construction of earth berms from local soil material. In both methods, topsoil is removed and vegetation is destroyed. In Ontario, the wetlands are designed to be shallow but continuously wetted (0.15–0.6 m deep) and large enough to store runoff of a 1-year 24-h storm for 1 day (Ontario Ministry of the Environment 2003). Stormwater ponds are usually simple in shape with a smooth, regular basin, but the sediment topography of habitat wetlands may be more complex as the construction team sometimes excavates multiple pits in search of suitable soil for berming. During the excavation process, soils are compacted, smoothed, and homogenized, which alters their porosity, hydraulic conductivity, and the availability of microtopographic sites (Stolt et al. 2000; Bruland and Richardson 2006), which may reduce subsequent plant and microbial diversity (Stottmeister et al. 2003). Both stormwater and habitat wetlands may have simple margins or may be designed with a more convoluted shoreline, featuring bars, small bays, and islands that improve some of the functions of the wetland, particularly habitat provision (Ontario Ministry of the Environment 2003). Stormwater wetland basins are lined with clay, which isolates the wetland from the groundwater. Habitat wetlands may or may not feature a liner, particularly if they are sited to take advantage of the natural hydrology, such as local topographic depressions or groundwater discharge zones (Bob Clay, personal communication). After construction, the wetland is seeded with mulch or topsoil to provide a substrate for aquatic organisms and then flooded. Plants may be installed during construction, or they may colonize the site through natural dispersal (Ontario Ministry of the Environment 2003). Initial vegetation is usually dominated by disturbance-exploiting species and may include many non-native or invasive species (Balcombe et al. 2005; Weaver et al. 2012). Although some sites eventually transition towards vegetation more similar to nearby natural wetlands (Balcombe et al. 2005), others do not (Campbell et al. 2002). Artificial wetlands under 10 years old are generally poor in organic matter (Stolt et al. 2000; Wolf et al. 2011) and have higher bulk densities (Campbell et al. 2002), which has implications for vegetation and microbial community structure (Campbell et al. 2002; Peralta et al. 2013), as well as mercury sorption dynamics.

Once established, both stormwater and habitat wetlands are very effective at removing a range of contaminants, including suspended solids and metals, hydrocarbons, and particulate phosphorus (Greenway 2004; Wadzuk et al. 2010). These functions derive from both the physical structure of an artificial wetland and the biotic processes it supports. Mechanically, an artificial wetland is designed to slow water flow, allowing suspended matter and particulate pollutants to settle out of solution (Kadlec and Wallace 2009). Biotic processes in artificial wetlands are equally important to their effective function. Aquatic macrophytes provide hydraulic drag, prevent erosion and resuspension, insulate the sediment, and directly absorb some pollutants (Brix 1997; Scholz and Lee 2005). Microbial activities in the sediment, water column, and periphyton mediate many of the valuable biogeochemical transformations that occur in artificial wetlands, particularly degradation of nitrogenous pollutants (Vymazal 2007; Lee et al. 2009; Peralta et al. 2013), hydrocarbons (Stottmeister et al. 2003), pesticides (Kao et al. 2001), sulfate loads (Fortin et al. 2000), and precipitation of metals complexed to hydrogen sulfide (Kosolapov et al. 2004).

The management of artificial wetlands differs between those designed for habitat provision or stormwater control. Management of stormwater wetlands focuses on maintaining the pollutant reduction and flood control capabilities of the facility. A constructed stormwater wetland has a lifecycle of 5–15 years, determined by the rate of sediment accumulation. Sedimentation alters effective storage capacity of the wetland, which greatly reduces the removal efficiency of suspended solids and nutrients and impairs the effectiveness of the wetland for flood control (Heal and Drain 2003). Dredging and disposal of sediment is therefore an important part of stormwater wetland maintenance (Drake and Guo 2008). Dredging is usually carried out through water diversion, dry sediment removal, and reflooding (Ontario Ministry of the Environment 2003). The process takes several weeks and destroys most of the existing vegetation (David Kenth, personal communication).

Habitat wetlands, by contrast, are rarely dredged (Bob Clay, personal communication) and management activities in these sites focus on improving species diversity and habitat value. The establishment of typical aquatic and littoral vegetation, sometimes accompanied by additional management interventions such as predator control, provision of nest boxes, removal of invasive plants, or stocking with desirable flora and fauna (Bromley et al. 1985) results in high habitat value and increased landscape level species diversity (Scholz and Lee 2005; Moore and Hunt 2012). Wetlands managed for habitat, particularly provision of waterbird habitat, are sometimes managed through manipulation of the water table to provide suitable water depth, forage, and nesting habitat (Ma et al. 2010).

Mercury Biogeochemistry

There are multiple sources of mercury—both anthropogenic and natural—that result in the release of this element into the environment. While accidental or careless point source releases of mercury to soils and water occur in some highly contaminated areas, the most globally important emissions are in gaseous and particulate-bound forms released into the atmosphere (Driscoll et al. 2013). The main anthropogenic sources of mercury to the atmosphere include artisanal scale gold mining, coal-fired electricity generation, industrial scale non-ferrous mining, cement production, and waste management, while important natural sources of mercury to the atmosphere include volcanic emissions, forest fires, and leaching from mercury-rich minerals such as cinnabar (Pacyna et al. 2006; Pirrone et al. 2010). Lake sediment and ice core records suggest that the atmospheric deposition of mercury has increased by two to five times since the industrial revolution (Schuster et al. 2002; Lindberg et al. 2007; Biester et al. 2007). This has led to greater re-emission of mercury from the terrestrial environment to the atmosphere and an overall rising global atmospheric pool (Lindberg et al. 1998; Driscoll et al. 2013). Owing to the relatively long residence time of gaseous elemental mercury in the atmosphere, mercury in the atmosphere is globally distributed and significant deposition to the Earth's surface occurs in areas far removed from the original sources (Fitzgerald et al. 1998). Mercury emissions have declined over the past few decades in some parts of the world (e.g., North America, Europe), but have increased significantly in others (e.g., China and India) (Pacyna et al. 2006, 2016). The newly ratified Minamata Convention on Mercury, a global treaty negotiated by the United Nations Environment Program (United Nations Environment Programme (UNEP) 2017) aims to reduce global mercury emissions to the environment, but it is too early to assess the effectiveness of the agreement.

For urban artificial wetlands, it is probable that the main sources of mercury are direct atmospheric deposition and runoff, especially for mercury bound to suspended particulate matter, such as street dust (Eckley and Branfireun 2008). However, comprehensive mass balance studies in urban artificial wetlands are lacking. Mercury has a high affinity for particles (Domagalski 1998; Schuster et al. 2007; Balogh et al. 2008) and thus, the strong sediment deposition characteristics of most wetlands result in significant settling of particles, which should include mercury attached to particles. Indeed, multiple studies in constructed treatment wetlands have found these systems to be significant sinks for particle-bound metals, including mercury (e.g., Gustin et al. 2006a; Nelson et al. 2006). Phytoremediation of mercury in artificial wetlands via absorption and/or uptake by specific aquatic plants species has also been successful, especially in mining areas and other highly mercury-contaminated systems (e.g. Gomes et al. 2014; Marrugo-Negrete et al. 2017). Fluvial total mercury outputs from wetlands are mostly in dissolved form and in strong association with dissolved organic matter (Driscoll et al. 1995), most likely bound to reduced sulfur functional groups on dissolved organic matter (Skyllberg et al. 2003). Photochemical reduction and other chemical reduction processes for mercury in wetlands have not been studied adequately. Data from natural wetlands suggest that reduction and volatilization may be an important pathway for mercury release from wetlands (e.g., Haynes et al. 2017), although the relatively high dissolved organic matter content in wetland surface waters may provide some shielding from photoreduction, as has been observed in tea-stained lake systems (O'Driscoll et al. 2004).

Methylmercury Biogeochemistry

The great majority of mercury transported through the environment is in the inorganic form. Methylmercury is formed from these inputs of IHg, with wetlands, including peatlands, ponds, lake margins, salt and freshwater marshes, being the most important sites of mercury methylation on a landscape scale (St. Louis et al. 1994; Grigal 2002; Hall et al. 2008; Chumchal and Drenner 2015; Fleck et al. 2016). The production of MeHg in these sites is a complex process mediated by the activity and identity of the microbial community, mercury bioavailability, the dynamic balance of mercury methylation and demethylation and the physicochemical and ecological parameters that influence these processes (Ullrich et al. 2001; Munthe et al. 2007; Driscoll et al. 2013). These topics are briefly reviewed below, before a discussion of the occurrence, geographical distribution, and potential controls on mercury methylation and MeHg accumulation in stormwater and habitat wetlands.

Mercury Methylation

In freshwater habitats, mercury methylation is carried out by anaerobic prokaryotes (Jensen and Jernelöv 1969; Robinson and Tuovinen 1984; Choi et al. 1994; Gilmour et al. 2013; Podar et al. 2015), and therefore, many of the controls on MeHg production are related to the ecology of mercury methylators. Methylmercury is formed from divalent mercury (Hg^{2+}) via the intracellular two-step *hgcAB* pathway (Parks et al. 2013; Schaefer et al. 2014), orthologues of which have been found in every confirmed mercury methylator to date (Gilmour et al. 2013). The discovery of this mercury methylation pathway has greatly expanded our understanding of the taxonomic diversity of mercury methylators, which are now confirmed to include members of the sulfate reducing (Compeau and Bartha 1985), iron reducing (Fleming et al. 2006; Kerin et al. 2006) and syntrophic (Bae et al. 2014) *Deltaproteobacteria*, metabolically diverse *Firmicutes* (Gilmour et al. 2013), methanogenic archaea (Hamelin et al. 2011), syntrophic *Chloroflexi* (Bae et al. 2014) and a marine nitrite oxidizer (Gionfriddo et al. 2016).

All confirmed mercury methylating strains for which ecological data is available are microaerophiles, facultative anaerobes, or obligate anaerobes (Gilmour et al. 2013; Gionfriddo et al. 2016). Furthermore, an extensive *in silico* survey of the ecological distribution of *hgcAB* has found examples in a very broad range of anaerobic, but not aerobic, habitats (Podar et al. 2015). Therefore, the established consensus is that mercury methylation is confined to anaerobic or only transiently aerobic environments. Evidence from freshwater environmental studies supports this paradigm, where mercury methylation in freshwater is limited to anoxic habitats and microsites, including stratified water masses, periphyton biofilms, aquifers and groundwater, and benthic sediments (Gilmour et al. 1998; Desrosiers et al. 2006; Guimarães et al. 2006; Correia et al. 2012; Podar et al. 2015). Benthic sediments and saturated soils are the largest and most consistently anoxic compartments in most wetland systems, and the location of most of the mercury load (Gilmour et al. 1992), but mercury methylation can also occur in the stratified, anoxic regions of the lacustrine water column (Eckley and Hintelmann 2006).

Controls on Mercury Methylation

As mercury methylation is an intracellular process, one of the most important controls on the production of methylmercury is the availability of Hg^{2+} substrate to methylators (Schaefer et al. 2011; Graham et al. 2012b). However, only a fraction of the total Hg^{2+} present in a system is bioavailable for methylation (Hsu-Kim et al. 2013). This bioavailable fraction is both variable among environments, and controlled by factors that are incompletely understood, but may include the size and quality of organic matter ligands, with an apparently increased bioavailability of Hg^{2+} bound to thiol-containing cysteine residues (Schaefer and Morel 2009; Schaefer et al. 2011; Jonsson et al. 2014) and competition with other trace elements

(Schaefer et al. 2014). In particular, sulfide (H_2S) is an aggressive ligand of Hg^{2+}, with concentrations above 0.3–3 mg/L greatly limiting MeHg production (Gilmour et al. 1998; Langer et al. 2001; Jay et al. 2002; Drott et al. 2007; Bailey et al. 2017). Mercury bioavailability also relates to the age of the mercury itself, with older Hg^{2+} being significantly less bioavailable for methylation (Hintelmann et al. 2002; Schwesig and Krebs 2003; Harris et al. 2007; Peng et al. 2012; Mao et al. 2013; Strickman and Mitchell 2016), possibly as a result of the agglutination of HgS particles into large colloids or crystalline structures, which are likely less easily taken up by mercury methylators (Graham et al. 2012a, b; Zhang et al. 2012). As a result of this complexity, neither the Hg^{2+} concentration nor the total mercury concentration of an environment is a reliable predictor of its methylation capacity (Hsu-Kim et al. 2013; Fig. 7.1).

In the absence of a robust understanding of mercury bioavailability, an empirically defined suite of environmental parameters that correlate with the efficiency of methylation has been identified (Fig. 7.1). Some of these relate to the metabolic needs of mercury methylating organisms, including a positive interaction with anoxia and higher temperatures (Ullrich et al. 2001; Benoit et al. 2003; Munthe et al. 2007). Mercury methylation is also enhanced by the supply of electron acceptors (i.e. primarily sulfate), which may reach the system through atmospheric deposition, internal cycling, groundwater flow, or surface water inputs (Fortin et al. 2000; Bates et al. 2002; Mitchell et al. 2008a; Åkerblom et al. 2013), or be regenerated by transiently oxic conditions (Feng et al. 2014; Coleman Wasik et al. 2015; Mueller et al. 2016; Oswald and Carey 2016). For this reason, environments subjected to repeated cycles of wetting and drying also have increased rates of mercury methylation (Compeau and Bartha 1985; Gustin et al. 2006b; Eckley et al. 2017). Mercury methylation can also be reduced by competitive inhibition of mercury methylators, such as stimulation of nitrate reducers by high nitrate concentrations, with resulting redirection of carbon flow away from sulfate reducers (Todorova et al. 2009; Shih et al. 2011). Finally, there are at least two environmental factors that regulate MeHg production apparently as a result of reduced Hg^{2+} bioavailability. Environments with a pH of 5 to <7 host higher rates of net mercury methylation (Ullrich et al. 2001), which may be due to liberation of Hg^{2+} from dissolved organic matter ligands (Amirbahman et al. 2002). Conversely, in high-sulfide environments where this solute exceeds 0.3–3 mg/L (Gilmour et al. 1998; Langer et al. 2001; Jay et al. 2002; Drott et al. 2007; Bailey et al. 2017), sulfide removes mercury from solution by precipitation as HgS_s (Benoit et al. 1999; Mitchell and Gilmour 2008) and also shifts the speciation of mercury compounds to charged, likely less-bioavailable species (Liu et al. 2012). This dual role of sulfate and its reaction product, sulfide, as stimulants and inhibitors of methylation has been termed the "goldilocks concept", with the greatest methylation capacity observed in environments where sulfate concentrations are high enough that the metabolism of sulfate reducers is not substrate-limited, but where the concentrations of sulfide are not sufficiently elevated to reduce the bioavailability of Hg^{2+} (Gilmour et al. 1998), as illustrated in Fig. 7.1.

Sediment organic matter is usually related positively to mercury and MeHg concentrations. Both inorganic mercury and MeHg have a high affinity for thiol binding

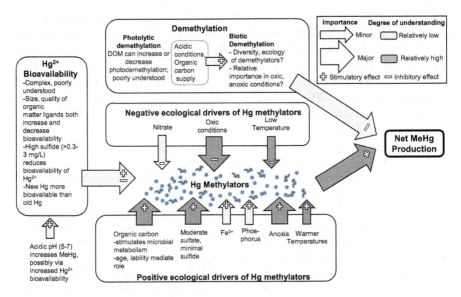

Fig. 7.1 Controls on the net production of methylmercury (MeHg). Large arrows represent a major control, while smaller arrows represent a more minor control. The degree to which the mechanism is understood is indicated by shading, with dark shading indicating a relatively well-understood process and lighter shading indicating a less well-understood process. Mechanisms that stimulate net MeHg production are indicated with a plus sign on the head of the relevant arrow; a minus sign indicates an inhibitory effect

sites on organic matter, resulting in greater adsorption and storage in high-organic sediments and soils (Skyllberg et al. 2000; He et al. 2007; Obrist et al. 2011; Meng et al. 2016). Transformation of Hg^{2+} to MeHg is also often enhanced in environments high in organic matter, which may be related to the metabolic stimulation of mercury methylators (King et al. 2000, 2002; Acha et al. 2005; Mitchell et al. 2008a; Windham-Myers et al. 2013) or, possibly, improved bioavailability of Hg^{2+} for methylation (Schaefer and Morel 2009; Graham et al. 2012b; Schaefer et al. 2014; Mazrui et al. 2016), although further research is needed on how organic matter affects mercury bioavailability.

Demethylation

Methylmercury can also be demethylated, through biotic and abiotic processes. The net accumulation of MeHg therefore integrates both methylation and demethylation, making demethylation an important modulator of MeHg concentrations in a range of environments (Avramescu et al. 2011; Hines et al. 2012; Kronberg et al. 2012; Zhao et al. 2016). In the photic zone, photolysis is an important degradation pathway for MeHg (Sellers et al. 1996), while biotic demethylation dominates outside this region. Dissolved organic matter (DOM) appears to be an important, but not fully understood control on photo-demethylation, with different studies demonstrating that DOM binding can either promote or inhibit photo-demethylation (Li et al. 2010; Zhang and Hsu-Kim 2010). Other researchers found that DOM levels had little impact on the photolytic degradation of MeHg (Black et al. 2012; Fleck et al. 2014), possibly because prior photobleaching of DOM may be an important co-correlate with photodemethylation (Klapstein et al. 2017) or because characteristics like hydrological residence times and shading by vegetation vary widely in natural systems, obscuring subtle relationships between DOM and photo-demethylation (Fleck et al. 2014). Despite the role of wetlands as important sites of DOM production, there is still a paucity of research on the DOM-photo-demethylation link, particularly in constructed wetland systems.

Biotic demethylation can be divided into two categories: reductive demethylation, which produces methane and Hg^0, and oxidative demethylation, with CO_2 and Hg^{2+} as the end products (Barkay et al. 2003). Demethylation abilities are broadly distributed among sulfate reducers, methanogens, at least one iron reducer, and a wide diversity of aerobes (Marvin-DiPasquale and Oremland 1998; Ullrich et al. 2001; Barkay et al. 2003; Avramescu et al. 2011; Lu et al. 2016). Demethylation proceeds most robustly in oxic environments, although it also occurs under anaerobic conditions (Compeau and Bartha 1984; Ullrich et al. 2001; Lin et al. 2012; Lu et al. 2016). Demethylation is more challenging to measure than mercury methylation (Hintelmann et al. 2000), and as a result, the ecology of biotic MeHg degradation remains relatively unknown. Positive correlations have been identified between demethylation and labile organic carbon (Marvin-DiPasquale and Oremland 1998; Marvin-DiPasquale et al. 2000; Li and Cai 2012; Hamelin et al. 2015) and more acidic pH (Miskimmin et al. 1992; Zhao et al. 2016) but there is a paucity of information on the responsiveness of MeHg demethylation to other standard biogeochemical variables, impairing efforts to reduce net MeHg concentrations by manipulations that would upregulate demethylation.

Variability in Concentrations

Given the complexity of controls on MeHg production and concentrations, it is not surprising that there is considerable variation in MeHg concentrations and production among different wetlands, as well as variations over fine spatial scales within an individual site and over the lifetime of a particular wetland (Mitchell et al. 2008b; Sinclair et al. 2012; Tjerngren et al. 2011; Hoggarth et al. 2015). This variability highlights the need to understand MeHg concentrations and production in an individual wetland, rather than extrapolating from sites that are superficially similar. Temporal variations in MeHg production and concentrations over the lifetime of a wetland are particularly relevant for understanding MeHg dynamics in artificial wetlands. In particular, MeHg concentrations are greatly elevated in new aquatic environments created by flooding terrestrial areas. This early-life spike in MeHg production has been observed in many types of new aquatic environments, especially reservoirs, where it is accompanied by marked increases in fish MeHg loads and MeHg efflux (St. Louis et al. 2004; Mailman et al. 2006). In a survey of beaver impoundments, which might be described as naturally created wetlands, researchers also observed higher production of MeHg in younger ponds (Roy et al. 2009). This phenomenon appears to result from the decay of submerged organic mass and subsequent stimulation of microbial activity (Hall et al. 2005; Mailman et al. 2006).

Methylmercury in Artificial Wetlands

Geographic Distribution of MeHg Contamination

Like other aquatic environments, artificial wetlands support net MeHg accumulation in sediments, overlying water, porewater and biota (King et al. 2002; St. Louis et al. 2004; Stamenkovic et al. 2005; Gustin et al. 2006a, b; Rumbold and Fink 2006; Chavan et al. 2007; Sinclair et al. 2012; Strickman and Mitchell 2017). Table 7.1 summarizes data from the peer-reviewed literature on the concentrations of methylmercury (MeHg) in sediment, porewater, surface water and biota, as well as the methylation rate potentials (K_{meth}) of artificial wetlands managed for stormwater control or habitat provision. Methylmercury in artificial wetlands has been detected over wide geographical ranges, including stormwater wetlands in arid areas of Nevada (Stamenkovic et al. 2005; Gustin et al. 2006a, b; Chavan et al. 2007) and the southeastern USA (King et al. 2002; Rumbold and Fink 2006), and in habitat and stormwater wetlands in temperate Ontario, Canada (Sinclair et al. 2012; Strickman and Mitchell 2017). The global nature of mercury contamination, and the occurrence of MeHg and mercury methylation in other types of managed wetlands on other continents (Wang and Zhang 2012; Gentès et al. 2013a; Rothenberg et al. 2014) strongly suggests that MeHg contamination of artificial wetlands occurs wherever these installations are built. Direct measurements of active mercury methylation in artificial wetlands have only been assessed in southern Ontario, where the authors of this review have used enriched mercury isotope assays to measure the potential methylation rate constants (K_{meth}) in the sediment. In a comparison of stormwater and habitat wetlands, K_{meth} values averaged 0.006 ± 0.004 per day in stormwater wetlands, and 0.014 ± 0.010 per day in habitat wetlands (Strickman and Mitchell 2017), which are values similar to those obtained from other freshwater wetland systems (Drott et al. 2008; Tjerngren et al. 2011; Marvin-DiPasquale et al. 2014; Hoggarth et al. 2015). While additional studies are needed to characterize the MeHg production capacity of artificial wetlands in other geographic areas, or those receiving runoff with differing biogeochemistry, it is likely that *in situ* methylation of Hg^{2+} is the main source of MeHg to artificial wetlands, and that their MeHg production rates are not markedly higher or lower than those of better-studied natural wetlands.

Table 7.1 Inventory of peer-reviewed studies of methylmercury (MeHg) concentrations in the sediment, porewater, surface water, and biota, and methylation rate potentials (K_{meth}) of artificial wetlands managed for stormwater control or habitat provision

Wetland type	Location	Reference	Notes	Sediment MeHg (ng/g)	Porewater MeHg (ng/L)	Surface water MeHg (ng/L)	K_{meth} (day^{-1})	Biota MeHg (ng/g)	Biota
Experimental stormwater mesocosms	Savannah, Georgia, USA	King et al. (2002)		–	0.2–3.5	0.1–2.4	–	–	–
Experimental stormwater mesocosms	Savannah, Georgia, USA	Harmon et al. (2004)	Control	1.4–1.5	0.5	0.22	–	–	–
			Low-sulfate treatment	1.5–2.4	1.7	0.17	–	–	–
			High-sulfate treatment	0.9–1.5	1.7	0.26	–	–	–
Experimental stormwater wetland	Reno, Nevada, USA	Stamenkovic et al. (2005)	Effluent from mesocosms receiving clean water		–	0.5–5	–	–	–
			Effluent from mesocosms receiving contaminated water	–	–	0.1–2	–	–	–
Experimental stormwater wetland	Reno, Nevada, USA	Gustin et al. (2006a)	Inflow water/sediment at inlet	0.5–2.3	–	0.91	–	–	–
			Effluent/sediment at outlet with contaminated sediment-contaminated water	0.5	–	1.41	–	–	
			Clean sediment, contaminated water	0.3	–	0.99	–	–	–
			Contaminated sediment, clean water	0.9	–	1.60	–	–	–
			Clean sediment, clean water	0.5	–	0.81	–	–	–

(continued)

Table 7.1 (continued)

Wetland type	Location	Reference	Notes	Sediment MeHg (ng/g)	Porewater MeHg (ng/L)	Surface water MeHg (ng/L)	K_{meth} (day^{-1})	Biota MeHg (ng/g)	Biota
Working stormwater wetland	Florida, USA	Rumbold and Fink (2006)	Problematic cell of larger system	–	–	Up to 20	–	c. 10–346	Mosquitofish
				–	–	–	–	Median range 10–580	Sunfish
				–	–	–	–	Median range 20–850	Bass
Experimental stormwater wetland	Reno, Nevada, USA	Chavan et al. (2007)	Five effluents from different sources of water and sediment	–	–	0.5–5.5	–	–	–
Habitat wetlands	Greater Toronto Area, Ontario, Canada	Sinclair et al. (2012)	One year old wetland	8	2	0.5	–	Up to 1700	Invertebrates (Diptera, Hemiptera, Coleoptera, and Odonata)
			Two year old wetlands	2.8–3	1.7–2.8	0.1–0.3	–	Up to 600	
			Three year old wetlands	2	1	0.4	–	Up to 350	
			Nine year old wetland	1.5	1.5	0.1	–	Up to 100	
			Urban control wetland	2.8	0.7	0.1	–	Up to 150	
			Rural control wetland	2	0.4	0.1	–	Up to 100	

7 Methylmercury in Managed Wetlands

Setting	Location	Reference	Description					
Working stormwater wetland	Florida Everglades	Zheng et al. (2013)	Outflow concentrations of problematic cell	—	—	0.11–0.95	—	—
Working stormwater wetland	Florida Everglades	Feng et al. (2014)	Cell experiencing two severe dryouts	—	—	—	7.8–321	Mosquitofish
			Cell experiencing one severe dryout	—	—	—	3–101	
			Cell experiencing no severe dryouts	—	—	—	4–33	
Stormwater and habitat wetlands	Greater Toronto Area	Strickman and Mitchell (2017)	Stormwater wetlands	0.29–0.88	—	—	0–0.016	—
			Habitat wetlands	0.83–2.77	—	—	0–0.031	—
Working stormwater wetlands	Greater Toronto Area	Strickman (2017)	New	0.04–0.34	—	—	0.020–0.22	—
			Dredged	0.12–1.97	—	—	0.004–0.096	—
			Mature	0.34–1.25	—	—	0.033–0.069	—

Unless otherwise indicated, numbers represent the mean or range of means of the relevant variable. A dash indicates that data were not available

Biogeochemical Controls on Mercury Methylation

The biogeochemical controls on mercury methylation in artificial wetlands have been investigated mainly through correlations between environmental variables with MeHg concentrations in different environmental compartments. Based on this work, sulfate reducing mercury methylators appear to be most important in MeHg production in artificial wetlands. This interpretation is based on observations of stimulation of MeHg production in sulfate-enriched mesocosms, or immediately after the addition of gypsum, a sulfate-containing mineral (King et al. 2002; Harmon et al. 2004), and co-correlations between sulfate in overlying water and MeHg export (Gustin et al. 2006b). Evidence from both the sediment and the overlying water compartments support this interpretation.

Sediment

In the sediment, patterns of mercury methylation are consistent with the ecology of sulfate reducing mercury methylators. More anoxic (Sinclair et al. 2012) and sulfate-rich sediments and porewaters (King et al. 2002; Harmon et al. 2004) have been related to higher MeHg concentrations. Sediment organic matter is also correlated to increased MeHg concentrations and production, with greater increases resulting from organic matter that is newer or more labile (King et al. 2002; Sinclair et al. 2012; Strickman 2017; Strickman and Mitchell 2017). This suggests that higher organic matter contents likely directly stimulated MeHg production (Meng et al. 2016) by supporting a more abundant and more active microbial community (Schallenberg and Jacob 1993).

The supply of Hg^{2+} is an important control on MeHg production, but a significant relationship between these variables is not usually observed in environmental studies because simple measurements of Hg^{2+} are rarely a robust proxy for the bioavailable fraction (Benoit et al. 2003; Hsu-Kim et al. 2013). In habitat wetlands, studies comparing Hg^{2+} and MeHg concentrations in sediments found no relation between the two variables (Sinclair et al. 2012; Strickman and Mitchell 2017). In stormwater wetlands, by contrast, stronger correlations have been found between sediment MeHg and Hg^{2+}, suggesting that the bioavailable fraction of Hg^{2+} was unusually consistent in this wetland type. Methylation rates were also dampened in these wetlands, suggesting that the ongoing inputs of Hg^{2+} in stormwater runoff had a consistent, but low bioavailability for methylation (Strickman 2017; Strickman and Mitchell 2017).

Overlying Water

Several authors have suggested that incoming water quality may affect the potential of a wetland to support mercury methylation. Zheng et al. (2013), in a comparison of inflow and outflow MeHg concentrations and water quality in a Florida stormwater wetland, found statistical support for the hypothesis that inflowing water quality conditions conducive to the activity of sulfate-reducing mercury methylators result in higher outflow MeHg concentrations. These patterns included a positive relationship between MeHg in the outflow water and DOC in the inflow water, as well as negative relationships with dissolved oxygen. Gustin et al. (2006b) similarly found positive correlations between sulfate, TOC, lower pH, and temperature in incoming water, and the concentration of MeHg in outflowing water, although a subsequent experimental manipulation of sulfate concentrations resulted in no difference between MeHg export from experimental and control mesocosms (Gustin et al. 2006b). Characteristics of the incoming water which alter the bioavailability of Hg^{2+} may also affect the production of MeHg. Zheng et al. (2013) found negative relationships between MeHg concentrations in outflowing water and high concentrations of chloride and sulfate in inflowing water. At low to moderate concentrations, sulfate enhances the activity of sulfate-reducing mercury methylators and results in higher MeHg concentrations. However, at higher concentrations the sulfide produced by sulfate-reducers and other biogeochemical processes reduces the bioavailability of mercury for methylation (Fig. 7.1). Total suspended solids in incoming water may also complex Hg^{2+} and reduce its bioavailability, as observed by Stamenkovic et al. (2005), who identified a negative correlation between total suspended solids and MeHg in outflows in experimental stormwater wetlands in Nevada. However, these studies did not mechanistically link endogenous MeHg production in the sediment or water column to inflowing water quality. Given that the majority of mercury methylation presumably occurs in the sediment of these shallow water and unstratified aquatic systems, rather than the oxic water column, these patterns could also be explained by differences in partitioning of MeHg from benthic sediments and porewaters to the overlying waters.

Nitrate concentrations in overlying water and porewater have been linked to reduced MeHg production in artificial wetlands. Working in experimental stormwater wetlands in Nevada, Stamenkovic et al. (2005) identified a significant negative relationship between nitrate in surface waters and MeHg concentrations in outflowing waters. Chavan et al. (2007) similarly found that mesocosms receiving low-nitrate inflow water functioned as MeHg sources year round and had high peak MeHg concentrations in outflowing water (>5 ng/L) in comparison to mesocosms fed with high nitrate water, which were sources only in the summer with peak outflow MeHg concentrations of 3.5 ng/L. Strickman (2017) mechanistically linked high nitrate porewater concentrations to lower MeHg production via measurements of K_{meth} in a series of working stormwater wetlands in Southern Ontario. Further research is needed, but together these studies suggest that competition between mercury methylators and nitrate reducers can limit the production of MeHg in stormwater wetlands, as has been observed in other systems (Shih et al. 2011). If MeHg production in stormwater wetlands is indeed dampened by inputs of nitrate, a common contaminant in stormwater runoff, then the generally low MeHg production in stormwater wetlands relative to other wetlands may be a widespread and serendipitously beneficial phenomenon (Strickman 2017). This would imply that nitrate additions to stormwater ponds upstream of high-risk environments might reduce

MeHg contamination, a technique successfully applied in an urban lake (Matthews et al. 2013). Conversely, reductions in nitrate pollution on a landscape scale may unintentionally increase MeHg production and export from artificial wetlands.

There has been very little research on the possible relationship between phosphorus inputs and mercury methylation in constructed wetlands, which is surprising, given that most constructed wetlands are freshwater environments and that phosphorus, more than nitrogen, is most often the limiting nutrient for benthic production in freshwater lakes and ponds (Elser et al. 2007). In wetlands where methylation by periphyton is important, increased phosphorus loading may augment methylation (Lazaro et al. 2016). Correlative evidence of increased dissolved phosphorus with greater MeHg concentrations has been reported, but this is generally attributed to a co-correlation with anoxia or possibly sulfate reduction, which both increase phosphorus solubility (Roden and Edmonds 1997; Balogh et al. 2006). Possible nutrient (nitrogen or phosphorus) limitation or stimulation of methylating microbes is overall not well understood and is an important area for future research, particularly in constructed wetlands. In many aquatic systems, increased primary productivity as a function of phosphorus inputs can result in the dilution of mercury in biomass at both the base of the food web and in consumers (Pickhardt et al. 2002; Simoneau et al. 2005), although significant unexplained variation exists between biomagnification and primary productivity across systems (Lavoie et al. 2013). Similar investigations have not yet been published on small-scale constructed wetland systems.

Temporal Patterns in MeHg Production

Early-life spikes in MeHg concentrations are especially relevant for managers of artificial wetlands, particularly those discharging to environments sensitive to MeHg contamination, such as the Florida Everglades (Liu et al. 2010). However, small artificial wetlands do not seem to consistently experience this early-life spike, with some systems showing clearly elevated MeHg concentrations or export soon after startup, while others do not (Table 7.1). In a comparison of constructed habitat wetlands of different ages in Ontario, Canada, Sinclair et al. (2012) found that that the proportion of total mercury in sediment that was methylated (i.e. % MeHg) was 2.4 times higher in a new wetland than in wetlands three or more years old (Sinclair et al. 2012). Working in habitat wetlands in the same area, Strickman and Mitchell (2017) found slight, though non-significant, decreases in MeHg concentrations at older sites. Elevated MeHg concentrations and export have also been observed during the first months or years of the lives of individual artificial wetlands. In an experimental mesocosm in Savannah, Georgia, USA, porewater MeHg concentrations declined 75–80% over the first 2 months of the study (Harmon et al. 2004). In a series of wetlands treating stormwater entering the Florida Everglades in the USA, MeHg in outflowing surface waters exceeded concentrations in inflowing waters (Rumbold and Fink 2006) before declining 2 years later, although this change may also have been the result of management changes that eliminated extreme wet-dry cycles (Zheng et al. 2013). In other systems, MeHg concentrations or export were more responsive to additions of gypsum than to the age of the mesocosm (King et al. 2002), or showed no age-related pattern at all (Rumbold and Fink 2006). Furthermore, a study of stormwater wetlands in Southern Ontario observed increased, rather than decreased MeHg concentrations across age chronosequences (Strickman and Mitchell 2017).

These variations are likely related to the microbiological conditions created by different artificial wetland construction strategies. The stormwater wetlands in Southern Ontario that showed an increase in MeHg with age were excavated into the mineral soil, lined with clay and received minimal additions of organic matter prior to floodup. As a result, these sites had low organic matter when young, and higher levels of organic matter as they aged (Strickman and Mitchell 2017). This lack of sediment organic matter would have provided little substrate for microbial activity, thus limiting MeHg production (Meng et al. 2016). Furthermore, it is likely that sediments had not yet been abundantly populated by typical wetland mercury methylators or that their activity was depressed by suboptimal biogeochemical conditions; similar reductions in the rates of other microbially mediated biogeochemical processes have been observed in newly created wetlands (Wolf et al. 2011). The older stormwater wetlands surveyed by Strickman and Mitchell (2017) had likely developed a microbial community relatively similar to a natural wetland (Duncan and Groffman 1994). This interpretation is supported by observations of reduced mercury methylation in a young, low-organic-matter stormwater wetland vs. a mature, high-organic matter stormwater wetland studied in a separate project that compared new, mature, and dredged stormwater wetlands over one growing season (Strickman 2017). In contrast, the experimental wetlands in Nevada and Florida that experienced more typical early-life spikes in MeHg concentrations were constructed from natural wetland sediments that likely had established microbial communities, including the mercury methylators typical of aquatic environments. Similarly, the young habitat wetlands studied by Sinclair et al. (2012) were constructed through berming of marshy areas rather than excavation (Bob Clay, personal communication), and thus had both high organic matter sediments and, potentially, populations of mercury methylators in anoxic microsites in the pre-flood soils (Kronberg et al. 2016).

Influence of Wetland Management on MeHg Dynamics

The management of artificial wetlands, particularly stormwater wetlands, is a fundamental driver of their biogeochemical and hydrological functioning. Research on this topic has been limited, but suggests that management interventions may have important implications for MeHg biogeochemistry. Dryout-rewetting cycles—an undesirable circumstance that artificial wetlands are managed to avoid—have been best studied in this respect. The degree of impact on MeHg production of water level fluctuations is related to the duration and extent of the dryout, with more severe dryouts resulting in stimulation of MeHg production or export, while routine water level variations or minor dryouts appear to have little effect. Working in experimental stormwater wetlands, Gustin et al. (2006b) created a 4-month-long dryout event, punctuated by two brief influxes of water. This treatment, which simulated poor hydrological management, resulted in MeHg concentrations in outflows that were up to 5.3 times higher than in the same wetland at a similar season the previous year. Similarly, Feng et al. (2014) found that two severe dryout events in one cell of a Florida stormwater treatment wetland resulted in mosquitofish mercury concentrations ten times higher than in a cell that did not experience dryouts. In contrast, Strickman and Mitchell (2017) found no difference in MeHg production or concentrations between areas of individual wetlands that presumably experienced frequent, intermittent, or no minor dryout events related to water level fluctuation. Supporting this interpretation, a single, brief dryout followed by immediate rewetting resulted in only a modest and shortlived elevation in MeHg export from an experimental stormwater wetland (Gustin et al. 2006b).

The most disruptive management activity applied to stormwater wetlands is dredging, in which water is drained and the accumulated sediment, along with emergent vegetation and resident biota, is removed and disposed of off-site. Only one study (Strickman 2017) has investigated the impact of this management activity on MeHg biogeochemistry, and this investigation found that the mercury methylation capacity of the sediment microflora was initially very low post-dredging, but rapidly rebounded over a period of five months to levels more similar to those found in a nearby mature, undredged wetland. It was unclear if the rebound in MeHg production and concentrations was complete, or if dredging would in fact cause ongoing increases and a spike in MeHg production in this wetland. While further research is needed, the Strickman (2017) study suggests that dredging of stormwater wetlands has only a short-term effect on MeHg production, and that MeHg concentrations in sediments will rapidly return to, or even exceed, the levels present before dredging. Further research comparing MeHg production, concentrations, and export in multiple wetlands before and after dredging is needed to characterize the response of the sediment microflora to this management activity.

Impacts of MeHg in Artificial Wetlands

Artificial wetlands may be important components of the landscape-level MeHg budget due to their spatial prevalence and high hydrologic connectivity. Annual MeHg export from small artificial wetlands may be much higher than that from natural environments, based on a study that estimated the export of MeHg from experimental stormwater wetlands constructed from sediments of differing contamination levels and receiving either clean or mercury-contaminated water (Gustin et al. 2006a). In this project, the researchers found that annual MeHg exports ranged from 6 to 143 μg/m^2/year. These rates are much higher than the 0.1–0.5 μg/m^2/year emissions from natural, relatively uncontaminated small northern wetlands (Krabbenhoft et al. 1995; St. Louis et al. 1996; Driscoll et al. 1998; Galloway and Branfireun 2004). It is also likely that the high level of Hg^{2+} contamination in the wetland studied by Gustin et al. (2006a) was a major factor in this difference. Additional studies are needed on the net export of MeHg from artificial wetlands under different climatic regions and with different levels of mercury contamination to confirm if artificial wetlands do export more MeHg than similar natural wetlands.

Seasonal differences in MeHg export within individual artificial wetlands have also been found, with the same wetland or mesocosm generally serving as a source in warmer months and as a sink in the cold/cool season (King et al. 2002; Gustin et al. 2006a; Chavan et al. 2007; Stamenkovic et al. 2005). These patterns may be linked to temperature-related stimulation of the microbiota or changing releases of labile organic carbon from plant roots over the course of the growing season (King et al. 2002; Harmon et al. 2004), although these interpretations remain speculative. The relative concentrations of MeHg in sediment and overlying water also influence the function of a wetland as a source or sink, with wetlands having more contaminated sediment and receiving relatively clean overlying water functioning as a source regardless of season (Gustin et al. 2006a; Stamenkovic et al. 2005). This suggests that factors controlling the efflux of MeHg from sediments to the water column—including MeHg concentrations, redox conditions at the sediment-water interface, and the relative concentrations of sulfides in porewater and overlying water (Holmes and Lean 2006; Bailey et al. 2017)—may help modulate MeHg efflux in this, and other systems. However, factors controlling the aqueous efflux of MeHg from artificial wetlands have never been investigated directly, and there is a particular dearth of information on how methylation and demethylation may control this process. Improved understanding of the extent, seasonality, and drivers of aqueous export of MeHg from artificial wetlands is needed to help characterize the position of these environments in the landscape-level MeHg cycle, and provide practical guidance to managers of artificial wetlands upstream of environments with existing MeHg contamination issues (Rumbold and Fink 2006).

The movement of MeHg from the sediments and waters of an artificial wetland to resident biota represents the entry point of MeHg to human and animal food webs. Overall, concentrations of MeHg in the biota of artificial wetlands suggest that there is greater MeHg exposure for subsequent animal consumers compared to at least terrestrial systems. In a Florida stormwater wetland, median concentrations of MeHg in fish tissue ranged from extremely low to up to 850 ng/g, depending on the species, season, and cell of the wetland in which the fish were collected (Rumbold and Fink 2006; Feng et al. 2014). Fish at the upper part of this range could represent a risk of reproductive harm to piscivorous animals or to human consumers (World Health Organization 2017). Only one study reports invertebrate MeHg concentrations in artificial wetlands (Sinclair et al. 2012). This study found that in wetlands with lower sediment MeHg concentrations, averaging c. 1.5 ng/g, the concentrations of MeHg in invertebrates were between c. 10 and 400 ng/g, which is below the thresholds for negative reproductive effects in insectivorous birds (Brasso and Cristol 2007). In the youngest wetland, however, invertebrate

MeHg concentrations averaged up to 1600 ng/g, which is well above the levels associated with reproductive effects in swallows consuming a diet of similar invertebrates (Brasso and Cristol 2007). Furthermore, while the MeHg burdens of most of the invertebrates examined by Sinclair et al. (2012) were below values associated with toxicity, these levels were nonetheless higher than loads in trophically similar taxa in comparable natural wetlands, which averaged 195 ng/g in other small Canadian wetlands (Bates and Hall 2012; Hall et al. 1998). This suggests that, while the overall exposure to MeHg in artificial wetlands is relatively low, some artificial wetlands may be important sources of MeHg to both aquatic and terrestrial food webs. Certainly, the small number of studies investigating MeHg in artificial wetland biota suggests a need for further work across diverse artificial wetland types. Specific work is needed to characterize the concentrations and drivers of MeHg in the biota of artificial wetlands, particularly the small fish species that frequently colonize older municipal stormwater ponds (Strickman, personal observation). Evidence from experimental ponds and mesocosms suggests that the presence of fish may reduce the total MeHg efflux from small wetlands by reducing the total biomass of the larger, higher-trophic larvae that are the prey of most fish species and shifting the overall biomass composition to smaller, low-trophic species that have lower MeHg burdens (Chumchal and Drenner 2015). Although this approach would increase exposure of piscivores to MeHg, in some circumstances it might be a valuable tool for controlling MeHg exposure for insectivorous species, including threatened or declining taxa such as little brown bats or rusty blackbirds (Edmonds et al. 2012; Little et al. 2015).

Plant tissues also bioaccumulate MeHg from saturated sediments (Strickman and Mitchell 2016), with positive bioaccumulation factors between sediment and aboveground photosynthetic tissue ranging between 1.1 and 8.1 (Cosio et al. 2014). In wild and white rice plants, the highest concentrations of MeHg are found in the grain, with concentrations in white rice in uncontaminated sites ranging between 0.86 and 5.8 ng/g (Rothenberg et al. 2014) and an average of 6 ng/g in wild rice. Windham-Myers et al. (2014b) estimated that a migratory waterfowl overwintering for 5 months in a fallow wild rice field could be exposed to up to 1.097 mg (1,097,000 ng) of THg, of which 40–60% is likely in the methylated form (Rothenberg et al. 2014). This represents exposure to four to five times as much MeHg as waterfowl overwintering in natural wetlands (Windham-Myers et al. 2014b). If wild rice or other similar plant species in artificial wetlands accumulate MeHg to a similar degree, then substantial MeHg may be transferred to granivorous as well as insectivorous and piscivorous birds, but no such published research yet exists.

MeHg Risks for Other Types of Artificial Wetlands

The work summarized in this review has focused on the MeHg dynamics of stormwater and habitat wetlands, the most-studied types of small artificial wetlands. However, based on our understanding of the ecology and drivers of mercury methylation and demethylation in these wetlands and in natural wetlands, there are a number of other types of artificial wetlands that may have high MeHg production or export. We explore possible MeHg-related issues in two additional artificial wetland types here for which no direct, published studies currently exist.

Floating Treatment Wetlands

Floating treatment wetlands are a relatively new approach to water treatment, that use created floating islands to mimic the unique conditions of natural floating wetlands. These relatively widespread natural systems are characterized by a dense upper layer of living plants rooted in a buoyant, self-supporting mat of peaty organic matter up to 60 cm thick. Living plant roots may extend into the water column beneath the floating island, supporting a low-density layer of decaying organic matter suspended over an area of open water. These dense mats of floating vegetation alter the conditions of the underlying water column and sediment, creating low-oxygen, high organic matter, low pH and dark conditions (Mallison et al. 2001). The sediment beneath a floating island is greatly enriched in low-density, sludgelike organic matter (Alam et al. 1996). Floating treatment wetlands mimic most of these conditions and show promise for managing agricultural runoff, acid mine drainage, and dairy and distillery wastewater (Strosnider et al. 2017).

Although MeHg production has never been investigated in either floating treatment wetlands or natural floating wetlands, previous work has found that the rhizoflora of floating tropical plants support active mercury methylation (Acha et al. 2005; Guimarães et al. 2006), as does the periphyton attached to submerged plants in several temperate regions (Gentès et al. 2013b; Hamelin et al. 2015). In one study, the MeHg production rate of root periphyton was more than ten times higher than the production rate of the underlying sediment (Gentès et al. 2013b). These findings, together with the known ecological requirements of major mercury methylating groups strongly suggest that floating treatment wetlands may be sites of elevated mercury methylation, a possibility that should be explored before they are used to remediate waters enriched in mercury or upstream of environments with pre-existing MeHg contamination problems.

Wetlands Amended with Gypsum or Sulfate

Artificial wetlands receiving waters enriched in metals are sometimes amended with gypsum ($CaSO_4 \cdot 2H_2O$) or sulfate, which are readily reduced to sulfide (HS^-). Sulfide forms insoluble complexes with a wide variety of metal cations, and removes them from the water column as precipitates, representing a powerful and simple remediation tool. Additions of sulfate and gypsum are also used to reduce methane emissions from wetlands, particularly rice paddies, an effect created by the competitive inhibition of methanogens by sulfate reducers (Gauci et al. 2004.; Lovley and Klug 1983; Pangala et al. 2010). However, if the system is not carefully managed to keep sulfide concentrations high enough to significantly reduce Hg^{2+} bioavailability, these amendments are likely to increase mercury methylation by stimulating the activities of mercury methylators (Jeremiason et al. 2006). This pattern has been observed in the northern areas of the Florida Everglades, where runoff from gypsum-amended agricultural fields has elevated MeHg production (Gilmour et al. 1998; Orem et al. 2011). Evidence from artificial wetlands suggests that a typical passive management approach is insufficient to prevent the formation of sulfate reducing conditions in at least some areas of the wetland. In a series of wetland mesocosms, King et al. (2002) found that additions of gypsum resulted in sharp spikes in MeHg concentrations in the effluent, rising from c. 0.2 ng/L pre-addition to 1.5 ng/L post-addition. Similarly, Harmon et al. (2004) found that sulfate amendments stimulated MeHg production and resulted in elevated MeHg concentrations in porewater (0.5–1.6 ng/L vs. > 0.2–0.5 ng/L), as well as greater proportions of mercury present as MeHg (18.5 vs. 9%). The extent of MeHg production in small wetlands directly amended with sulfate or gypsum should be assessed.

Future Directions

The low level of investment in quantifying mercury methylation in artificial wetlands, and particularly in stormwater management systems, is at odds with the large and growing significance of these manmade structures in our urban and suburban environments. Existing research on MeHg production and concentrations in artificial wetlands has been limited to only three geographic areas (i.e. Nevada, the southeastern USA, and southern Ontario, Canada), and does not explore the full range of treatment wetland types, water characteristics, or variations among different ecoregions or climate zones. In addition, most studies of artificial wetlands have neglected one or more environmental compartments, with a particular dearth of information on methylation rates, estimates of the net efflux of MeHg from the wetland, and the concentrations in biota. Further basic characterization is also needed on the extent of several processes that have not been robustly investigated in the existing literature. These include nitrate-related reductions in MeHg production and export by artificial wetlands, basic investigations on the impact of phosphate on MeHg production and concentrations, the net annual export of MeHg in artificial wetlands relative to comparable natural wetlands, the biotic risk associated with MeHg produced in artificial wetlands, and the impact of dredging in stormwater wetlands. The water-table manipulations used to manage habitat wetlands have never been investigated in terms of their impact on MeHg biogeochemistry. There is also a need to quantitatively balance the many benefits of artificial wetlands with the production of MeHg (Stamenkovic et al. 2005). In most cases, it is likely that the benefits greatly outweigh the relatively modest risk of MeHg production and export, particularly in regions with low mercury contamination. However, the location of artificial wetlands in MeHg sensitive landscapes is of more concern. These include areas with pre-existing MeHg contamination problems, (Rumbold and Fink 2006;

Feng et al. 2014), wetlands surrounding mercury contaminated ports (Marvin-DiPasquale and Agee 2003; Figueiredo et al. 2014), or wetlands providing significant food supplies to especially sensitive or valuable aquatic and terrestrial species (Scheuhammer et al. 2007; Edmonds et al. 2012; Little et al. 2015).

There is also a need for more research on the microbial ecology of mercury methylation and demethylation in artificial wetlands. In particular, the limited understanding of demethylation reduces our ability to characterize complete MeHg budgets for individual systems and to identify potential control mechanisms. In some natural wetlands, degradation of MeHg significantly affects the total amount of MeHg exported from the system (Windham-Myers et al. 2014a). Demethylation has never been directly measured in any artificial wetland, although some studies suggested that photodemethylation (Gustin et al. 2006a) or stimulation of biotic demethylation (King et al. 2002) may have reduced net MeHg exports. Investigations of both biotic and abiotic demethylation in the water column and sediment of artificial wetlands are needed to help fill these gaps and identify potential opportunities for managing MeHg accumulation. Similarly, the taxonomic diversity of important mercury methylators in artificial wetlands should be more thoroughly explored. While there is ample evidence that sulfate reducers contribute to mercury methylation in most, if not all artificial wetlands surveyed, in other environments iron reducers and methanogens dominate MeHg production (Kerin et al. 2006; Hamelin et al. 2011). Recently, primers optimized for the environmental detection and identification of the *hgcAB* gene pair have been developed (Christensen et al. 2016), which enable the rapid and relatively inexpensive identification of the mercury methylating flora, and semi-quantification of its activity. A better understanding of the diversity and activity of mercury methylators in artificial wetlands will reveal new insights about their ecology, and may further inform practical management strategies.

Acknowledgements The authors would like to thank David Kenth, an engineer at the City of Brampton, for professional information on the practical management of stormwater ponds in Southern Ontario. We would also like to thank Bob Clay, of the Friends of the Rouge Watershed in Toronto, Ontario, who provided information on the construction of habitat wetlands in the Rouge National Park. Funding for our work in Southern Ontario was provided by a Natural Sciences and Engineering Research Council of Canada Discovery Grant.

References

Acha D, Iniguez V, Roulet M, Guimarães JRD, Luna R, Alanoca L, Sanchez S (2005) Sulfate-reducing bacteria in floating macrophyte rhizospheres from an Amazonian floodplain lake in Bolivia and their association with Hg methylation. Appl Environ Microbiol 71:7531–7535. https://doi.org/10.1128/AEM.71.11.7531-7535.2005

Åkerblom S, Bishop K, Björn E, Lambertsson L, Eriksson T, Nilsson MB (2013) Significant interaction effects from sulfate deposition and climate on sulfur concentrations constitute major controls on methylmercury production in peatlands. Geochim Cosmochim Acta 102:1–11. https://doi.org/10.1016/j.gca.2012.10.025

Alam SK, Ager LA, Rosegger TM, Lange TR (1996) The effects of mechanical harvesting of floating plant tussock communities on water quality in Lake Istokpoga, Florida. Lake Reservoir Manage 12:455–461. https://doi.org/10.1080/07438149609354285

Amirbahman A, Reid AL, Haines TA, Kahl S, Arnold C (2002) Association of methylmercury with dissolved humic acids. Environ Sci Technol 36:690–695. https://doi.org/10.1021/es011044qAmir

Avramescu M-L, Yumvihoze E, Hintelmann H, Ridal J, Fortin D, Lean DRS (2011) Biogeochemical factors influencing net mercury methylation in contaminated freshwater sediments from the St. Lawrence River in Cornwall, Ontario, Canada. Sci Total Environ 409:968–978. https://doi.org/10.1016/j.scitotenv.2010.11.016

Bae H-S, Dierberg FE, Ogram A (2014) Syntrophs dominate sequences associated with the mercury methylation-related gene hgcA in the water conservation areas of the Florida Everglades. Appl Environ Microbiol 80:6517–6526. https://doi.org/10.1128/AEM.01666-14

Bailey LT, Mitchell CPJ, Engstrom DR, Berndt ME, Coleman Wasik JK, Johnson NW (2017) Influence of porewater sulfide on methylmercury production and partitioning in sulfate-impacted lake sediments. Sci Total Environ 580:1197–1204. https://doi.org/10.1016/j.scitotenv.2016.12.078

Balcombe CK, Anderson JT, Fortney RH, Rentch JS, Grafton WN, Kordek WS (2005) A comparison of plant communities in mitigation and reference wetlands in the mid-Appalachians. Wetlands 25:130–142

Balogh SJ, Swain EB, Nollet YH (2006) Elevated methylmercury concentrations and loadings during flooding in Minnesota rivers. Sci Total Environ 368:138–148

Balogh SJ, Swain EB, Nollet YH (2008) Characteristics of mercury speciation in Minnesota rivers and streams. Environ Pollut 154:3–11. https://doi.org/10.1016/j.envpol.2007.11.014

Barkay T, Miller SM, Summers AO (2003) Bacterial mercury resistance from atoms to ecosystems. FEMS Microbiol Rev 27:355–384. https://doi.org/10.1016/S0168-6445(03)00046-9

Bates LM, Hall BD (2012) Concentrations of methylmercury in invertebrates from wetlands in the Prairie Pothole Region of North America. Environ Pollut 160:153–160

Bates AL, Orem WH, Harvey JW, Spiker EC (2002) Tracing sources of sulfur in the Florida Everglades. J Environ Qual 31:287–299. https://doi.org/10.2134/jeq2002.0287

Benoit JM, Gilmour CC, Mason RP, Heyes A (1999) Sulfide controls on mercury speciation and bioavailability to methylating bacteria in sediment pore waters. Environ Sci Technol 33:951–957. https://doi.org/10.1021/es9808200

Benoit JM, Gilmour CC, Heyes A, Mason RP, Miller CL (2003) Geochemical and biological controls over methylmercury production and degradation in aquatic ecosystems. In: Cai Y, Braids OC (eds) Biogeochemistry of environmentally important trace elements. American Chemical Society, San Diego, CA, pp 262–297

Biester H, Bindler R, Martinez-Cortizas A, Engstrom DR (2007) Modeling the past atmospheric deposition of mercury using natural archives. Environ Sci Technol 41:4851–4860. https://doi.org/10.1021/es0704232

Black FA, Poulin BA, Flegal AR (2012) Factors controlling the abiotic photo-degradation of monomethylmercury in surface waters. Geochim Cosmochim Acta 84:492–507. https://doi.org/10.1016/j.gca.2012.01.019

Boucher O, Jacobson SW, Plusquellec P, Dewailly E, Ayotte P, Forget-Dubois N, Jacobson JL, Muckle G (2012) Prenatal methylmercury, postnatal lead exposure, and evidence of Attention Deficit/Hyperactivity Disorder among Inuit Children in Arctic Québec. Environ Health Perspect 120:1456–1461. https://doi.org/10.1289/ehp.1204976

Brasso RL, Cristol DA (2007) Effects of mercury exposure on the reproductive success of tree swallows (Tachycineta bicolor). Ecotoxicology 17:133–141. https://doi.org/10.1007/s10646-007-0163-z

Brix H (1997) Do macrophytes play a role in constructed treatment wetlands? Water Sci Technol 35:11–17. https://doi.org/10.1016/S0273-1223(97)00047-4

Bromley PT, Buhlmann KA, Helfrich LA (1985) Management of wood ducks on private lands and waters. Virginia Cooperative Extension Service, Virginia State University, Petersburg, VA

Bruland GL, Richardson CJ (2006) Comparison of soil organic matter in created, restored and paired natural wetlands in North Carolina. Wetl Ecol Manage 14:245–251. https://doi.org/10.1007/s11273-005-1116-z

Campbell DA, Cole CA, Brooks RP (2002) A comparison of created and natural wetlands in Pennsylvania, USA. Wetl Ecol Manage 10:41–49

Chavan PV, Dennett KE, Marchand EA, Gustin MS (2007) Evaluation of small-scale constructed wetland for water quality and Hg transformation. J Hazard Mater 149:543–547

Choi S-C, Chase T, Bartha R (1994) Enzymatic catalysis of mercury methylation by Desulfovibrio desulfuricans LS. Appl Environ Microbiol 60:1342–1346

Christensen GA, Wymore AM, King AJ, Podar M, Hurt RA Jr, Santillan EU, Soren A, Brandt CC, Brown SD, Palumbo AV, Wall JD, Gilmour CC, Elias DA (2016) Development and validation of broad-range qualitative and clade-specific quantitative molecular probes for assessing mercury methylation in the environment. Appl Environ Microbiol 82(19):6068–6078. https://doi.org/10.1128/AEM.01271-16

Chumchal MM, Drenner RW (2015) An environmental problem hidden in plain sight? Small human-made ponds, emergent insects, and mercury contamination of biota in the Great Plains: an environmental problem hidden in plain sight. Environ Toxicol Chem 34:1197–1205. https://doi.org/10.1002/etc.2954

Coleman Wasik JK, Engstrom DR, Mitchell CPJ et al (2015) The effects of hydrologic fluctuation and sulfate regeneration on mercury cycling in an experimental peatland: drought increases mercury in peatlands. J Geophys Res Biogeo 120:1697–1715. https://doi.org/10.1002/2015JG002993

Compeau G, Bartha R (1984) Methylation and demethylation of mercury under controlled redox, pH and salinity conditions. Appl Environ Microbiol 48:1203–1207

Compeau GC, Bartha R (1985) Sulfate-reducing bacteria: principal methylators of mercury in anoxic estuarine sediment. Appl Environ Microbiol 50:498–502

Correia RRS, de Oliveira DCM, Guimarães JRD (2012) Total mercury distribution and volatilization in microcosms with and without the aquatic macrophyte Eichhornia crassipes. Aquat Geochem 18:421–423

Cosio C, Flück R, Regier N, Slaveykova VI (2014) Effects of macrophytes on the fate of mercury in aquatic systems: Hg and macrophytes. Environ Toxicol Chem 33:1225–1237. https://doi.org/10.1002/etc.2499

de Wit HA, Kainz MJ, Lindholm M (2012) Methylmercury bioaccumulation in invertebrates of boreal streams in Norway: effects of aqueous methylmercury and diet retention. Environ Pollut 164:235–241. https://doi.org/10.1016/j.envpol.2012.01.041

Desrosiers M, Planas D, Mucci A (2006) Total mercury and methylmercury accumulation in periphyton of Boreal Shield Lakes: influence of watershed physiographic characteristics. Sci Total Environ 355:247–258. https://doi.org/10.1016/j.scitotenv.2005.02.036

Domagalski J (1998) Occurrence and transport of total mercury and methyl mercury in the Sacramento River Basin, California. J Geochem Explor 64:277–291

Drake J, Guo Y (2008) Maintenance of Wet Stormwater Ponds in Ontario. Canadian Water Resources Journal 33(4):351–368. https://doi.org/10.4296/cwrj3304351

Driscoll CT, Blette V, Yan C, Schofield CL, Munson Rm Holsapple J (1995) The role of dissolved organic carbon in the chemistry and bioavailability of mercury in remote Adirondack lakes. In: Porcella DB, Huckabee JW, Wheatley B (eds) Proceedings of the third international conference on mercury as a global pollutant. Springer, New York, pp 499–508

Driscoll CT, Holsapple J, Schofield CL, Munson R (1998) The chemistry and transport of mercury in a small wetland in the Adirondack region of New York, USA. Biogeochemistry 40:137–146

Driscoll CT, Mason RP, Chan HM et al (2013) Mercury as a global pollutant: sources, pathways, and effects. Environ Sci Technol 47:4967–4983. https://doi.org/10.1021/es305071v

Drott A, Lambertsson L, Björn E, Skyllberg U (2008) Do potential methylation rates reflect accumulated methyl mercury in contaminated sediments? Environ Sci Technol 42:153–158. https://doi.org/10.1021/es0715851

Drott A, Lambertsson L, Björn E, Skyllberg U (2007) Importance of Dissolved Neutral Mercury Sulfides for Methyl Mercury Production in Contaminated Sediments. Environ Sci Technol 41:2270–2276. https://doi.org/10.1021/es061724z

Duncan CP, Groffman PM (1994) Comparing microbial parameters in natural and constructed wetlands. J. J Environ Qual 23:298–305

Eckley CS, Branfireun B (2008) Mercury mobilization in urban stormwater runoff. Sci Total Environ 403:164–177. https://doi.org/10.1016/j.scitotenv.2008.05.021

Eckley CS, Hintelmann H (2006) Determination of mercury methylation potentials in the water column of lakes across Canada. Sci Total Environ 368:111–125. https://doi.org/10.1016/j.scitotenv.2005.09.042

Eckley CS, Luxton TP, Goetz J, McKernan J (2017) Water-level fluctuations influence sediment porewater chemistry and methylmercury production in a flood-control reservoir. Environ Pollut 222:32–41. https://doi.org/10.1016/j.envpol.2017.01.010

Edmonds ST, O'Driscoll NJ, Hillier NK et al (2012) Factors regulating the bioavailability of methylmercury to breeding rusty blackbirds in northeastern wetlands. Environ Pollut 171:148–154. https://doi.org/10.1016/j.envpol.2012.07.044

Elser JJ, Bracken MES, Cleland EE et al (2007) Global analysis of nitrogen and phosphorus limitation of primary producers in freshwater, marine and terrestrial ecosystems. Ecol Lett 10:1135–1142. https://doi.org/10.1111/j.1461-0248.2007.01113.x

Feng S, Ai Z, Zheng S, Binhe G, Yuncong L (2014) Effects of dryout and inflow water quality on mercury methylation in a constructed wetland. Water Air Soil Pollut 255:1929. https://doi.org/10.1007/s11270-014-1929-6

Figueiredo NLL, Areias A, Mendes R, Canário J, Duarte A, Carvalho C (2014) Mercury-resistant bacteria from salt marsh of Tagus Estuary: the influence of plants presence and mercury contamination levels. J Toxicol Environ Health A 77:959–971. https://doi.org/10.1080/15287394.2014.911136

Fitzgerald WF, Engstrom DR, Mason RP, Nater EA (1998) The case for atmospheric mercury contamination in remote areas. Environ Sci Technol 32:1–7

Fleck JA, Gill G, Bergamaschi BA, Kraus TEC, Downing BD, Alpers CN (2014) Concurrent photolytic degradation of aqueous methylmercury and dissolved organic matter. Sci Total Environ 484:263–275. https://doi.org/10.1016/j.scitotenv.2013.03.107

Fleck JA, Marvin-DiPasquale M, Eagles-Smith CA, Ackerman JT, Lutz MA, Tate M, Alpers CN, Hall BD, Krabbenhoft DP, Eckley CS (2016) Mercury and methylmercury in aquatic sediment across western North America. Sci Total Environ 568:727–738. https://doi.org/10.1016/j.scitotenv.2016.03.044

Fleming EJ, Mack EE, Green PG, Nelson DC (2006) Mercury methylation from unexpected sources: molybdate-inhibited freshwater sediments and an iron-reducing bacterium. Appl Environ Microbiol 72:457–464. https://doi.org/10.1128/AEM.72.1.457-464.2006

Fortin D, Goulet R, Roy M (2000) Seasonal cycling of Fe and S in a constructed wetland: the role of sulfate-reducing bacteria. Geomicrobiol J 17:221–235

Galloway M, Branfireun B (2004) Mercury dynamics of a temperate forested wetland. Sci Total Environ 325:239–254. https://doi.org/10.1016/j.scitotenv.2003.11.010

Gauci V, Matthews E, Dise N, Walter B, Koch D, Granberg G, Vile M (2004) Sulfur pollution suppression of the wetland methane source in the twentieth and twenty-first centuries. Proc Natl Acad Sci U S A 101(34):12583–12587. https://doi.org/10.1073/pnas.0404412101

Gentès S, Maury-Brachet R, Guyoneaud R, Monperrus M, André J-M, Davail S, Legeay A (2013a) Mercury bioaccumulation along food webs in temperate aquatic ecosystems colonized by aquatic macrophytes in south western France. Ecotoxicol Environ Saf 91:180–187. https://doi.org/10.1016/j.ecoenv.2013.02.001

Gentès S, Monperrus M, Legeay A, Maury-Brachet DS, André J-M, Guyoneaud R (2013b) Incidence of invasive macrophytes on methylmercury budget in temperate lakes: central role of bacterial periphytic communities. Environ Pollut 172:116–123. https://doi.org/10.1016/j.envpol.2012.08.004

Gilmour CC, Henry EA, Mitchell R (1992) Sulfate stimulation of mercury methylation in freshwater sediments. Environ Sci Technol 26:2281–2287

Gilmour CC, Riedel GS, Ederington MC, Bell JT, Gill GA, Stordal MC (1998) Methylmercury concentrations and production rates across a trophic gradient in the northern Everglades. Biogeochemistry 40:327–345

Gilmour CC, Podar M, Bullock AL, Graham AM, Brown SD, Somenahally A, Johs A, Hurt RA Jr, Bailey KL, Elias DA (2013) Mercury methylation by novel microorganisms from new environments. Environ Sci Technol 47:11810–11820. https://doi.org/10.1021/es403075t

Gionfriddo CM, Tate MT, Wick RR, Schultz MB, Zemla A, Thelen MP, Schofield R, Krabbenhoft DP, Holt KE, Moreau JW (2016) Microbial mercury methylation in Antarctic sea ice. Nat Microbiol 1:16127. https://doi.org/10.1038/nmicrobiol.2016.127

Gomes MVT, de Souza RR, Teles VS, Araújo Mendes É (2014) Phytoremediation of water contaminated with mercury using *Typha domingensis* in constructed wetland. Chemosphere 103:228–233. https://doi.org/10.1016/j.chemosphere.2013.11.071

Graham AM, Aiken GR, Gilmour CC (2012a) Dissolved organic matter enhances microbial mercury methylation under sulfidic conditions. Environ Sci Technol 46:2715–2723. https://doi.org/10.1021/es203658f

Graham AM, Bullock AL, Maizel AC, Elias DA, Gilmour CC (2012b) Detailed assessment of the kinetics of Hg-cell association, Hg methylation, and methylmercury degradation in several Desulfovibrio species. Appl Environ Microbiol 78:7337–7346. https://doi.org/10.1128/AEM.01792-12

Greenway M (2004) Constructed wetlands for water pollution control processes—parameters and performance. Asia Pac J Chem Eng 12:491–504

Grigal DF (2002) Inputs and outputs of mercury from terrestrial watersheds: a review. Environ Rev 10:1–39. https://doi.org/10.1139/a01-013

Guimarães JRD, Mauro JBN, Meili M, Sundbom M, Haglund AL, Coelho-Souza SA, Hylander LD (2006) Simultaneous radioassays of bacterial production and mercury methylation in the periphyton of a tropical and a temperate wetland. J Environ Manag 81:95–100. https://doi.org/10.1016/j.jenvman.2005.09.023

Gustin MS, Chavan PV, Dennett KE, Donaldson S, Marchand E, Feranadez G (2006a) Use of constructed wetlands with four different experimental designs to assess the potential for methyl and total Hg outputs. Appl Geochem 21:2023–2035

Gustin MS, Chavan PV, Dennett KE, Marchand EA, Donaldson S (2006b) Evaluation of wetland methyl mercury export as a function of experimental manipulations. J Environ Qual 35:2352–2359

Hall BD, Bodaly RA, Fudge RJP, Rudd JWM, Rosenberg DM (1997) Food as the dominant pathway of methylmercury uptake by fish. Water Air Soil Pollut 100:13–24

Hall BD, Rosenberg DM, Wiens AP (1998) Methyl mercury in aquatic insects from an experimental reservoir. Can J Fish Aquat Sci 55(9):2036–2047

Hall BD, Louis VLS, Rolfhus KR, Bodaly RA, Beaty KG, Paterson MJ, Peech Cherewyk KA (2005) Impacts of reservoir creation on the biogeochemical cycling of methyl mercury and total mercury in boreal upland forests. Ecosystems 8:248–266. https://doi.org/10.1007/s10021-003-0094-3

Hall BD, Aiken GR, Krabbenhoft DP, Marvin-Dipasquale M, Swarzenski CM (2008) Wetlands as principal zones of methylmercury production in southern Louisiana and the Gulf of Mexico region. Environ Pollut 154:124–134. https://doi.org/10.1016/j.envpol.2007.12.017

Hamelin S, Amyot M, Barkay T, Wang Y, Planas D (2011) Methanogens: principal methylators of mercury in lake periphyton. Environ Sci Technol 45:7693–7700. https://doi.org/10.1021/es2010072

Hamelin S, Planas D, Amyot M (2015) Mercury methylation and demethylation by periphyton biofilms and their host in a fluvial wetland of the St. Lawrence River (QC, Canada). Sci Total Environ 512–513:464–471. https://doi.org/10.1016/j.scitotenv.2015.01.040

Harmon SM, King JK, Gladden JB, Chandler GT, Newman LA (2004) Methylmercury formation in a wetland mesocosm amended with sulfate. Environ Sci Technol 38:650–656. https://doi.org/10.1021/es030513g

Harris RC, Rudd JWM, Amyot M, Babiarz CL, Beaty KG, Blanchfield PJ, Bodaly RA, Branfireun BA, Gilmour CC, Graydon JA, Hetes A, Hintelmann H, Hurley JP, Kelly CA, Krabbenhoft DP, Lindberg SE, Mason RP, Paterson MJ, Podemski CL, Robinson A, Sandlands KA, Southworth GR, St. Louis VL, Tate MT (2007) Whole-ecosystem study shows rapid fish-mercury response to changes in mercury deposition. Proc Nat Acad Sci U S A 104:16586–16591. https://doi.org/10.1073/pnas.0704186104

Haynes KM, Kane ES, Potvin L, Lilleskov EA, Kolka R, Mitchell CPJ (2017) Mobility and transport of mercury and methylmercury in peat as a function of changes in water table regime and plant functional groups: climate change and peat pore water Hg. Glob Biogeochem Cycles. https://doi.org/10.1002/2016GB005471

He T, Lu J, Yang F, Feng X (2007) Horizontal and vertical variability of mercury species in pore water and sediments in small lakes in Ontario. Sci Total Environ 386:53–64. https://doi.org/10.1016/j.scitotenv.2007.07.022

Heal K, Drain SJ (2003) Sedimentation and sediment quality in sustainable urban drainage systems. In: Pratt CJ, Davies JW, Newman AP, Perry JL (eds) Proceedings of the 2nd national conference on sustainable drainage, Coventry, UK

Hines ME, Poitras EN, Covelli S, Faganeli J, Emili A, Žižek S, Horvat M (2012) Mercury methylation and demethylation in Hg-contaminated lagoon sediments (Marano and Grado Lagoon, Italy). Estuar Coast Shelf Sci 113:85–95. https://doi.org/10.1016/j.ecss.2011.12.021

Hintelmann H, Keppel-Jones K, Evans RD (2000) Constants of mercury methylation and demethylation rates in sediments and comparison of tracer and ambient mercury availability. Environ Toxicol Chem 19:2204–2211

Hintelmann H, Harris R, Heyes A, Hurley JP, Kelly CA, Krabbenhoft DP, Lindberg S, Rudd JWM, Scott KJ, St. Louis VL (2002) Reactivity and mobility of new and old mercury deposition in a boreal forest ecosystem during the first year of the METAALICUS study. Environ Sci Technol 36:5034–5040. https://doi.org/10.1021/es025572t

Hoggarth CGJ, Hall BD, Mitchell CPJ (2015) Mercury methylation in high and low-sulphate impacted wetland ponds within the prairie pothole region of North America. Environ Pollut 205:269–277. https://doi.org/10.1016/j.envpol.2015.05.046

Holmes J, Lean D (2006) Factors that influence methylmercury flux rates from wetland sediments. Sci Total Environ 368:306–319. https://doi.org/10.1016/j.scitotenv.2005.11.027

Hsu-Kim H, Kucharzyk KH, Zhang T, Deshusses MA (2013) Mechanisms regulating mercury bioavailability for methylating microorganisms in the aquatic environment: a critical review. Environ Sci Technol 47:14385–14394. https://doi.org/10.1021/es304370g

Jay JA, Murray KJ, Gilmour CC, Mason RP, Morel FMM, Roberts AL, Hemond HF (2002) Mercury Methylation by Desulfovibrio desulfuricans ND132 in the Presence of Polysulfides. Appl Environ Microbiol 68:5741–5745. https://doi.org/10.1128/AEM.68.11.5741-5745.2002

Jennifer Drake, Yiping Guo, (2008) Maintenance of Wet Stormwater Ponds in Ontario. Canadian Water Resources Journal 33 (4):351-368

Jensen S, Jernelöv A (1969) Biological methylation of mercury in aquatic organisms. Nature 223:753–754. https://doi.org/10.1038/223753a0

Jonsson S, Skyllberg U, Nilsson MB, Lundberg E, Andersson A, Björn E (2014) Differentiated availability of geochemical mercury pools controls methylmercury levels in estuarine sediment and biota. Nat Commun 5:4624–4631. https://doi.org/10.1038/ncomms5624

Jeremiason JD, Engstrom DR, Swain EB, Nater EA, Johnson BM, Almendinger JE, Monson BA, Kolka RK (2006) Sulfate addition increases methylmercury production in an experimental wetland. Environ Sci Technol 40:3800–3806. https://doi.org/10.1021/es0524144

Kadlec RH, Wallace S (2009) Treatment wetlands, 2nd edn. CRC Press, Taylor & Francis, New York

Kao CM, Wang JY, MJ W (2001) Evaluation of atrazine removal processes in a wetland. Water Sci Technol 44:539–544

Karagas MR, Choi AL, Oken E, Horvat M, Schoeny R, Kamai E, Cowell W, Grandjean P, Korrick S (2012) Evidence on the human health effects of low-level methylmercury exposure. Environ Health Perspect 120:799–806. https://doi.org/10.1289/ehp.1104494

Kerin EJ, Gilmour CC, Roden E, Suzuki MT, Coates JD, Mason RP (2006) Mercury methylation by dissimilatory iron-reducing bacteria. Appl Environ Microbiol 72:7919–7921. https://doi.org/10.1128/AEM.01602-06

King JK, Kostka JE, Frischer ME, Saunders FM (2000) Sulfate-reducing bacteria methylate mercury at variable rates in pure culture and in marine sediments. Appl Environ Microbiol 66:2430–2437. https://doi.org/10.1128/AEM.66.6.2430-2437.2000

King JK, Harmon SM, TT F, Gladden JB (2002) Mercury removal, methylmercury formation, and sulfate-reducing bacteria profiles in wetland mesocosms. Chemosphere 46:859–870

Kivaisi AK (2001) The potential for constructed wetlands for wastewater treatment and reuse in developing countries: a review. Ecol Eng 16:545–560

Klapstein SJ, Ziegler SE, Risk DA, O'Driscoll NJ (2017) Quantifying the effects of photoreactive dissolved organic matter on methylmercury photodemethylation rates in freshwaters. Environ Toxicol Chem 36(6):1493–1502. https://doi.org/10.1002/etc.3690

Kosolapov DB, Kuschk P, Vainshtein MB, Vatsourina AV, Wießner A, Kästner M, Müller RA (2004) Microbial processes of heavy metal removal from carbon-deficient effluents in constructed wetlands. Eng Life Sci 4:403–411. https://doi.org/10.1002/elsc.200420048

Krabbenhoft DP, Benoit JM, Babiarz CL, Hurley JP, Andren AW (1995) Mercury cycling in the Allequash Creek watershed, northern Wisconsin. In: Porcella DB, Huckabee JW, Wheatley B (eds) Mercury as a global pollutant. Springer, Dordrecht, The Netherlands, pp 425–433

Kronberg R-M, Tjerngren I, Drott A, Björn E, Skyllberg U (2012) Net degradation of methyl mercury in alder swamps. Environ Sci Technol 46:13144–13151. https://doi.org/10.1021/es303543k

Kronberg R-M, Jiskra M, Wiederhold JG, Björn E, Skyllberg U (2016) Methyl mercury formation in hillslope soils of boreal forests: the role of forest harvest and anaerobic microbes. Environ Sci Technol 50:9177–9186. https://doi.org/10.1021/acs.est.6b00762

Langer CS, Fitzgerald WF, Visscher PT, Vandal GM (2001) Biogeochemical cycling of methylmercury at Barn Island salt marsh, Stonington, CT, USA. Wetl Ecol Manage 9:295–310

Lavoie RA, Jardine TD, Chumchal MM, Kidd KA, Campbell LM (2013) Biomagnification of mercury in aquatic food webs: a worldwide meta-analysis. Environ Sci Technol 47:13385–13394

Lazaro WL, Diez S, da Silva CJ, Ignacio ARA, Guimaraes JRD (2016) Waterscape determinants of net mercury methylation in a tropical wetland. Environ Res 150:438–445

Lee C, Fletcher TD, Sun G (2009) Nitrogen removal in constructed wetland systems. Eng Life Sci 9:11–22. https://doi.org/10.1002/elsc.200800049

Li Y, Mao Y, Liu G, Tachiev G, Roelant D, Feng X, Cai Y (2010) Degradation of methylmercury and its effects on mercury distribution and cycling in the Florida Everglades. Environ Sci Technol 44:6661–6666. https://doi.org/10.1021/es1010434

Li Y, Cai Y (2012) Progress in the study of mercury methylation and demethylation in aquatic environments. Chin Sci Bull 58:177–185. https://doi.org/10.1007/s11434-012-5416-4

Lin C-C, Yee N, Barkay T (2012) Microbial transformations in the mercury cycle. In: Lui G, Cai Y, O'Driscoll N (eds) Environmental chemistry and toxicology of mercury. Wiley-Blackwell, Hoboken, pp 155–191

Lindberg SE, Hanson PJ, Meyers TP, Kim K-H (1998) Air/surface exchange of mercury vapor over forests—the need for a reassessment of continental biogenic emissions. Atmos Environ 32:895–908. https://doi.org/10.1016/S1352-2310(97)00173-8

Lindberg S, Bullock R, Ebinghaus R, Engstrom D, Feng X, Fitzgerald W, Pirrone N, Prestbo E, Seigneur C (2007) A synthesis of progress and uncertainties in attributing the sources of mercury in deposition. AMBIO J Hum Environ 36:19–33

Little ME, Burgess NM, Broders HG, Campbell LM (2015) Mercury in little brown bat (*Myotis lucifugus*) maternity colonies and its correlation with freshwater acidity in Nova Scotia, Canada. Environ Sci Technol 49:2059–2065. https://doi.org/10.1021/es5050375

Liu G, Naja GM, Kalla P, Scheidt D, Gaiser E, Cai Y (2010) Legacy and fate of mercury and methylmercury in the Florida Everglades. Environ Sci Technol 45:496–501

Liu G, Li Y, Cai Y (2012) Adsorption of mercury on solids in the aquatic environment. In: Liu G, Cai Y, O'Driscoll N (eds) Environmental chemistry and toxicology of mercury, 1st edn. Wiley, Hoboken, NJ, pp 367–387

Lovley DR, Klug MJ (1983) Sulfate reducers can outcompete methanogens at freshwater sulfate concentrations. Appl Environ Microbiol 45:187–192

Lu X, Liu Y, Johs A, Zhao L, Wang T, Yang Z, Lin H, Elias D, Pierce EM, Liang L, Barkay T, Gu B (2016) Anaerobic mercury methylation and demethylation by *Geobacter bemidjiensis* Bem. Environ Sci Technol 50:4366–4373. https://doi.org/10.1021/acs.est.6b00401

Ma Z, Cai Y, Li B, Chen J (2010) Managing wetland habitats for waterbirds: an international perspective. Wetlands 30:15–27. https://doi.org/10.1007/s13157-009-0001-6

Mahaffey KR, Clickner RP, Bodurow CC (2003) Blood organic mercury and dietary mercury intake: national health and nutrition examination survey, 1999 and 2000. Environ Health Perspect 112:562–570. https://doi.org/10.1289/ehp.6587

Mailman M, Stepnuk L, Cicek N, Bodaly RA (2006) Strategies to lower methyl mercury concentrations in hydroelectric reservoirs and lakes: a review. Sci Total Environ 368:224–235. https://doi.org/10.1016/j.scitotenv.2005.09.041

Malaviya P, Singh A (2012) Constructed wetlands for management of urban stormwater runoff. Crit Rev Environ Sci Technol 42:2153–2214. https://doi.org/10.1080/10643389.2011.574107

Mallison CT, Stocker RK, Cichra CE (2001) Physical and vegetative characteristics of floating islands. J Aquat Plant Manag 39:107–111

Mao Y, Li Y, Richards J, Cai Y (2013) Investigating uptake and translocation of mercury species by sawgrass (*Cladium jamaicense*) using a stable isotope tracer technique. Environ Sci Technol 47:9678–9684. https://doi.org/10.1021/es400546s

Marrugo-Negrete J, Enamorado-Montes G, Durango-Hernández J, Díez S (2017) Removal of mercury from gold mine effluents using *Limnocharis flava* in constructed wetlands. Chemosphere 167:188–192. https://doi.org/10.1016/j.chemosphere.2016.09.130

Marvin-DiPasquale MC, Oremland RS (1998) Bacterial methylmercury degradation in Florida Everglades peat sediment. Environ Sci Technol 32:2556–2563

Marvin-DiPasquale M, Agee J, McGowan C, Oremland RS, Thomas M, Krabbenhoft D, Gilmour CC (2000) Methyl-mercury degradation pathways: a comparison among three mercury-impacted ecosystems. Environ Sci Technol 34:4908–4916. https://doi.org/10.1021/es0013125

Marvin-DiPasquale M, Agee JL (2003) Microbial mercury cycling in sediments of the San Francisco Bay-Delta. Estuaries 26:1517–1528

Marvin-DiPasquale M, Windham-Myers L, Agee JL, Kakouros E, Kieu LH, Fleck JA, Alpers CN, Stricker CA (2014) Methylmercury production in sediment from agricultural and non-agricultural wetlands in the Yolo Bypass, California, USA. Sci Total Environ 484:288–299. https://doi.org/10.1016/j.scitotenv.2013.09.098

Matthews DA, Babcock DB, Nolan JG, Prestigiacomo AR, Effler SW, Driscoll CT, Todorova SG, Kuhr KM (2013) Whole-lake nitrate addition for control of methylmercury in mercury-contaminated Onondaga Lake, NY. Environ Res 125:52–60. https://doi.org/10.1016/j.envres.2013.03.011

Mazrui NM, Jonsson S, Thota S, Zhao J, Mason RP (2016) Enhanced availability of mercury bound to dissolved organic matter for methylation in marine sediments. Geochim Cosmochim Acta 194:153–162. https://doi.org/10.1016/j.gca.2016.08.019

Meng B, Feng X, Qiu G, Li Z, Yao H, Shang L, Yan H (2016) The impacts of organic matter on the distribution and methylation of mercury in a hydroelectric reservoir in Wujiang River, Southwest China: the influence of organic matter on mercury cycling. Environ Toxicol Chem 35:191–199. https://doi.org/10.1002/etc.3181

Mergler D, Anderson HA, Chan LHM, Mahaffey KR, Murray M, Sakamoto M, Stern AH (2007) Methylmercury exposure and health effects in humans: a worldwide concern. Ambio 36:3–11. https://doi.org/10.1579/0044-7447(2007)36[3:MEAHEI]2.0.CO;2

Miskimmin BM, Rudd JW, Kelly CA (1992) Influence of dissolved organic carbon, pH, and microbial respiration rates on mercury methylation and demethylation in lake water. Can J Fish Aquat Sci 49:17–22

Mitchell CPJ, Gilmour CC (2008) Methylmercury production in a Chesapeake Bay salt marsh. Journal of Geophysical Research. Biogeosciences 113:G00C04

Mitchell CPJ, Branfireun BA, Kolka RK (2008a) Assessing sulfate and carbon controls on net methylmercury production in peatlands: an *in situ* mesocosm approach. Appl Geochem 23:503–518. https://doi.org/10.1016/j.apgeochem.2007.12.020

Mitchell CPJ, Branfireun BA, Kolka RK (2008b) Spatial characteristics of net methylmercury production hot spots in peatlands. Environ Sci Technol 42:1010–1016. https://doi.org/10.1021/es0704986

Moore TLC, Hunt WF (2012) Ecosystem service provision by stormwater wetlands and ponds—a means for evaluation? Water Res 46:6811–6823. https://doi.org/10.1016/j.watres.2011.11.026

Mueller P, Jensen K, Megonigal JP (2016) Plants mediate soil organic matter decomposition in response to sea level rise. Glob Chang Biol 22:404–414. https://doi.org/10.1111/gcb.13082

Munthe J, Bodaly RA, Branfireun BA, Driscoll CT, Gilmour C, Harris R, Horvat M, Lucotte M, Malm O (2007) Recovery of mercury-contaminated fisheries. Ambio 36:33–44. https://doi.org/10.1579/0044-7447(2007)36[33:ROMF]2.0.CO;2

Nelson EA, Specht WL, Knox AS (2006) Metal removal from water discharges by a constructed treatment wetland. Eng Life Sci 6:26–30. https://doi.org/10.1002/elsc.200620112

Obrist D, Johnson DW, Lindberg SE, Luo Y, Hararuk O, Bracho R, Battles JJ, Dail DB, Edmonds RL, Monson RK, Ollinger SV, Pallardy SG, Pregitzer KS, Todd DE (2011) Mercury distribution across 14 U.S. forests. Part I: spatial patterns of concentrations in biomass, litter, and soils. Environ Sci Technol 45:3974–3981. https://doi.org/10.1021/es104384m

O'Driscoll NJ, Lean DRS, Loseto LL, Carignan R, Siciliano SD (2004) Effect of dissolved organic carbon on the photoproduction of dissolved gaseous mercury in lakes: potential impacts of forestry. Environ Sci Technol 38:2664–2672. https://doi.org/10.1021/es034702a

Ontario Ministry of the Environment (2003) Stormwater management planning and design manual. MOE, Toronto, ON, Canada

Orem W, Gilmour C, Axelrad D, Krabbenhoft D, Scheidt D, Kalla P, McCormick P, Gabriel M, Aiken G (2011) Sulfur in the south Florida ecosystem: distribution, sources, biogeochemistry, impacts, and management for restoration. Crit Rev Environ Sci Technol 41:249–288. https://doi.org/10.1080/10643389.2010.531201

Oswald CJ, Carey SK (2016) Total and methyl mercury concentrations in sediment and water of a constructed wetland in the Athabasca Oil Sands Region. Environ Pollut 213:628–637. https://doi.org/10.1016/j.envpol.2016.03.002

Pacyna EG, Pacyna JM, Steenhuisen F, Wilson S (2006) Global anthropogenic mercury emission inventory for 2000. Atmos Environ 40:4048–4063. https://doi.org/10.1016/j.atmosenv.2006.03.041

Pacyna JM, Travnikov O, De Simone F, Hedgecock IM, Sundseth K, Pacyna EG, Steenhuisen F, Pirrone N, Munthe J, Kindbom K (2016) Current and future levels of mercury atmospheric pollution on a global scale. Atmos Chem Phys 16:12495–12511. https://doi.org/10.5194/acp-16-12495-2016

Pangala SR, Reay DS, Heal KV (2010) Mitigation of methane emissions from constructed farm wetlands. Chemosphere 78:493–499. https://doi.org/10.1016/j.chemosphere.2009.11.042

Parks JM, Johs A, Podar M, Bridou R, Hurt RA Jr, Smith SD, Tomanicek SJ, Qian Y, Brown SD, Brandt CC, Palumbo AV, Smith JC, Wall JD, Elias DA, Liang L (2013) The genetic basis for bacterial mercury methylation. Science 339:1332–1335. https://doi.org/10.1126/science.1230667

Peng X, Liu F, Wang W-X, Ye Z (2012) Reducing total mercury and methylmercury accumulation in rice grains through water management and deliberate selection of rice cultivars. Environ Pollut 162:202–208. https://doi.org/10.1016/j.envpol.2011.11.024

Peralta RM, Ahn C, Gillevet PM (2013) Characterization of soil bacterial community structure and physicochemical properties in created and natural wetlands. Sci Total Environ 443:725–732. https://doi.org/10.1016/j.scitotenv.2012.11.052

Pickhardt PC, Folt CL, Chen CY, Klaue B, Blum JD (2002) Algal blooms reduce the uptake of toxic methylmercury in freshwater food webs. Proc Natl Acad Sci U S A 99:4419–4423

Pirrone N, Cinnirella S, Feng X, Finkelman RB, Friedli HR, Leaner J, Mason R, Mukherjee AB, Stracher GB, Streets DG, Telmer K (2010) Global mercury emissions to the atmosphere from anthropogenic and natural sources. Atmos Chem Phys 10:5951–5964. https://doi.org/10.5194/acp-10-5951-2010

Podar M, Gilmour CC, Brandt CC, Soren A, Brown SD, Crable BR, Palumbo AV, Somenahally AC, Elias DA (2015) Global prevalence and distribution of genes and microorganisms involved in mercury methylation. Sci Adv 1:e1500675–e1500675. https://doi.org/10.1126/sciadv.1500675

Robinson JB, Tuovinen OH (1984) Mechanisms of microbial resistance and detoxification of mercury and organomercury compounds: physiological, biochemical, and genetic analyses. Microbiol Rev 48(2):95–124

Roden EE, Edmonds JW (1997) Phosphate mobilization in iron-rich anaerobic sediments: microbial Fe(III) oxide reduction versus iron-sulfide formation. Arch Hydrobiol 139:347–378

Rothenberg SE, Windham-Myers L, Creswell JE (2014) Rice methylmercury exposure and mitigation: a comprehensive review. Environ Res 133:407–423. https://doi.org/10.1016/j.envres.2014.03.001

Roy V, Amyot M, Carignan R (2009) Beaver ponds increase methylmercury concentrations in Canadian Shield streams along vegetation and pond-age gradients. Environ Sci Technol 43:5605–5611. https://doi.org/10.1021/es901193x

Rumbold DG, Fink LE (2006) Extreme spatial variability and unprecedented methylmercury concentrations within a constructed wetland. Environ Monit Assess 112:115–135

Schaefer JK, Morel FM (2009) High methylation rates of mercury bound to cysteine by *Geobacter sulfurreducens*. Nat Geosci 2:123–126

Schaefer JK, Rocks SS, Zheng W, Liang L, Gu B, Morel FMM (2011) Active transport, substrate specificity, and methylation of Hg(II) in anaerobic bacteria. Proc Natl Acad Sci U S A 108:8714–8719. https://doi.org/10.1073/pnas.1105781108

Schaefer JK, Szczuka A, Morel FMM (2014) Effect of divalent metals on Hg(II) uptake and methylation by bacteria. Environ Sci Technol 48:3007–3013. https://doi.org/10.1021/es405215v

Schallenberg M, Jacob K (1993) The ecology of sediment bacteria in lakes and comparison with other aquatic ecosystems. Ecology 74(3):919–934. https://doi.org/10.2307/1940816

Scheuhammer AM, Meyer MW, Sandheinrich MB, Murray MW (2007) Effects of environmental methylmercury on the health of wild birds, mammals, and fish. Ambio 36:12–19

Scholz M, Lee B (2005) Constructed wetlands: a review. Int J Environ Stud 62:421–447. https://doi.org/10.1080/00207230500119783

Schuster PF, Krabbenhoft DP, Naftz DL, Cecil D, Olson ML, Dewild JF, Susong DD, Green JR, Abbott ML (2002) Atmospheric mercury deposition during the last 270 years: a glacial ice core record of natural and anthropogenic sources. Environ Sci Technol 36:2303–2310. https://doi.org/10.1021/es0157503

Schuster PF, Shanley JB, Marvin-Dipasquale M, Reddy MM, Aiken GR, Roth DA, Taylor HE, Krabbenhoft DP, DeWild JF (2007) Mercury and organic carbon dynamics during runoff episodes from a northeastern USA watershed. Water Air Soil Pollut 187:89–108. https://doi.org/10.1007/s11270-007-9500-3

Schwesig D, Krebs O (2003) The role of ground vegetation in the uptake of mercury and methylmercury in a forest ecosystem. Plant Soil 253:445–455

Sellers P, Kelly CA, Rudd JWM, MacHutchon AR (1996) Photodegradation of methylmercury in lakes. Nature 380:694–697

Shih R, Robertson WD, Schiff SL, Rudolph DL (2011) Nitrate controls methyl mercury production in a streambed bioreactor. J Environ Qual 40:1586. https://doi.org/10.2134/jeq2011.0072

Simoneau M, Lucotte M, Garceau S, Laliberte D (2005) Fish growth rates modulate mercury concentrations in walley (Sander vitreus) from eastern Canadian lakes. Environ Res 98:73–82

Sinclair KA, Xie Q, Mitchell CPJ (2012) Methylmercury in water, sediment, and invertebrates in created wetlands of Rouge Park, Toronto, Canada. Environ Pollut 171:207–215. https://doi.org/10.1016/j.envpol.2012.07.043

Skyllberg U, Qian J, Frech W, Xia K, Bleam WF (2003) Distribution of mercury, methyl mercury and organic sulphur species in soil, soil solution and stream of a boreal forest catchment. Biogeochemistry 64:53–76

Skyllberg U, Xia K, Bloom PR, Nater EA, Bleam WF (2000) Binding of mercury(II) to reduced sulfur in soil organic matter along upland-peat soil transects. J Environ Qual 29:855–865

Smith SV, Renwick WH, Bartley JD, Buddemeier RW (2002) Distribution and significance of small, artificial water bodies across the United States landscape. Sci Total Environ 299:21–36

St. Louis VL, Rudd JW, Kelly CA, Beaty KG, Bloom NS, Flett RJ (1994) Importance of wetlands as sources of methyl mercury to boreal forest ecosystems. Can J Fish Aquat Sci 51:1065–1076

St. Louis VL, Rudd JW, Kelly CA, Beaty KG, Flett RJ, Roulet NT (1996) Production and loss of methylmercury and loss of total mercury from boreal forest catchments containing different types of wetlands. Environ Sci Technol 30:2719–2729

St. Louis VL, Rudd JWM, Kelly CA, Bodaly RA, Paterson MJ, Beaty KG, Hesslein RH, Heyes A, Majewski AR (2004) The rise and fall of mercury methylation in an experimental reservoir. Environ Sci Technol 38:1348–1358. https://doi.org/10.1021/es034424f

Stamenkovic J, Gustin MS, Dennett KE (2005) Net methyl mercury production versus water quality improvement in constructed wetlands: trade-offs in pollution control. Wetlands 25:748–757

Stolt MH, Genthner MH, Daniels WL, Groover VA, Nagle S, Haering KC (2000) Comparison of soil and other environmental conditions in constructed and adjacent palustrine reference wetlands. Wetlands 20:671–683

Stottmeister U, Wießner A, Kuschk P, Kappelmeyer U, Kästner M, Bederski O, Müller RA, Moormann H (2003) Effects of plants and microorganisms in constructed wetlands for wastewater treatment. Biotechnol Adv 22:93–117. https://doi.org/10.1016/j.biotechadv.2003.08.010

Strickman RJ (2017) Methylmercury in managed wetlands. Ph.D. dissertation, University of Toronto, Toronto, ON, Canada

Strickman RJ, Mitchell CPJ (2016) Accumulation and translocation of methylmercury and inorganic mercury in Oryza sativa: an enriched isotope tracer study. Sci Total Environ 574:1415–1423. https://doi.org/10.1016/j.scitotenv.2016.08.068

Strickman RJ, Mitchell CPJ (2017) Methylmercury production and accumulation in urban stormwater ponds and habitat wetlands. Environ Pollut 221:326–334. https://doi.org/10.1016/j.envpol.2016.11.082

Strosnider WH, Schultz SE, Strosnider KAJ, Nairn RW (2017) Effects on the underlying water column by extensive floating treatment wetlands. J Environ Qual 46:201. https://doi.org/10.2134/jeq2016.07.0257

Tjerngren I, Karlsson T, Björn E, Skyllberg U (2011) Potential Hg methylation and MeHg demethylation rates related to the nutrient status of different boreal wetlands. Biogeochemistry 108:335–350. https://doi.org/10.1007/s10533-011-9603-1

Todorova SG, Driscoll CT, Matthews DA, Effler SW, Hines ME, Henry EA (2009) Evidence for regulation of monomethyl mercury by nitrate in a seasonally stratified, eutrophic lake. Environ Sci Technol 43:6572–6578

Trasande L, Landrigan PJ, Schechter C (2005) Public health and economic consequences of methyl mercury toxicity to the developing brain. Environ Health Perspect 113:590–596. https://doi.org/10.1289/ehp.7743

Ullrich SM, Tanton TW, Abdrashitova SA (2001) Mercury in the aquatic environment: a review of factors affecting methylation. Crit Rev Environ Sci Technol 31:241–293. https://doi.org/10.1080/20016491089226

United Nations Environment Programme (UNEP) (2017) Minimata Convention on Mercury. Available via www.mercuryconvention.org. Accessed 23 Mar 2017

Vymazal J (2007) Removal of nutrients in various types of constructed wetlands. Sci Total Environ 380:48–65. https://doi.org/10.1016/j.scitotenv.2006.09.014

Wadzuk BM, Rea M, Woodruff G, Flynn K, Traver RG (2010) Water-quality performance of a constructed stormwater wetland for all flow conditions. J Am Water Resour Assoc 46:385–394. https://doi.org/10.1111/j.1752-1688.2009.00408.x

Wang F, Zhang J (2012) Mercury contamination in aquatic ecosystems under a changing environment: implications for the Three Gorges Reservoir. Chin Sci Bull 58:141–149. https://doi.org/10.1007/s11434-012-5490-7

Weaver MA, Zablotowicz RM, Krutz LJ, Bryson CT, Locke MA (2012) Microbial and vegetative changes associated with development of a constructed wetland. Ecol Indic 13:37–45. https://doi.org/10.1016/j.ecolind.2011.05.005

Wiener JG, Krabbenhoft DP, Heinz GH, Scheuhammer AM (2003) Ecotoxicology of mercury. In: Hoffman DJ, Rattner BA, Burton GA, Cairns J (eds) Handbook of ecotoxicology, 2nd edn. CRC Press, Boca Raton, FL, pp 409–463

Windham-Myers L, Marvin-DiPasquale M, Stricker CA, Agee JL, Kieu LH, Kakourous E (2013) Mercury cycling in agricultural and managed wetlands of California, USA: experimental evidence of vegetation-driven changes in sediment biogeochemistry and methylmercury production. Sci Total Environ 484:300–307. https://doi.org/10.1016/j.scitotenv.2013.05.028

Windham-Myers L, Fleck JA, Ackerman JT, Marvin-DiPasquale M, Striker CA, Heim WA, Bachand PA, Eagles-Smith CA, Gill G, Stephenson M, Alpers CN (2014a) Mercury cycling in agricultural and managed wetlands: a synthesis of methylmercury production, hydrologic export, and bioaccumulation from an integrated field study. Sci Total Environ 484:221–231. https://doi.org/10.1016/j.scitotenv.2014.01.033

Windham-Myers L, Marvin-DiPasquale M, Kakouros E, Agee JL, Kieu LH, Stricker CA, Fleck JA, Ackerman JT (2014b) Mercury cycling in agricultural and managed wetlands of California, USA: seasonal influences of vegetation on mercury methylation, storage, and transport. Sci Total Environ 484:308–318. https://doi.org/10.1016/j.scitotenv.2013.05.027

Wolf KL, Ahn C, Noe GB (2011) Development of soil properties and nitrogen cycling in created wetlands. Wetlands 31:699–712. https://doi.org/10.1007/s13157-011-0185-4

World Health Organization (2017) Technical information on development of fish consumption advice—FDA/EPA advice on what women and parents should know about eating fish. Available via FDA. www.fda.gov/Food/FoodborneIllnessContaminants/Metals/ucm531136.htm. Accessed 15 Mar 2017

Yuan Y (2012) Methylmercury: a potential environmental risk factor contributing to epileptogenesis. Neurotoxicology 33:119–126. https://doi.org/10.1016/j.neuro.2011.12.014

Zhao L, Qiu G, Anderson CWN, Meng B, Wang D, Shang L, Yan H, Feng X (2016) Mercury methylation in rice paddies and its possible controlling factors in the Hg mining area, Guizhou province, Southwest China. Environ Pollut 215:1–9. https://doi.org/10.1016/j.envpol.2016.05.001

Zhang T, Hsu-Kim H (2010) Photolytic degradation of methylmercury enhanced by binding to natural organic ligands. Nat Geosci 3:473–476. https://doi.org/10.1038/ngeo892

Zhang T, Kim B, Levard C, Reinsch BC, Lowry GV, Deshusses MA, Hsu-Kim H (2012) Methylation of mercury by bacteria exposed to dissolved, nanoparticulate, and microparticulate mercuric sulfides. Environ Sci Technol 46:6950–6958. https://doi.org/10.1021/es203181m

Zheng S, Gu B, Zhou Q, Li Y (2013) Variations of mercury in the inflow and outflow of a constructed treatment wetland in south Florida, USA. Ecol Eng 61:419–425. https://doi.org/10.1016/j.ecoleng.2013.10.015

Chapter 8
Ornamental Flowers and Fish in Indigenous Communities in Mexico: An Incentive Model for Pollution Abatement Using Constructed Wetlands

Marco A. Belmont, Eliseo Cantellano, and Noe Ramirez-Mendoza

Introduction

Like many other countries, Mexico has challenges with water scarcity, as well as pollution of existing water resources. Most of the watersheds in Mexico have water deficits and there are over one hundred underground water reservoirs that are being overexploited (Aboites et al. 2010). Furthermore, of the 243 m^3/s of municipal wastewater produced in Mexico, only 40% of this volume is treated. For non-municipal wastewater, only 16% is treated of the 188.7 m^3/s volume that is produced (i.e. 29.9 m^3/s), and the other 84% is used for irrigation (Arreguín et al. 2010). It is well known that untreated wastewater is a health and ecological hazard. The use of untreated wastewater in irrigation carries high risks of adverse health effects due to exposure to parasites and waterborne diseases (Keraita et al. 2015). Also, using untreated wastewater for irrigation may damage the soils by increasing salinity and the accumulation of pollutants (Qadir et al. 2015a, b).

The Mezquital Valley, located in the State of Hidalgo is an arid region in central Mexico that receives millions of cubic meters annually of untreated wastewater from Mexico City and the surrounding metropolitan area. The volume of wastewater received varies from 50 to 300 m^3/s, depending on the season, with an annual average of 60 m^3/s (Romero-Álvarez 1997). The wastewater is transported to the Mezquital Valley (MV) through an 80 km network of channels, tunnels and storage

---

M.A. Belmont (✉)
Toronto Public Health, City of Toronto, Toronto, ON, Canada
e-mail: marco.belmont@toronto.ca

E. Cantellano
FES Zaragoza, Universidad Nacional Autónoma de México, México D.F., Mexico

N. Ramirez-Mendoza
Cooperativa La Coralilla, Ixmiquilpan, Hidalgo, Mexico

© Springer International Publishing AG 2018
N. Nagabhatla, C.D. Metcalfe (eds.), *Multifunctional Wetlands*,
Environmental Contamination Remediation and Management,
DOI 10.1007/978-3-319-67416-2_8

ponds that irrigate an area of approximately 83,000 ha (Romero-Álvarez 1997). This wastewater irrigation system, which is one of the largest in the world, increases agricultural production due to input of water and nutrients, but creates a variety of health and environmental problems. Past studies have shown that the wastewater entering the MV has very high loads of fecal coliforms and helminth eggs (Cortés 1989; Cifuentes et al. 1994), and this has led to high rates of waterborne diseases in wastewater irrigated areas (Heinz et al. 2011). The VM also receives large amounts of nitrogen, phosphorous, calcium, potassium and heavy metals, which has been accumulating in the soil and in aquifers (Jiménez and Chávez 2004; Cajuste et al. 1991). Contaminants of emerging concern enter in the wastewater as well. For example, the pharmaceuticals, ibuprofen and naproxen are found in the wastewater, although the concentrations of these compounds are reduced by almost 100% through the irrigation network (Navarro et al. 2015). Antibiotic-resistant bacteria have also been detected in wastewater irrigated soils in the region (Broszat et al. 2014).

Concerns about human health risks has motivated the development of legislation to ban the cultivation of vegetables and to control the use of the land by the local population (Cifuentes et al. 1994; Downs et al. 2000; Hernández-Acosta et al. 2014; Navarro et al. 2015). There is an urgent need for treatment of the wastewater before it is used for irrigation. Unfortunately, there are insufficient resources for communities in the MV to construct and operate conventional wastewater treatment plants. However, constructed treatment wetlands may be a suitable option for wastewater treatment in small rural communities because of the low costs of construction and operation, and their simple technology. Reclaimed wastewater provides the advantage of retaining high concentrations of nutrients for irrigation, once the pollutants and pathogens have been removed (Edwards 1980). Furthermore, the availability of land for locating constructed wetlands is usually not a problem in remote areas (Denny 1997; Kivaisi 2001; Belmont et al. 2004).

However, the efficient operation of a constructed wetland requires continuous maintenance to control flows, remove debris and sludge, and monitor water quality parameters (Wu et al. 2015a, b). Small communities do not have the resources to pay municipal workers to maintain these systems. Therefore, incentives are required to compensate individuals within the community for their time and labour for maintaining a treatment wetland. The "La Corallila" cooperative located in the MV is an interesting case study in a rural Indigenous community where water treatment is coupled with the production of ornamental flowers with high commercial value and the production of fish for human consumption. This model promotes the creation of employment and economic development for the community as an incentive for maintaining a constructed wetland that improves the quality of wastewater. This model has been successfully replicated in other small communities in Mexico, and could also be implemented in other arid regions of the world.

Fig. 8.1 Location of the Mezquital Valley (*shaded*) within Hidalgo State in central Mexico

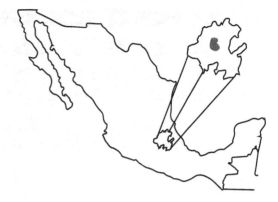

Table 8.1 Stages of development of the constructed wetland project in the community of Villagrán, in the Municipality of Ixmiquilpan, Mexico

Stage	Period	Activities
I. Start up	1995–1997	Pilot testing of the treatment wetland
II. Maturation	1998–2000	Expansion of wetlands and fish ponds; Testing of pisciculture with several fish species
III. Production	2001–2004	Beginning of profitability of the project
IV. Expansion	2005–present	Ornamental flower cultivation; Pisciculture and fish restaurant

Study Location

The MV is situated in the high plateau of central Mexico, 60 km north of Mexico City (Fig. 8.1), with an altitude between 1700 and 2100 m above sea level. About 75,000 farmers in the valley irrigate approximately 90,000 ha of land with mostly untreated wastewater (Qadir et al. 2015a, b). Historically, the area was a semi-desert, with very poor soil and was not suitable for agriculture. However, with the transport to the region of wastewater for irrigation, beginning at the end of the nineteenth century, agricultural production increased rapidly.

The constructed treatment wetland described in this study is located in the community of Villagrán in the Municipality of Ixmiquilpan in central Mexico. This community is located close to the distal margin of the irrigation system (Navarro et al. 2015; Martínez and Bandala 2015). The development of the project took place in four stages, dating back to 1995 (Table 8.1). The development of the project through these stages is described below in detail, but the description of improvements in water quality is only described for Stage IV of the operation, once cultivation of ornamental flowers was initiated.

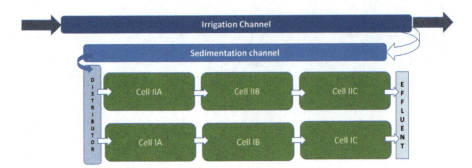

Fig. 8.2 Schematic of the Phase IV constructed wetland for treatment of wastewater

Treatment Wetland and Pisciculture

The treatment system presently in place in the community is based upon a pilot-scale system previously described by Belmont et al. (2004). It consists of a sedimentation channel and a six-cell subsurface flow wetland (Fig. 8.2). The system takes wastewater from a large channel at a rate of approximately 1 L/s. This water passes through a sedimentation channel to remove large solids (34 m length × 0.4 m width), and then the wastewater passes through two treatment trains consisting of 3 subsurface flow wetlands (SSFW) in series (Fig. 8.2), each with dimensions of 10 m length × 5 m width. In each train, there is a 0.8 m-drop between cells, which helps with aeration of the water before entering the next cell. The SSFW cells were filled with gravel (2 cm diameter approximately) and planted with Calla lilly (*Zantedeschia aethiopica*). The plants were 8–12 cm tall at the time of planting and were planted at 0.6 m distance between them. Calla lily was used in order to produce flowers of high commercial value and because this species was shown in a previous pilot-scale wetland to thrive in a subsurface flow wetland with untreated wastewater (Belmont et al. 2004). After treatment, the treated water is used in a series of tanks used for pisciculture.

The fish farm system consists of a stabilization pond that promotes algae growth and increases the amount of dissolved oxygen in the water, four cement ponds for hatching and growing the fry of tilapia (*Oreochromis niloticus*), three cement ponds for growing juvenile tilapia to market-size, and two ponds that are not lined with concrete for growing other species of fish.

Water Quality Monitoring

Since construction of the wetland beginning in 2005 to the present, there have been several monitoring periods to evaluate the performance of the wetland and the quality of the treated water, but there has not been continuous monitoring over this entire period. The measurements taken at various times included temperature, dissolved

Table 8.2 Methods used for analysis of water and wastewater

Parameter	Method	Method
Total solids	Gravimetric (NMX-AA-034-SCFI-2001)	2
Total suspended solids	Photometric (HACH 8006)	2
Oil and grease	Soxhlet extraction (NMX-AA-005-SCFI-2000)	4
Hardness	EDTA titration (HACH 8213)	2
Alkalinity	Titration, fenolftalein (HACH 8203)	2
COD	Closed reflux/spectrometry ((NMX-AA-030-SCFI-2001)	4
Total chlorine	DPD (HACH 10102)	1
Free chlorine	DPD (HACH 8167)	1
Sulfate	Sulfa Ver4 (HACH 8051)	3
Sulfide	Methylene blue (HACH 8131)	3
Orthophosphate	Fosfo Ver3 with ascorbic acid (HACH 8048)	1
Total nitrogen	Kjeldhal (HACH 10072)	3
Ammonium	Salicylate (HACH 8155)	2
Organic nitrogen	Kjeldhal (HACH 10072)	4
Nitrate	Cadmium reduction (HACH 8039)	2
Nitrite	Colorimetric (HACH 8507)	2
$cBOD_5$	Dilution method (NMX-AA-028-SCFI-2001)	4
Total coliforms	Membrane filtration (HACH Caldo m-ColiBlue24)	3
E. coli	Membrane filtration (HACH Caldo m-ColiBlue24)	3

(*1*) US EPA method for wastewater and standard methods, (*2*) US EPA method for wastewater, (*3*) Standard Methods for the Examination of Water and Wastewater, (*4*) Normas Oficiales Mexicanas

oxygen, pH, conductivity, hardness, alkalinity, total solids, total suspended solids, total coliforms, E. *coli*, chemical oxygen demand (COD), biological oxygen demand (BOD), oil and grease, and concentrations of total and organic nitrogen, ammonium, nitrate, nitrite and orthophosphate. The analytical techniques used are described in previous reports (López 2013). These methods are listed in Table 8.2.

Project Development

The project started as a proposal to use treatment wetlands to solve pollution problems faced by an organization of milk producers from the Ixmiquilpan municipality belonging to the Indigenous group, Hñahñu or Otomí. The collaboration with this group was favourable because they already were using the wastewater for growing cattle forage and vegetables. However, because of uncertainty over ownership of the land, a plan to establish a treatment wetland was abandoned. Fortunately, the community group, "La Coralilla", from the town Ex-Hacienda de Ocotza, Villagrán in Ixmiquilpan municipality remained interested and started experimental work with a treatment wetland. This stage of the work lasted 2 years (1995–1997). This initial work demonstrated that water quality was improved by treatment in the wetland, including a 90–95% reduction in total coliforms. Initially, the project aimed to reuse the reclaimed wastewater for agriculture, but the community group decided to test the use of the treated wastewater for pisciculture.

In Stage II of the project (1998–2000), external funding allowed for an increase in the surface area of the treatment wetlands and the construction of fish ponds. Six cells of wetlands and six fish ponds were built. Three plants were tested in the wetland, *Typha angustifolia*, *Phragmites australis* and *Canna flaccida*. The consortium also tested different fish species (grass carp, bagre and tilapia) and different fish feed (food waste, agricultural waste, and mesquite flour). During Stage II, the project operated in deficit due to the high capital expenses. However, during Stage III, the project produced small profits, ranging from $1400 to $7000 USD per year. At this time, the academic collaboration with the UNAM increased as well and the project was used for field trips and practical courses with students. At the end of 2000, it was decided to expand the project to include the sale of cooked fish in a restaurant. This change produced a higher participation of the women in the project, and the role of women from the community was fundamental to the development of the project.

During the first 4 years of the project, all funds were provided by the community group and accounted for 60% of the total investment. The remaining 40% of the total project investment came from government grants over subsequent years. The project did not start to become profitable until the fifth year, making the cost/benefit ratio for the first 8 years only 0.8, while the yield rate was only 7.1. It is important to note that the project was first conceived as an academic exercise for demonstrating the suitability of treatment wetlands for wastewater treatment, while the business model was adopted later.

Stage IV of the project (i.e. 2005 to the present) involves large scale cultivation of Calla lily in the treatment wetlands and tilapia in the fish ponds (Fig. 8.3). These flowers have a high market value in Mexico. During this stage, there has been a large increase in the production of both flowers and fish. The group officially formed a cooperative (i.e. Sociedad Cooperativa La Coralilla) which has been recognized by the municipal and state governments. Since 2005, the project operates with a team of nine people on average, with some seasonal support from another five workers. Since 2005, the La Coralilla cooperative produces approximately 3500 kg of fish (tilapia) and 10,000 flowers (Calla lily) annually. This production generates an approximate annual net profit of $14,000 USD, which is a very important revenue stream for the families involved in the operation and maintenance of the facility.

From the academic side, there have been ten professors involved in a variety of studies and many students have done practical courses and theses at the site. Currently the farm receives about 60 visitors a month, and most of these visitors are undergraduate students. Throughout the project, both the academic staff and students learned a great deal about the processes and operations of the treatment wetland, as well as the production, preparation and sales of the farmed fish.

Fig. 8.3 Cultivation of Calla lily in one cell of the La Coralilla treatment wetland during Stage II of the project. Photo by Eliseo Cantellano

Wastewater Treatment

The analysis of wastewater in the wetland has shown that the quality of wastewater in the influent is very variable and that the level of some of the pollutants can be relatively low. Results from a first sampling campaign performed in 2005 and 2006 showed a mean concentration of $N-NO_3$ in the influent of 4.5 mg/L. A second sampling campaign in 2010–2011 showed a mean $N-NO_3$ concentration in influent of 8.0 mg/L and extreme values of nitrate of up to 15 mg/L. The mean nitrate values in wastewater over these monitoring periods were below the level of 10 mg/L recognized by the USEPA as the maximum for rivers, and are low when compared with the usual level in wastewater of 40 mg/L (Metcalf and Eddy Inc. 1991). Therefore, it is likely that some denitrification is occurring within the wastewater distribution system after discharge from Mexico City and the surrounding municipality. However, these values for nitrate are similar to the levels reported for industrial pre-treated wastewater (Rodríguez-González et al. 2013). Ward et al. (1996) reported that levels of nitrate as low as 4 mg/L can produce damage to ecosystems and human health. There is also concern about contamination of groundwater with nitrates in the study area, as Downs et al. (1999) measured nitrate in groundwater in the range of 47–69 mg/L.

The average concentration of ammonium ($N-NH_4$) in the influent was 1.8 mg/L in the first sampling campaign (2005–2006), and 0.66 mg/L in the second sampling campaign (2010–2011). Even though these levels are relatively low when compared

Table 8.3 Mean, median, minimum and maximum values for water quality parameters measured between September 2010 and August 2011 in influent (untreated) and effluent (treated) from the treatment wetland (n = 12)

Parameter	Site	Mean	Median	Min.	Max.	% removal
Hardness (mg/L $CaCO_3$)	Influent	91.4	96.5	39	106	–
	Effluent	84.8	91.5	38	108	
Alkalinity (mg/L $CaCO_3$)	Influent	509	511	492	528	–
	Effluent	508	515	492	528	
Total nitrogen (mg-N/L)	Influent	16	18	8	24	27
	Effluent	12	11	6	20	
Organic nitrogen (mg/L)	Influent	0.7	0.6	0.2	1.2	77
	Effluent	0.2	BDL	BDL	1.2	
Ammonium (mg-N/L)	Influent	0.66	0.15	0.22	1.2	77
	Effluent	0.15	BDL	BDL	1.0	
Nitrate (mg-N/L)	Influent	8.2	8.2	2.9	15	23
	Effluent	6.5	5.8	1.0	13	
Nitrite (mg-N/L)	Influent	1.6	1.3	0.5	2.7	73
	Effluent	0.3	0.4	BDL	0.9	
Orthophosphate (mg/L)	Influent	2.4	2.7	1.2	3.0	21
	Effluent	1.8	2.1	1.0	2.4	
BOD (mg/L)	Influent	41	44	11	97	40
	Effluent	26	21	6	48	
COD (mg/L)	Influent	120	128	50	212	27
	Effluent	89	90	24	176	
Total solids (mg/L)	Influent	1.5	1.3	0.1	5.1	27
	Effluent	0.9	1.2	0.1	1.3	
Total suspended solids (mg/L)	Influent	33	36	23	41	78
	Effluent	7.8	7.1	4	13	
Oil and grease (mg/L)	Influent	23	23	12	41	21
	Effluent	17	19	8	33	
Total coliforms (CFU/100 mL)	Influent	592	660	232	1130	79
	Effluent	38	130	10	93	
E. coli (CFU/100 mL)	Influent	558	550	223	1100	70
	Effluent	45	130	9	100	

BDL below detection limit

with levels that have been reported for untreated wastewater, they can have toxic effects, as short term exposure to ammonium at concentrations between 0.6 and 2.0 mg N-NH_4/L are toxic to fish. In fact, some authors recommend a maximum concentration of 0.1 mg/L (Pillay 2008).

In general, the treatment wetland improved the quality of the water, as shown in the summary of water quality parameters for influent and effluent in the wetland from a sampling campaign between September 2010 and August 2011 (Table 8.3). On average, over the monitoring period, the treatment system removed >70% of coliform bacteria, 78% of suspended solids, 73% of nitrite, 77% of ammonium and a high proportion of biochemical and chemical oxygen demand (Table 8.3). Removals of orthophosphate (21%) and nitrate (23%) were lower, but poor removal of these nutrients is not unusual in subsurface flow wetlands (White et al. 2006; Vymazal 2007). Hybrid systems with vertical and horizontal flow constructed wetlands (Kabelo Gaboutloeloe et al. 2009; Vymazal 2013) or other wetland configurations (Wu et al. 2015a, b) may more effectively remove nitrogen and phosphorus from the wastewater.

The treatment wetland removed ammonium efficiently, as the 2010–2011 data showed reductions of mean ammonium concentrations of 0.66 mg/L in the influent to mean levels of 0.15 mg/L in the wetland effluent (Table 8.3). The highest ammonium concentrations in effluent were recorded in January and February, and this is probably due to the lower microbial activity in the wetland substrate during the winter (Yan and Xu 2014). However, none of the effluent samples that were collected showed ammonium concentrations higher than the US EPA guideline of 1.5 mg/L. Even though the ammonium concentration in the effluent can occasionally be elevated to levels close to the limits for protection of aquatic life, there have not been adverse effects observed on the tilapia. It is probable that the ammonium levels are further reduced by nitrification because of the aeration of the effluent when discharged into the stabilization pond, and the use of air injectors operating in the aquaculture ponds.

After the water passes through the treatment wetlands, it is suitable for pisciculture since the microbiological, physical and chemical parameters are within recommended values (Fig. 8.4). The World Health Organization established a limit for

Fig. 8.4 Stabilization ponds (foreground) and holding tanks for pisciculture at the La Coralilla treatment wetland during Stage IV of the project. Photo by Eliseo Cantellano

fecal coliforms of 1000 CFU/100 mL for aquaculture applications (WHO 1989; Blumenthal et al. 2000a, b). This value is set in order to avoid bacterial infection in the fish when using treated wastewater for aquaculture. All water quality parameters are typically within the range for tilapia culture in Mexico described by García and Calvario (2008), which includes the following lower and upper limits: pH (4–11), alkalinity (11–200 mg/L $CaCO_3$), hardness (20–350 mg/L), nitrite (up to 0.45 mg/L), nitrate (no more than 10^3 mg/L), orthophosphate (0.6–1.5 mg/L) and temperature (25–36 °C). Every year, the aquaculture system has been recognized for good aquaculture practices for the culture of tilapia established by the "Sistema Nacional de Sanidad, Inocuidad y Calidad Agroalimentaria" (i.e., National System for Agro-food Safety), which monitors the adequate processing of fish to avoid contamination (García and Calvario 2008).

The levels of alkalinity are very high in the influent water and were not reduced in the treatment wetland (Table 8.3). This is consistent with previous studies done by Jiménez (2005) on the wastewater originating from Mexico City that is diverted to the MV. This author reported alkalinity in the range of 709–779 mg/L and did not observe alkalinity reduction using different water treatment systems. However, these levels of alkalinity averaging 508 mg/L are not a problem for pisciculture. Boyd and Tucker (2012) reported alkalinity in natural waters up to 500 mg/L as $CaCO_3$. Furthermore, alkalinity can be beneficial for the fish since the dissolved anions can buffer pH changes, especially diurnal variations, and can reduce the toxicity of metals. The water quality of the system has been sufficiently good to allow culturing of 12,000 tilapia fry in three ponds with a total volume of 550 m^3, for an average density of 20 fish/m^3.

Conclusions

The La Coralilla cooperative in the Mezquital Valley of central Mexico operates a constructed treatment wetland that produces Calla lily flowers that are sold in a local market and generates treated water that is of sufficient quality for commercial pisciculture of tilapia. The community has organized a business that sells approximately 1000 Calla lily flowers and approximately 300 kg of tilapia per month. This operation contributes to water pollution abatement and also generates employment and economic benefits for the community. The revenue stream from the sale of fish and flowers provides an incentive to maintain and operate the wetland. This approach can be used in other small, rural communities around the world that do not have the resources to build and operate conventional wastewater treatment plants.

References

Aboites AL, Birrichaga GD, Garay TJ (2010) El Manejo de las Aguas Mexicanas en el Siglo XX. In: Jiménez B, Torregrosa MI, Aboites AL (eds) El Agua en México: cauces y encauces. Academia Mexicana de Ciencias, México, pp 21–50

Arreguín CF, Alcocer YV, Marengo MH, Cervantes JC, Albornoz GP, Salinas JG (2010) Los retos del agua. In: Jiménez B, Torregrosa ML, Aboites AL (eds) El agua en México: cauces y encauces. Academia Mexicana de Ciencias, Tlalpan, Mexico, pp 51–77

Belmont MA, Cantellano E, Thompson S, Williamson M, Sánchez A, Metcalfe CD (2004) Treatment of domestic wastewater in a pilot-scale natural treatment system in central Mexico. Ecol Eng 23:299–311

Blumenthal UJ, Mara DD, Peasey A, Ruiz-Palacios G, Stott R (2000a) Guidelines for the microbiological quality of treated wastewater used in agriculture: recommendations for revising WHO guidelines. Bull World Health Organ 78:1104–1116

Blumenthal UJ, Peasey A, Ruiz-Palacios G, Mara D (2000b) Guidelines for wastewater reuse in agriculture and aquaculture: recommended revisions based on new research evidence. WELL Study, Task No.: 68 Part 1. Water and Environmental Health at London and Loughboroug, London, UK

Boyd CE, Tucker CS (2012) Pond aquaculture water quality management. Springer, New York, 123 p

Broszat M, Nacke H, Blasi R, Siebe C, Huebner J, Daniel R, Grohmann E (2014) Wastewater irrigation increases the abundance of potentially harmful Gammmaproteobacteria in soils in Mezquital Valley, Mexico. Appl Environ Microbiol 80:5282–5291

Cajuste LJ, Carrillo R, Cota E, Laird RJ (1991) The distribution of metals from wastewater in the Mexican valley of Mezquital. Water Air Soil Pollut 57-58:763–771

Cifuentes E, Blumenthal U, Ruiz-Palacios G, Bennett S (1994) Escenario epidemiológico del uso agrícola del agua residual: El Valle del Mezquital, México. Revista Salud Pública de México 36(1):3–9

Cortés J (1989) Caracterización microbiológica de las aguas residuales con fines agrícolas. Informe de estudío realizado en el Valle del Mezquital. Instituto Mexicano de Tecnología Acuática, Jiutepec, México, 46 p

Denny P (1997) Implementation of constructed wetlands in developing countries. Water Sci Technol 35:27–34

Downs TJ, Cifuentes-Garcia E, Suffet IM (1999) Risk screening for exposure to groundwater pollution in a wastewater irrigation district of the Mexico City region. Environ Health Perspect 107:553

Downs T, Cifuentes E, Edward R, Suffett IM (2000) Effectiveness of natural treatment in a wastewater irrigation district of the Mexico City region: asynoptic field survey. Water Environ Res 72(1):4–21

Edwards P (1980) A review of recycling organic wastes into fish, with emphasis on the tropics. Aquaculture 21:261–279

García A, Calvario O (2008) Manual de buenas prácticas de producción acuícola de tilapia para la inocuidad alimentaria. Centro de Investigación en Alimentos y Desarrollo (CIAD), Mazatlán, Sinaloa, México, 104 p

Heinz I, Salgot M, Koo-Oshima S (2011) Water reclamation and inter-sectoral water transfer between agriculture and cities—a FAO economic wastewater study. Water Sci Technol 63:1068–1074

Hernández-Acosta E, Quiñones-Aguilar EE, Cristóbal-Acevedo D, Rubiños-Panta JE (2014) Calidad biológica de aguas residuales utilizadas para riego de cultivos forrajeros en Tulancingo, Hidalgo, México. Revista Chapingo Serie Ciencias Forestales y del Ambiente 20:89–100

Jiménez B, Chávez A (2004) Quality assessment of an aquifer recharged with wastewater for its potential use as drinking source: "El Mezquital Valley" case. Water Sci Technol 50:269–276

Jiménez B (2005) Treatment technology and standards for agricultural wastewater reuse: a case study in Mexico. Irrig Drain 54(S1):32–45

Kabelo Gaboutloeloe G, Chen S, Barber ME, Stöckle CO (2009) Combination of horizontal and vertical flow constructed wetlands to improve nitrogen removal. Water Air Soil Pollut Focus 9:279–286

Keraita B, Medlicott K, Dreschel P, Mateo-Sagasta J (2015) Health risks and cost-effective health risk management in wastewater use systems. In: Dreschel P, Qadir M, Wichelns D (eds) Wastewater: economic asset in an urbanizing world. Springer, pp 39–54

Kivaisi AK (2001) The potential for constructed wetlands for wastewater treatment and reuse in developing countries: a review. Ecol Eng 16:545–560

López CN (2013). Evaluación de la calidad del agua en un humedal construido de flujo subsuperficial en el municipio de Ixmiquilpan, Hidalgo. Ph.D. dissertation, Universidad Nacional Autónoma de México, DF, México

Martínez PF, Bandala ER (2015) Issues and challenges for water supply, storm water drainage and wastewater treatment in the Mexico City metropolitan area. In: Aguilar-Barajas I, Mahlknecht J, Kaledin J, Kjellén M, Mejía-Betancourt A (eds) Water and cities in Latin America: challenges for sustainable development. Routledge, Abingdon, UK, pp 109–125

Metcalf and Eddy Inc. (1991) Wastewater engineering. treatment, disposal and reuse, 3rd edn. McGraw-Hill Inc., New York, 1333 p

Navarro I, Chavez A, Barrios JA, Maya C, Becerril E, Lucario S, Jimenez B (2015) Wastewater reuse for irrigation—practices, safe reuse and perspectives. In: Javaid MS (ed) Irrigation and drainage—sustainable strategies and systems. InTech, Rijeka. ISBN: 978-953-51-2123-7. https://doi.org/10.5772/59361

Rodríguez-González MR, Molina-Burgos J, Jácome-Burgos A, Suárez-López J (2013) Humedal de flujo vertical para tratamiento terciario del efluente físico-químico de una estación depuradora de aguas residuales domésticas. Ingeniería, Investigación y Tecnología 14:223–235

Romero-Álvarez H (1997) Case Study VII—the Mezquital Valley, Mexico. In: WHO-UNEP, Water Pollution Control: a guide to the use of water quality management principles. Helmer Hespanhol, Suffolk, UK, pp 23–38

Pillay TVR (2008) Aquaculture and the environment. Wiley, New York, p 239

Qadir M, Boelee E, Amerasinghe P, Danso G (2015a) Costs and benefits of using wastewater for aquifer recharge. In: Dreschel P, Qadir M, Wichelns D (eds) Wastewater: economic asset in an urbanizing world. Springer, Amsterdam, pp 153–167

Qadir M, Mateo-Sagasta J, Jiménez B, Siebe C, Siemens J, Hanjra MA (2015b) Environmental risks and cost-effective risk management in wastewater use systems. In: Dreschel P, Qadir M, Wichelns D (eds) Wastewater: economic asset in an urbanizing world. Springer, Amsterdam, pp 55–74

Vymazal J (2007) Removal of nutrients in various types of constructed wetlands. Sci Total Environ 380:48–65

Vymazal J (2013) The use of hybrid constructed wetlands for wastewater treatment with special attention to nitrogen removal: a review of a recent development. Water Res 47:4795–4811

Ward MH, Mark SD, Cantor KP, Weisenburger DD, Correa-Vilasenore V, Zahm SH (1996) Drinking water nitrate and the risk of non-Hodgkin's lymphoma. Epidemiology 7:465–471

White JR, Reddy KR, Newman JM (2006) Hydrologic and vegetation effects on water column phosphorus in wetland mesocosms. Soil Sci Soc Am J 70:1242–1251

WHO (1989) Guidelines for the safe use of wastewater and excreta in agriculture and aquaculture. World Health Organization, Geneva, Switzerland

Wu H, Fan J, Zhang J, Hao Ngo H, Guo W, Liang S, Hu Z, Liu H (2015a) Strategies and techniques to enhance constructed wetland performance for sustainable wastewater treatment. Environ Sci Pollut Res 22:14637–14650

Wu H, Zhang J, Hao Ngo H, Guo W, Hu Z, Liang S, Fan J, Liu H (2015b) A review of sustainability of constructed wetlands for wastewater treatment: design and operation. Bioresour Technol 175:594–601

Yan Y, Xu J (2014) Improving winter performance of constructed wetlands for wastewater treatment in northern China: a review. Wetlands 34:243–253

Chapter 9
Phytoremediation Eco-models Using Indigenous Macrophytes and Phytomaterials

Kenneth Yongabi, Nidhi Nagabhatla, and Paula Cecilia Soto Rios

Introduction

Water is an important natural resource and key to the implementation of the sustainable development agenda (United Nations 2016). Access to safe drinking water and sanitation was included in the Millennium Development Goals (UNICEF/WHO 2017) and is currently identified in Goal 6 of the 2030 Agenda for Sustainable Development (United Nations 2015). However, water is considered a finite resource, even though it is renewable, and therefore, it is imperative to ensure effective management of water resources. For long term water security, urban areas must adopt strategies to secure and protect the water resources on which individuals and nature depend (Abell et al. 2017). For this reason, the Rio+20 document, "The Future We Want", emphasized the importance of water in the 2030 Agenda, and specifically stated in passage 122, "*We recognize the key role that ecosystems play in maintaining water quantity and quality and support actions within respective national boundaries to protect and sustainably manage these ecosystems*" (UNEP 2013).

K. Yongabi (✉)
Phytobiotechnology Research Foundation Institute, Catholic University of Cameroon, Bamenda, Bamenda, Cameroon
e-mail: yongabika@yahoo.com

N. Nagabhatla
United Nations University—Institute for Water, Environment and Health, Hamilton, ON, Canada

School of Geography and Earth Science, McMaster University, Hamilton, ON, Canada

P.C.S. Rios
United Nations University—Institute for Water, Health and Environment, Hamilton, ON, Canada

Graduate School of Life Sciences, Tohoku University, Sendai, Miyagi, Japan

© Springer International Publishing AG 2018
N. Nagabhatla, C.D. Metcalfe (eds.), *Multifunctional Wetlands*, Environmental Contamination Remediation and Management, DOI 10.1007/978-3-319-67416-2_9

Water pollution resulting from metals is widespread in the industrialized and the developing world. While contamination of surface and ground water systems may occur via circulation of natural geological deposits in the crust, the main drivers of metal pollution of water resources are anthropogenic sources, such as discharges of untreated municipal and industrial wastewater. This is especially true in countries with emerging economies because of an expanding their industrial presence (Obodo 2002). Often, threats to human health from metals are associated with exposure to lead, cadmium, mercury and arsenic (Järup 2003). Arsenic, although a metalloid, is usually included in discussions related to metal contamination (Aziz-Abraham and Al-Hajjaji 1984; Mortor and Dunette 1994).

Most countries have national guidelines for controlling the levels of toxic metals in water and sediments, as well as limits on consumption of metal contaminants in fish, shellfish and other foods (WHO 1997). Although there are existing guidelines for metal related pollution, there is little international coordination in regulation and enforcement of these guidelines in various regions of the globe (Bontoux 1998). Advanced technologies for removing metals from water and wastewater are beyond the resources of many countries, and especially in communities in rural or remote regions. As an alternative, researchers are investigating mitigation techniques using vegetation based approaches in studies conducted both in microcosms and in natural/field conditions (Rai 2008; Sukumaran 2013).

Wetland ecosystems serve as "natural water infrastructure", as they have the capacity to remove a variety of pollutants, including nutrients, pathogens, and metals. In recent years, there have been significant developments on the use of "Nature-based Solutions" for water treatment (Nesshöver et al. 2017; Díaz et al. 2015; TEEB 2010). Nature-based Solutions (NbS) are defined by The International Union for the Conservation of Nature (IUCN) and the World Bank as, "the actions to protect, sustainably manage and restore natural or modified ecosystems, which address societal challenges in an efficient and adaptable manner while simultaneously providing human well-being and biodiversity benefits" (Cohen-Schacham et al. 2016; Nesshöver et al. 2017). With its inclusive goal to address social, environmental and economic dimensions in tandem, NbS natural processes and structures closely align with the agenda of Sustainable Development (European Commission 2015). In that sense, the NbS framework is clearly applicable to the management of wetlands.

Wetlands have been for some time perceived as key elements of a green infrastructure approach to water treatment that contributes to human water security (Mahmood et al. 2013; Faulwetter et al. 2009; Li et al. 2015; Sun et al. 2015; Yabe and Nakamura 2010). Constructed wetlands are engineered systems that are developed to replicate the processes or functions of natural wetlands (Vymazal and Březinová 2016). Plants are a key component of wetland ecosystem functioning, both in constructed or natural wetlands (Jan and Tereza 2016). Phytoremediation approaches are commonly used in constructed wetlands for water and wastewater treatment. However, wetlands with even the same vegetation type and composition often show notable differences in hydrology and sediment characteristics (Williams 2002), making it crucial to test and select specific plant species for use in constructed wetlands that demonstrate high potential to remove targeted contaminants.

Phytomaterials harvested from plant species often show good capacity to treat water and in that sense, represent a cost-effective, safe, efficient and sustainable option for avoiding the outbreaks of waterborne diseases that frequently occur both in rural and urban areas of developing countries (Zhang et al. 2006). Treatment of drinking water at the household level, rather than centralized treatment systems may be an effective approach for reducing waterborne diseases in developing countries (Megersa et al. 2016). However, the availability of these phytomaterials is dependent on biogeography. Identification of candidate materials that are both effective and abundant in a specific region is a task requiring scientific expertise and investigation.

In this chapter, we review case studies from Nigeria that describe the accumulation of metals in aquatic macrophytes (Ogunkunle et al. 2016), the use of plants as bioindicators of metal contamination (Shuaibu and Nasiru 2011/ Ogunkunle et al. 2016), and the use of phytomaterials harvested from a terrestrial plant as a phytocoagulant and disinfectant for water treatment (Caceres et al. 1991; Bina 1991; Montakhab et al. 2010). These approaches are discussed in relation to the development of an integrated eco-model with potential for community engagement that fits well with the NbS approach for water and wastewater treatment. The case studies from Nigeria were implemented in an urban center in the north-central part of the country and in the Bauchi-guinea Savannah region.

Phytoremediation and Bioindicators

In recent years, various phytoremediation technologies have been implemented to remove pollutants from the environment (Cunningham and Berti 1993; Raskin et al. 1994; Salt et al. 1995; Leon Romero et al. 2017). These plant based remediation technologies are often inexpensive when compared to engineered solutions (Niemi and McDonald 2004; Wamelink et al. 2005). Typical plant species employed for metal pollution abatement include *Typha* sp., *Striga* sp., algae and water hyacinths, as well as terrestrial plants such as rice, and species of *Vetiver*, *Colocasia*, *Moringa* and *Jatropha* (Audu and Lawal 2006; Yongabi KA, 2012). However, essential cations (e.g. copper, iron, zinc) and toxic metals (e.g. cadmium, mercury, lead) can cause detrimental effects in plants, if present in high enough concentration (Bruins et al. 2000). Therefore, comprehensive studies are needed to assess which plants can be appropriately used for phytoremediation.

"Bioindicators" are living organisms which are utilized to screen the health of the natural ecosystem in the environment (Parmar et al. 2016), including the presence of contaminants in environmental media (e.g. water, sediments, soil, air). Species that are used to monitor pollution should have certain important features (Gadzała-Kopciuch et al. 2004). The most important characteristics of bioindicator plants were summarized by Kabata-Pendias and Pendias (1992) and include:

- High accumulation of contaminants
- Availability and ease of identification in the field. They should have a wide distribution, with no seasonal differences in abundance

- "Toxi-tolerance", meaning that the species should have low sensitivity to pollutants
- Existence of a correlation between accumulation and inputs to the ecosystem

Situation Analysis: Africa

Short-term economic development goals, degradation of natural habitats, water scarcity, poor sanitation and escalating human populations are among the key challenges in many African nations, as well as many other developing countries (Godfrey 2003; Muyibi et al. 2002c; Mahmood et al. 2013). Access to potable water at a reasonable cost is a key development problem in many parts of Africa (Masangwi et al. 2008). Sources of drinking water in this region, including rivers, ponds and dug wells are often contaminated with pathogens and metals (McConnachie et al. 1999; Muyibi et al. 2002a; Alkhatib et al. 2014), and very often the water is consumed without any type of treatment (Pritchard et al. 2009). Discharges of poorly treated wastewater; poor sanitation and contamination from agricultural runoff have resulted in deteriorating water quality in southwest Cameroon, Nigeria, Uganda and Benin (Yongabi 2012; Alkhatib et al. 2014). Expanding demand for bananas and cocoa is leading to large-scale production of these crops and extensive use of herbicides and pesticides, leading to contamination of surface waters through leaching and run-off (Asogwa and Dongo 2009). The pesticides used on rice fields in the northwest and southwest regions of Cameroon, as well as in rural areas of Nigeria are detected on a regular basis in water resources (Goufo 2008; Oluwole and Cheke 2011).

The Delimi, Badiko and Bindir rivers in the Gumau district of Bauchi State, Nigeria are impacted by the large human populations in the region, through wastewater contamination, bathing, laundry and fishing, and this region also reports high prevalence of diarrhea and other waterborne diseases (Akogun 1990). In Nigeria, children under 5 years of age are at high risk of diarrhea, reported as the main cause of fatalities in the country (Ahmed et al. 2007). Contamination of water resources by metals is also a potential key driver of health problems. Pollution problems include abandoned mine sites that threaten water resources due to leaching and runoff (Adamu et al. 2015). Metals can cause hyper-pigmentation, de-pigmentation, keratosis, dermatitis and skin cancer, as well as cardiovascular and neurological disorders (WHO 1997; Caerio et al. 2005). Studies in the Asa, Agba, Unilorin and Sobi wetlands in Nigeria reported high level of Pb and Cd in drinking water (Ogunkunle et al. 2016). Analysis by Alkhatib et al. (2014) also showed Pb contamination of drinking water in Nigeria, in addition to high Pb level in soil and vegetables (i.e. lettuce, spinach, and onion). Olokesus (1988) accessed a wide range of data and information sources to present an overview of the extent of water pollution in Nigeria, highlighting, the need to regularly monitor the levels of pollution, and to review and improve environmental protection legislations to reduce pollution. However, since that review 40 years ago, water pollution problems still persist in Nigeria.

To improve this situation, the scientific community in the region has conducted several studies to examine the potential of low-cost natural systems for treatment of domestic and industrial wastewater (Wood and Pybus 1992; Yongabi 2009; Yongabi et al. 2009; Mahmood et al. 2013). This work has built upon previous studies that demonstrated that natural plant systems can safeguard the sustainability of ecosystems, as well as protect human health (Burton and Peterson 1979; Rai et al. 1981; Fayed and Abdel-Shafy 1985; Van Straalen et al. 1987; Magalhaes et al. 1994; Alkhatib et al. 2014). As described in the present study, this has led to the development of an eco-model for an integrated approach to the use of plants to reduce risks from metal contamination.

Background and Objectives

In Nigeria, there are uncertainties about the ability of the nation to meet the Sustainable Development Goals for universal access to clean water and sanitation (i.e. SDG 6), due to the serious state of water pollution (Adedejia and Ako 2009). Pollution by metals is a major threat to the environment and to human health due to the persistence, and toxicity of these elements (Kabata-Pendias and Pendias 1992; Aktar et al. 2010; Singare et al. 2012). In the present review, the authors describe several case studies from Nigeria that collectively contribute to an eco-based model for using plant species for bioremediation of metal contamination, as bioindicators of contamination and as a low-cost treatment system for drinking water.

Case Study 1: Metal Accumulation in Urban Wetlands, Ilorin, Nigeria

This case study illustrates metal contamination in the urban center of Ilorin in the north-central part of Nigeria. In a recent study, Ogunkunle et al. (2016) monitored four lakes/wetlands in the region that are confined by earthen dams (i.e. Asa, Agba, Unilorin and Sobi). These lakes serve as sources of drinking water for the inhabitants of the Ilorin metropolis (Fig. 9.1). Ten water samples, as well as several macrophyte species were collected at each location. Macrophyte samples were dried and ground to a fine powder, then digested with a mixture of HNO_3 and $HClO_4$ acid. The concentrations of Pb, Cd, Ni and Mn in the digests, as well as in acidified water samples were measured by Atomic Absorption Spectrophotometry (AAS), as described previously (Ogunkunle et al. 2016). Analytical results were expressed as mg/L in water and mg/kg dry weight in macrophytes. Bioaccumulation factors (BAF) were calculated to establish the capacity of the macrophytes to accumulate metals (Nowell et al. 1999):

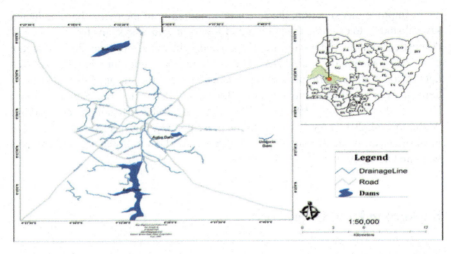

Fig. 9.1 (a) Location of the lake dams (b) Inset: map of Nigeria showing the states. Source: Ogunkunle et al. 2016, map not to scale

Table 9.1 Mean (±SD) concentrations and bioaccumulation factors (BAFs) of Mn in macrophyte species collected at selected lakes near Ilorin, Nigeria, and calculated concentration (%) in the plants

Macrophyte species	Concentration (mg/kg dw)	BAF	Site	Concentration in plants (%)[a]
Ceratophyllum demersum	79.8 ± 24.7	665	Asa	0.008
Pycreus lanceolatus	146.5 ± 81	862	Agba	0.0147
Eichhornia crassipes	70.5 ± 13.4	271	Sobi	0.0071
Pistia stratiotes	157.2 ± 40.4	605	Sobi	0.0157
Azolla Africana	37.7 ± 15.4	105	Unilorin	0.0038
Ceratophyllum demersum	277.0 ± 89.2	770	Unilorin	0.0277
Eclipta prostrata	133.2 ± 47.3	370	Unilorin	0.0133
Ludwigia abyssinica	170.5 ± 66.2	474	Unilorin	0.0171

[a]Mn threshold for hyper-accumulator is 1% (Xing et al. 2013)

$$BAF = \sum_{i-0}^{n} \frac{C_m}{Wi}$$

Where: C_m is the trace metal (mg/kg dry weight) in the macrophyte and Wi (mg/L) is the concentration of the trace metal in the water sample.

Table 9.1 summarizes the analytical results for Mn in macrophytes, as this element was detected at the highest concentration of 277 mg/kg dry weight in *Pycreus lanceolatus*. The BAFs for Mn were assessed to determine if they were high enough to qualify the macrophyte as a "hyper-accumulator" that bioconcentrates a metal by 1000 fold (Zayed and Gowthaman 1998; Xing et al. 2013). In a few samples, the BAF for Mn in the macrophytes approached this threshold (Table 9.1). For the other metals (i.e. Pb, Cd and Ni), all BAFs were well below values of 100. Therefore, three species of macrophytes (*Ceratophyllum demersum*, *Pycreus lanceolatus* and *Pistia stratiotes*) show potential as bioindicators for monitoring Mn pollution in aquatic ecosystems. Other studies have noted that hyper-accumulator plants can concentrate metals to levels as high as 1500 mg/kg without harmful effects (Pais and Jones 2000). None of the concentrations of Mn approached this threshold in the macrophytic plants (Table 9.1).

The analytical results for water samples also demonstrated that the four lakes are contaminated with Pb, with Asa and Unilorin lakes having the highest mean concentrations in water (i.e. 0.65 ± 0.21 mg/L and 0.53 ± 0.17 mg/L, respectively). The levels of Cd were relatively low, with the highest mean concentration of Cd in the surface water of Asa lake (i.e. 0.05 ± 0.01 mg/L). The levels of Ni were low at Agba and Sobi lakes.

Case Study 2: Metal Contamination in the Tourist Region of Bauchi, Nigeria

Bauchi State, located in the northeast region of Nigeria is the fifth largest state of Nigeria with a surface area of 49,119 km² and this region has been nicknamed as the "The Pearl of Tourism" (Fig. 9.2). Shadawanka stream/wetland is considered one of

Fig. 9.2 Study area in Bauchi, Nigeria (Source: Google Earth)

the major streams in the region that provides water for domestic consumption and irrigation and also regulates the aquatic habitat (Alkhatib et al. 2014; Adamu et al. 2015). The wetland is located in the commercial and industrial center of Bauchi, which contributes to the chronic metal contamination in the wetland (Bowen 1979; Burton and Peterson 1979; Magalhaes et al. 1994; Obodo 2002).

This case study investigates the potential of using an indigenous plant species as a bioindicator of metal contamination, with the goal of monitoring health risks related to water pollution in these communities. *Nymphaea lotus*, is a herbaceous wetland plant species that is widely distributed in wetland systems of Nigeria, including wetlands associated with streams, rivers and ponds (Welman 1948; White 1965; Imevbore 1971; Obot 1987). As illustrated in Fig. 9.3, this water lily has pink, white or yellow flowers and the leaves float or are partially submerged (Fayed and Abdel-Shafy 1985). The local community utilizes the species for medical treatments for circulatory system disorders, digestive system disorders, dysentery, genitourinary system disorders, as a diuretic, and for mental disorders and hyposomnia, as well as for treatment of leprosy, infections and inflammations (Burkill 1997). The seeds are also of value and frequently used in the Nigerian diet and for food preparations. This species is also used for medical purposes and as food in a number of other West African countries, for example, to make lotion in Sierra Leone (Holm-Nelson and Larsen 1979). The common occurrence, wide acceptance, popularity and most important its bio-accumulation potential makes this aquatic plant a good candidate for use as a bio indicator species.

Shuaibu and Nasiru (2011) collected *N. lotus* samples from ten different locations along Shadawanka wetlands to assess the levels of metals detected in the species. As shown in Fig. 9.4, concentrations varied, depending on the location, with the order of concentrations of the metals as Zn > Pb > Fe > Cd. Pais and Jones (2000) observed that hyper-accumulator plants can accumulate metals to levels as

Fig. 9.3 *Nymphaea lotus* growing in a wetland in Bauchi state, Nigeria

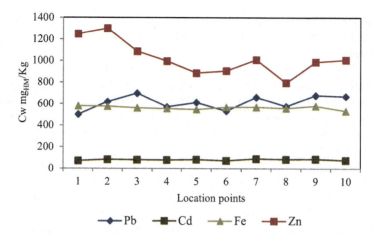

Fig. 9.4 Mean concentrations of Pb, Cd, Fe and Zn in *Nymphaea lotus* (mg/kg) collected from ten locations in Shadawanka wetlands, Nigeria. Adapted from Shuaibu and Nasiru (2011)

high as 1500 mg/kg without harmful effects (Pais and Jones 2000). In *N. lotus*, the concentrations of Zn approached 1400 mg/kg at one location, but concentrations of all other metals were well below this toxic threshold (Fig. 9.4). Overall, the study provided evidence that this aquatic species can be used as a bioindicator of metal contamination. Further investigation is required to study the potential health impact of contaminated vegetation either through direct consumption or through use in medical remedies (Khan et al. 2000).

The fundamental characteristic of a good bioindicator species is the ability to bioaccumulate several or selected elements to high levels (Wittig 1993). The case study in Bauchi demonstrates the potential for using three macrophytes species (*Ceratophyllum demersum, Pycreus lanceolatus* and *Pistia stratiotes*) harvested from local wetlands as bioindicators of metal pollution in aquatic ecosystems. In addition, the case study with *N. lotus*, showed that this aquatic plant also has a high potential for bioaccumulation of some metals. These species are all widely distributed in wetlands in Nigeria and in many other African countries (Welman 1948; White 1965), and may be useful for monitoring localized pollution (Ogunkunle et al. 2016). In Nigeria, the average annual maximum and minimum temperature varies from 35 °C and 18 °C in the North to 31 °C and 23 °C in the South (SGCBP 2008), so temperature is not expected to have a significance influence on the bioaccumulation capacity of the macrophytic plants.

Galadima et al. (2015) in a study situated in Nigeria demonstrated accumulation of metals by *Nymphaea* species and showed that the roots adsorbed high levels of Pb and Cd. Based on these results, the researchers concluded that this aquatic plant species served as a promising vector for removal of metals from wetland ecosystems. Shuaibu and Nasiru (2011) studied metal accumulation in different vegetative parts (i.e. flowers, leaves, stems, roots, seeds) of *Nymphaea* collected from localities where the plant is either used for medical purposes or consumed as food. There have been no reports of poisoning from direct consumption of *Nymphaea lotus*, but it can't be discounted that long term consumption of contaminated vegetation could pose a health risk.

Since *Nymphaea* species shows good potential for accumulation of metals, more investigation is required to establish the potential to use these macrophyte species in natural and constructed wetlands as an NbS for removal of contaminants. This may benefit the local communities both in terms of providing a safe potable water supply and through livelihood generation from the production of natural products with medicinal value.

The Millennium Development Report (2010) emphasized the importance of involving communities in identifying solutions for local level challenges as a prerequisite for sustainable solutions to development problems. Additionally, there is a critical need for dedicated management, technical and financial resources to deal with these challenges. Local, innovative solutions are needed that show potential for upscaling from pilot systems to the full-scale. While phytoremediation has been extensively analyzed in the laboratory and in microcosms, there have been only a few full-scale applications (Williams 2002). There have been even fewer models that have demonstrated community involvement. Examples of phytoremediation solutions are needed in a constructed or natural wetlands setting to demonstrate the effectiveness of low-cost NbS strategies.

Case Study 3: Locally Designed Water Treatment System in Nigeria Using a Moringa/Sand Filter

This case study illustrates a water treatment approach that employs locally sourced phytomaterials. The objective of this work was to assess the effectiveness of a filtration system that includes phytomaterials prepared from seeds of *Moringa oleifera* in combination with a sand filter to treat drinking water in a community in Bauchi, Nigeria. Commonly utilized coagulants for drinking water treatment include salts of aluminum and iron (Muyibi and Evison 1995; Kebreab et al. 2005). Chlorine is widely used for disinfection of drinking water to kill pathogens. Most of these coagulants and disinfectants are imported to Africa at a high cost. Negative impacts of typically employed water treatment solutions have been reported. For example, aluminum in water has been linked to Alzheimer's disease (Crapper and Krishnan 1973), although more recent scientific literature has largely discounted this connection. Chlorine treatment results in the formation of trichloromethanes, reported to be carcinogenic (Okuda and Baes 2001). Slow sand filters are popular systems for treating drinking water, as these systems can remove >96% of fecal coliforms, 100% of protozoa and helminths (Stauber et al. 2009), and 50–90% of organic and inorganic toxicants from water (Barth et al. 1997). However, there are challenges with cleaning these systems to avoid recontamination (Clark et al. 2012). Therefore, there is a need for locally sourced alternative solutions for water treatment that may also provide an opportunity for community engagement (Jahn 1988; Yin 2010; Deshmukh et al. 2013).

The high cost and possibility of health effects associated with conventional water treatment technologies has stimulated efforts to use coagulants and disinfectants derived from nature (Muyibi et al. 2002a, b; Kebreab et al. 2005; Yin 2010). Phytomaterials may provide natural disinfectants and coagulants as an alternative to chemical-based treatment solutions (Olsen 1987; Yongabi 2012). Various plants have been screened for their capacity as coagulants (Muyibi et al. 2002a; Kebreab et al. 2005; Yin 2010; Yongabi and Lewis 2011a), but not all coagulants have disinfectant properties. Phytomaterials prepared from *Moringa oleifera* have been shown to be efficient coagulants, as well as a disinfectant (Eilert 1978; Jahn 1979, 1988; Fewster et al. 2003; Yarahmadi and Hossieni 2009). Water treated with seed extracts prepared from this plant produces less sludge volume compared to alum (Ndabigengesere and Narasiah 1998). Also, the seeds were found to have antimicrobial effects against *Staphylococcus aureus* and *Bacillus subtilis* (Yongabi 2009). *Moringa* seeds are widely available in West Africa, and each tree produces approximately 15,000–25,000 seeds (Fig. 9.5) and 400–1000 seed pods per year (Jahn 1988). However, previous studies with water pre-treated with *M. oleifera* showed that there was bacterial re-growth within 48 h (Sutherland et al. 1990), so an integrated system with *Moringa* seed pre-treatment followed by sand filtration was investigated as a treatment technology (Yongabi and Lewis 2011b).

The innovative water treatment system was developed by sourcing locally available resources, such as the sand, gravel and charcoal used for construction of the sand filter. This process involved repeated washing and rinsing by clean water and

Fig. 9.5 *Moringa oleifera*, with pods containing seeds

Fig. 9.6 Sand filter drums. Source: Yongabi and Lewis (2011b)

layering the materials in the filter in the order of: gravel, charcoal, coarse sand and two layers of fine sand. Mature seeds of *M. oleifera*, a local terrestrial species, were obtained from households in Bauchi (Fig. 9.5). The seeds found in pods were shelled and ground in a clean mortar using a pestle. The nearby wetland in Gwallameji neighborhood of Bauchi, Nigeria was the source of contaminated water for pilot-scale testing. The powder prepared from the seeds was sprinkled into 100 L of untreated pond water at a dose of powder from one seed per litre of water in a 150 L capacity drum. The mixture was stirred using a clean wood stirrer and allowed to sit for 30 min and then filtered using muslin sack cloth. *Moringa* pretreated water was then passed through a sand filter and the filtered water was collected, as shown in Fig. 9.6. Water samples from the drum were analyzed for water quality parameters. To understand replicability of the process, sample water was sourced from three different wetlands/ponds (Yongabi and Lewis 2011b). Table 9.2 summarizes the mean values for a range of water quality parameters analyzed in samples of the pond water in three different scenarios; (a) water treated with *M. oleifera*;(b) water treated with sand filter; (c) and water treated with combination of *Moringa* pre-treatment and the sand filter.

Pre-treatment, the water had high level of aerobic bacteria, fecal indicator bacteria and other microbiological material, rendering it unfit for consumption as drinking water, or even a risk for bathing (Cheesbrough 1984). Post-treatment, the turbidity, suspended solids and numbers of various microbiological indicator species showed a significant decline. The filtered water compared to acceptable levels for drinking water according to threshold values recommended by the WHO (Table 9.2).

This case study demonstrates how integrating simple technologies based on phytomaterials and sand filters, can provide potable water for rural communities,

Table 9.2 Mean values for water quality parameters for untreated, *Moringa* treated, sand filter treated and *Moringa* pre-treated and sand filtered pond water reported in Yongabi and Lewis (2011b)

Parameter	Untreated pond water	Treatment with Moringa alone	Treatment with sand filter alone	Moringa and sand filter	WHO values (ranges)
Turbidity (NTU)	130.2	30.8	22.7	4.5	0–5 (25)
Total solids (mg/dm^3)	466.0	352.0	327	298	500 (1500)
Total aerobic mesophilc bacterial counts (cfu/mL)	TNTC	394.0	189.7	8	0–500
E. coli counts (cfu/mL)	5100	572.7	13.7	0.3	0–1
Coliform counts (cfu/mL)	7300	437.7	15.7	5.3	0–10
Yeast counts (cfu/mL)	2452.7	756.0	23.7	9.3	–
Pseudomonas counts (cfu/mL)	151	154.0	12.3	3	–

TNTC too numerous to count; Values in brackets indicate maximum permissible limit. Source: Yongabi and Lewis (2011b)

especially during dry season, when water shortages are acute. The benefit of applying coagulants and disinfectants derived from seeds of *M. oleifera* is also documented by Kebreab et al. (2005).

Population projections for Nigeria indicate that more than 60% of Nigerians will live in urban centers by 2025 (Millennium Development Report 2010). The need for nation-wide basic services to communities under the Community-based Urban Development Programme has been widely discussed (Adedejia and Ako 2009). This includes access to potable water and suitable sanitation (Gross et al. 2001). The SDG 6 sustainable development agenda clearly outlines the need for nations to provide universal access to clean drinking water and sanitation services by 2030. One solution is to utilize the multiple functions of natural and constructed wetlands as water management tools and to find incentives for communities to employ low-cost, NbS treatment systems (Mahmood et al. 2013).

NbS Eco-model in Africa

African countries will need to make significant investments to meet the objectives of the Sustainable Development Goals for 2030. Access to sanitation and safe drinking water is a continuing challenge. In Nigeria, more than 60% of the population lacks access to adequate supplies of safe drinking water, which is 14.8% less than the MDG's benchmark of 77% (Millennium Development Report 2015). In 2014, the urban areas of Nigeria recorded about 74.6% access to adequate drinking water, relative to the 57.6% in rural areas. The south west region of the country reported 70.6% access in rural areas, compared to other rural landscapes in the country that recorded proportions between 53 and 68% (Millennium Development Report 2015).

Inadequate financial and human resources are described among the key causes of inadequate treatment of drinking water or the lack of pollution mitigation efforts in many rural communities in Africa. The UN Human Development Report (2015) indicates about 40% of the population in Cameroon lives below the poverty line. A recent global report estimated that two thirds of the Cameroon population lacks access to safe water and nearly half of the population lacks access to adequate sanitation, which contributes to diarrhea being the third leading cause of death in the country (UNICEF/WHO 2017). Commonly applied approaches to drinking water disinfection with halogenated compounds such as chlorine (Muyibi et al. 2002a), do not guarantee safe drinking water and this treatment is not effective in removing heavy metals and other pollutants. Low-cost solutions for water treatment that are local and community managed may have better potential for adoption and impact. The concept of Nature based Solution (NbS) has emerged within the last decade as an approach to treatment of water that utilizes natural ecosystems rather than relying on conventional methods (Cohen-Schacham et al. 2016), as these solutions protect the environment and may also provide economic, social and health benefits for the community. However, the scaling up of NbS solutions from the bench to full scale may require an incentive-based approach or an entrepreneurial model to engage local communities and ensure long-term sustainability of the intervention.

The eco-model illustrated in Fig. 9.10 proposes using aquatic and terrestrial plants as bioindicators, remediation and local-scale water treatment technologies as an integrated approach to identifying and mitigating pollution in the region. Once sources of contamination are identified, the hyper-accumulator plant species can assist with the removal of metals and other contaminants from water in natural and constructed wetlands. Phytomaterials can be also used as coagulants and disinfectants to improve the water quality of community water sources.

The senior author of this chapter is currently undertaking a project to test this NbS approach at the community level to understand technical requirements for adoption of this approach by the community and to demonstrate the long term benefits for providing safe water, environmental protection and building the skills, capacity, and work opportunities for the community. This intervention may generate local-scale employment during the construction, operation and maintenance of the system. Development practitioners have for a long time argued that a key component of these types of projects is the involvement of local communities in the design and implementation of initiatives that aim to improve local conditions. The proposed eco-model can fulfil this objective by: (a) organizing community members to manage local biological extraction and plantation activities; (b) providing markets for local resources; (c) potentially including biogas and fertilizer for agriculture as products from the system; and (d) building capacity among local communities for developing a business model around this approach.

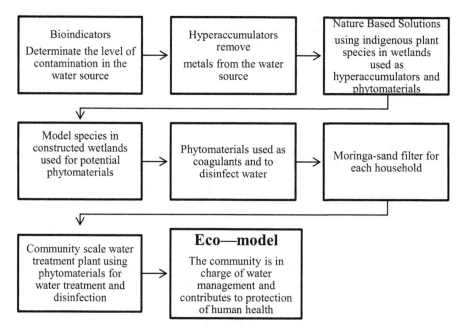

Fig. 9.10 Schematic representation of an eco-model employing plants and phytomaterials as elements in an integrated NbS approach

Challenges and Solutions for the NbS Approach

Various local and indigenous plant and algal species can be used in the NbS approach for addressing water quality issues and achieving the Sustainable Development Goals (i.e. SDG 6) that focuses on objectives for universal access to clean water and sanitation (UN Water 2016). However, in order to utilize the full potential of phytomaterials, studies are needed on the variability in bio-climatic and hydrological conditions, species diversity, seasonality, and water chemistry in natural and constructed wetland systems. Use of wetlands used for phytoremediation also requires building technical capacity of community members for the design and maintenance of the systems. For example, the plant species need to be managed after dieback to prevent recycling of accumulated contaminants.

The information provided in this chapter documents the effectiveness of various plant species and phytomaterials for removing contaminants from water and wastewater and contributes to the knowledge base on the use of phytoremediation as a practical and effective community based NbS. Among other services, the key benefits of the proposed eco-model include provision of clean water, and engagement of communities in building the local economy and improving community health. As countries prepare to address the goals and targets within SGD 6, the participating countries can aggregate their knowledge on using NbS to address water challenges in order to measure their progress.

Acknowledgements We are very grateful to external reviewers Hisae Nagashima, Ph.D. (Ecosystem Adaptability Center, Tohoku University Japan) and Marco Antonio Leon Romero, Ph.D. for the useful comments that helped to improve the message packaging in this chapter.

References

Abell R, Asquith N, Boccaletti G, Brehmer L, Chapin E, Erickson-Quiroz A, Higgins J, Johnson J, Kang S, Karres N, Lehner B, McDonald R, Raepple J, Shemie D, Simmons E, Sridhar A, Vigerstøl K, Vogl A, Wood S (2017) Beyond the source: the environmental, economic and community benefits of Source Water Protection. The Nature Conservancy, Arlington, VA

Adamu CI, Nganje TN, Edet A (2015) Heavy metal contamination and health risk assessment associated with abandoned barite mines in Cross River State, southeastern Nigeria. Environ Nanotechnol Monitor Manage 3:10–21

Adedejia A, Ako RT (2009) Towards achieving the United Nations' Millennium Development Goals: the imperative of reforming water pollution control and waste management laws in Nigeria. Desalination 248:642–649

Ahmed AA, Zezi AU, Yaro AH (2007) Antidiarrhoeal activity of the leaf extracts of *Daniella oliveri, Hutchutd dalz* (Fabaceae) and *Ficus sycomorus* (Moraceae) African Journal of Traditional Complementary and Alternative. Medicine 4:524–528

Akogun OB (1990) Water demand and schistosomiasis among the Gumau people of Bauchi State, Nigeri. Trans R Soc Trop Med Hyg 84(4):548–550

Aktar MW, Paramasivam M, Ganguly M, Purkait S, Sengupta D (2010) Assessment and occurrence of various heavy metals in surface water of Ganga River around Kolkata: a study for toxicity and ecological impact. Environ Monit Assess 160(1–4):207–213

Alkhatib AJ, Alhassan AJ, Ishaq M, Habib Y (2014) Tracking lead (Pb) in the environment of Jakara Kano State, Nigeria. Eur Sci J 10:7881–7431

Asogwa EU, Dongo LN (2009) Problems associated with pesticide usage and application in Nigerian cocoa production: a review. Afr J Agric Res 4(8):675–683

Audu AA, Lawal AO (2006) Variation in metal contents plants in vegetable garden sites in Kano Metropolis. Journal of Applied Science and. Environ Manag 10(2):105–109

Aziz-Abraham AM, Al-Hajjaji A-ZZL (1984) Environmental impact of heavy metals. J Environ Health 40:306–310

Barth G., Hall B, Chinnock S (1997) The use of slow sand filtration for disease control in recirculating hydroponics systems. In: Proceedings of the 4th national conference of the Australian Hydroponics Association

Bina B (1991) Investigation into the use of natural plant coagulants in the removal of bacteria and bacteriophage from turbid waters. Ph.D. dissertation, University of New Castle Upon Tyne, Newcastle, UK

Bontoux L (1998) The regulatory status of wastewater reuse in the European Union. In: Asano T (ed) Wastewater reclamation and reuse, water quality management library, vol 10. CRC Press, Boca Raton, FL, pp 1463–1475

Bowen HJM (1979) Environmental chemistry of the elements. Academic Press, London, UK, pp 105–120

Bruins MR, Kapil S, Oehme FW (2000) Microbial resistance to metals in the environment. Ecotoxicol Environ Saf 45(3):198–207

Burkill HM (1997) The useful plants of west tropical Africa. Royal Botanic Gardens Kew 2(4):264–267

Burton M, Peterson P (1979) Metal accumulation by aquatic bryophytes from polluted mine streams. Environ Pollut 19:439–461

Caceres A, Cabrera O, Morales O, Mollinedo P, Mendia P (1991) Pharmacological properties of *Moringa oleifera*: preliminary screening for antimicrobial activity. J Ethnopharmacol 33:213–216

Caerio S, Costa MH, Ramos TB, Fernandes F, Silveira N, Coimbra A, Painho M (2005) Assessing heavy metal contamination in Sado Estuary sediment: an index analysis approach. Ecol Indic 5:155–167

Cheesbrough M (1984) Medical laboratory manual for tropical countries. Butterworths, Boston, 79 p

Clark PA, Pinedo CA, Fadus M, Capuzzi S (2012) Slow-sand water filter: design, implementation, accessibility and sustainability in developing countries. Medical Science Monitor: International Medical Journal of Experimental and Clinical Research 18(7):105–117

Cohen-Schacham E, Walters G, Janzen C, Maginnis S (2016) Nature-based solutions to address global societal challenges. IUCN, Gland, Switzerland, p 97

Crapper DR, Krishnan SS (1973) Brain aluminum distribution in Alzheimer's disease and experimental neurofibrillary degeneration. Science 180(4085):511–513

Cunningham SD, Berti WR (1993) Remediation of contaminated soils with green plants: an overview. In Vitro Cell Dev Biol Plant 29(4):207–212

Deshmukh BS, Pimpalkar SN, Rakhunde RM, Joshi VA (2013) Evaluation performance of natural *Strychnos potatorum* over the synthetic coagulant alum, for the treatment of turbid water. Int J Innov Res Sci Eng Technol 2(11)

Díaz S, Demissew S, Carabias J, Joly C, Lonsdale M, Ash N (2015) The IPBES conceptual framework connecting nature and people. Curr Opin Environ Sustain 14:1–16

Eilert U (1978) Antibiotic principles of seeds of *Moringa oleifera* plant, Indian. Mater J 39:1013–1016

European Commission (2015) Towards an EU Research and innovation policy agenda for Nature-based Solutions and re-naturing cities, Final Report of the Horizon 2020 Expert Group on 'Nature-Based Solutions and Re-Naturing Cities, Directorate-General for Research and Innovation 2015 Climate Action, Environment, Resource Efficiency and Raw Materials EN

Faulwetter JL, Gagnon V, Sundeberg C, Chazarenc F, Burr M, Brisson J, Camper AK, Stein O (2009) Microbial processes influencing performance of treatment wetlands: a review. Ecol Eng 35:987–1004

Fayed SE, Abdel-Shafy H (1985) Accumulation of Cu, Zn, Cd and Pb by aquatic microphytes. Environ Int 11:77–87

Fewster E, Mol A, Wiessent-Brandsma C (2003) The bio-sand filter. Long term sustainability, user habits and technical performance evaluated. Presentation given at the 2003 international symposium on Household Technologies for Safe Water, June 16–17 2004, Nairobi, Kenya

Gadzała-Kopciuch R, Berecka B, Bartoszewicz J, Buszewski B (2004) Some considerations about bioindicators in environmental monitoring. Pol J Environ Stud 13(5):453–462

Galadima LG, Wasagu RS, Lawal M, Aliero A, Magajo UF, Suleman H (2015) Biosorption activity of *Nymphaea lotus* (water lily). The International Journal of Engineering and Science 4(3):66–70. ISSN (e): 2319-1813 ISSN (p): 2319-1805

Godfrey S (2003) Appropriate chlorination techniques for wells in Angola. Waterlines 21(4):6–8

Goufo P (2008) Rice production in cameroon: a review. Res J Agric Biol Sci 4(6):745–756. http://www.aensiweb.net/AENSIWEB/rjabs/rjabs/2008/745-756.pdf

Gross B, Wijk CV, Mukherjee N (2001) Linking sustainability with demand, gender and poverty: a study in community-managed water supply projects in 15 Countries. IRC International Water and Sanitation Centre, Delft, The Netherlands

Holm-Nelson B, Larsen IC (1979) Tropical botany geographical distribution of aquatic plants. Botanical Institute, University of Aahus, Denmark

Human Development Report 2015, 'Work For Human Development' (2015) Link to source: http://hdr.undp.org/sites/default/files/2015_human_development_report.pdf

Imevbore AMA (1971) Floating vegetation of Lake Kainji. Nature 230:599–600

Jahn SAA (1988) Using *Moringa* seeds as coagulants in developing countries. J Am Water Wastewater Assoc 80(6):43–50

Jahn SAA (1979) Studies on natural water coagulants in Sudan with special reference to *Moringa oleifera* seeds. Water SA 5-2:90–97

Jan V, Tereza B (2016) Accumulation of heavy metals in aboveground biomass of *Phragmites australis* in horizontal flow constructed wetlands for wastewater treatment: a review. Chem Eng J 290:232–242

Järup L (2003) Hazards of heavy metal contamination. Br Med Bull 68:167–182

Kabata-Pendias A, Pendias H (1992) Trace elements in soils and plants, 2nd edn. CRC Press, Boca Raton, FL, 315 p

Kebreab A, Gunaratna KR, Henriksson H, Brumer H, Gunnel D (2005) A simple purification and activity assay of the coagulant protein from *Moringa oleifera* seed. Water Res 39:2338–2344

Khan AG, Kuek C, Chaudhry TM, Khoo CS, Hayes WJ (2000) Role of plants, mycorrhizae and phytochelators in heavy metals contaminated land remediation. Chemosphere 41:197–207

Leon Romero MA, Soto-Rios PA, Fujibayashi M, Nishimura O (2017) Impact of NaCl solution pretreatment on plant growth and the uptake of multi-heavy metal by the model plant *Arabidopsis thaliana*. Water Air Soil Pollut 228:64

Li W, Li L, Qiu G (2015) Energy consumption and economics of typical wastewater treatment systems in Shenzhen. China J Clean Prod:1–5

Magalhaes VF, Karez CS, Pfeiffer W, Filho GM (1994) Trace metal accumulation by algae in Sepetiba Bay, Brazil. Environ Pollut 83:351–356

Mahmood Q, Pervez A, Saima B, Zaffar H, Yaqoob H, Waseem M, Afsheen S (2013) Natural treatment systems as sustainable ecotechnologies for the developing countries. Biomed Res Int 2013:796373

Masangwi SJ, Morse T, Ferguson G, Zawdie G, Grimason AM (2008) A preliminary analysis of the Scotland-Chikwawa health initiative project on morbidity. Environ Health Int 10(2):10–22

McConnachie GL, Folkard GK, Mtawali MA, Sutherland JP (1999) Field trials of appropriate hydraulic flocculation process. Water Res 33(6):1425–1434

Megersa M, Beyene A, Ambelu A, Triest L (2016) A preliminary evaluation of locally used plant coagulants for household water treatment. Water Conservation Science and Engineering 1:95–102

Millennium Development Report (2010) Nigeria. http://www.ng.undp.org/content/nigeria/en/home/library/mdg/nigeria-mdg-report/

Millennium Development Report (2015) Assessing Progress in Africa Toward the Millennium Development Goals. http://www.undp.org/content/undp/en/home/librarypage/mdg/mdg-reports/africa-collection.html

Montakhab A, Ghazali A, Johari M, Mohd M, Mohamed TA, Yusuf B (2010) Effects of drying and salt extraction of *Moringa oleifera* on its coagulation of high turbid water. J Am Sci 6(10):387–391

Mortor WE, Dunette DA (1994) Health effects of environmental arsenic. In: Arsenic in the environment, Part II, human health and ecosystem effects. Wiley, New York, NY, p 17

Muyibi SA, Evison LM (1995) Optimizing physical parameters affecting coagulation of turbid water with *Moringa oleifera*. Water Res 29(12):2689–2695

Muyibi SA, Megat J, Johari MMN, Ahmadun FR, Ameen ESM (2002a) Effects of oil extraction from *Moringa oleifera* seeds on coagulation of turbid water. Int J Environ Stud 59(2):243–254

Muyibi SA, Megat J, Loon L (2002b) Bench scale studies for pretreatment of sanitary landfill leachate with *Moringa oleifera* seeds extract. Int J Environ Stud 59(5):513–535

Muyibi SA, Noor MJMM, Leong TK, Loon LH (2002c) Effect of oil extraction from Moringa oleifera seeds on coagulation of turbid water. Environ. Studies, 59(2):243-254

Ndabigengesere A, Narasiah KS (1998) Quality of water treated by coagulation using *Moringa oleifera* seeds. Water Res 32(3):781–791.

Nesshöver C, Assmuth T, Irvine K, Rusch G, Waylen K, Delbaere B, Haase D, Jones-Walters L, Keune H, Kovacs EK, Külvik M, Rey F, Van D, Vistad O, Wilkinson M, Wittmer H (2017) The science, policy and practice of nature-based solutions: an interdisciplinary perspective. Sci Total Environ 579:1215–1227

Niemi GJ, McDonald ME (2004) Application of ecological indicators. Ann Rev Ecol Evol Syst 35:89–111

Nowell LH, Capel PD, Dileanis P (1999) Pesticides in stream sediment and aquatic biota: distribution, trends, and governing factors. CRC Press, Boca Raton, FL, 1040 p

Obot EA (1987) *Echinochloa stagnina*, a potential dry season livestock fodder for arid regions. J Arid Environ 12:175–177

Obodo GA (2002) Toxic metals in River Niger and its tributaries. Journal of the Indian Association of. Environ Manag 28:147–151

Ogunkunle C, Mustapha K, Oyedeji S, Fatoba P (2016) Assessment of metallic pollution status of surface water and aquatic macrophytes of earthen dams in Ilorin, north-central of Nigeria as indicators of environmental health. J King Saud Univ Sci 28:324–331

Okuda T, Baes AU (2001) Coagulation mechanism of salt solution extracted active component in *Moringa oleifera* seeds. Water Res 35(3):830–834

Olokesus F (1988) An overview of pollution in Nigeria and the impact of legislated standards on its abatement. Environmentalist 8(1):31–37

Olsen A (1987) Low technology water purification by bentonite clay and *Moringa oleifera* seed flocculation as performed in Sudanese village: effects on *Schistosoma mansoni* cercariae. Water Res 21(5):517–522

Oluwole O, Cheke RA (2011) Health and environmental impacts of pesticide use practices: a case study of farmers in Ekiti State, Nigeria. Int J Agric Sustain 7:153–163

Pais I, Jones JB (2000) The Handbook of Trace Elements. Luice Press, Florida, St

Parmar TK, Rawtani D, Agrawal YK (2016) Bioindicators: the natural indicator of environmental pollution. Fronties in Life Science 9. http://www.tandfonline.com/toc/tfls20/current

Pritchard M, Mkandawire T, Edmondson A, O'Neill JG, Kululanga G (2009) Potential of using plant extracts for purification of shallow well water in Malawi. Phys Chem Earth 34:799–805

Rai LC, Gour JP, Kumar HD (1981) Phycology and heavy metal pollution. Biol Rev 56:99–151

Rai PK (2008) Heavy metal pollution in aquatic ecosystems and its phytoremediation using wetland plants: an eco-sustainable approach. Int J Phytoremediation 10:133–160

Raskin I, Nanda P, Dushenkov S, Salt DE (1994) Bioconcentration of heavy metals by plants. Curr Opin Biotechnol 5:285–290

Salt DE, Blaylock M, Kumar NP, Dushenkov V, Ensley BE, Chet I, Raskin I (1995) Phytoremediation: a novel strategy for the removal of toxic metals from the environment using plants. Biotechnology 13:468–474

Shuaibu US, Nasiru AS (2011) Phytoremediation of trace metals in Shadawanka stream of Bauchi Metropolis, Nigeria. Univers J Environ Res Technol 2:176–181. http://www.environmentaljournal.org/1-2/ujert-1-2-10.pdf

Singare PU, Mishra RM, Trivedi M (2012) Sediment contamination due to toxic heavy metals in Mithi River of Mumbai. Adv Anal Chem 2(3):14–24

State Governance Capacity Building Project (SGCBP) (2008) Final report of the Environmental and Social Management Framework. Abuja Nigeria

Stauber CE, Ortiz GM, Loomis DP, Sobsey MD (2009) A randomized controlled trial of the concrete biosand filter and its impact on diarrheal disease in Banao, Dominican Republic. Am J Trop Med Hyg 80(2):286–293

Sukumaran D (2013) Phytoremediation of heavy metals from industrial effluent using constructed wetland technology. Appl Ecol Environ Sci 1(5):92–97

Sun X, Li Y, Zhu X, Cao K, Feng L (2015) Integrative assessment and management implications on ecosystem services loss of coastal wetlands due to reclamation. J Clean Prod:1–12

Sutherland JP, Folkard G, Grant WD (1990) Natural coagulants for appropriate water treatment: a novel approach. Water Lines 8:30–32

TEEB (2010) The Economics of Ecosystems and Biodiversity (TEEB): mainstreaming the economics of nature—a synthesis of the approach, conclusions and recommendations of TEEB. UNEP, Bonn, Germany

UNEP (2013) Biodiversity: Natural solutions for water security. https://www.cbd.int/doc/newsletters/development/news-dev-2015-2013-05-en.pdf

United Nations International Children's Emergency Fund (UNICEF)/World Health Organization (WHO) (2017) Progress for children with equity in the Middle East and North Africa. https://www.unicef.org/mena/Progress_for_Children_in_MENA_Web.pdf

United Nations (2015) UN General Assembly, Transforming our world: the 2030 Agenda for Sustainable Development, 21 October 2015, A/RES/70/1. http://www.refworld.org/docid/57b6e3e44.html

United Nations (2016) Global sustainable development report 2016. Department of Economic and Social Affairs, New York. https://sustainabledevelopment.un.org/content/documents/2328Global%20Sustainable%20development%20report%202016%20(final).pdf

UN Water (2016) Integrated monitoring guide for SDG 6 targets and global indicators, Version July 19, 2016. http://www.unwater.org/app/uploads/2017/03/SDG-6-targets-and-global-indicators_2016-07-19.pdf

Van Straalen NM, Burghouts T, Doornhoof MJ, Groot GM, Jansen MPM, Joossee EGG, Van Meerendonk JH, Theeuwen JJ, Verhoef HA, Zoomer HR (1987) Efficiency of lead and cadmium excretion in population of *Orchesella cinca* (collembola) from various contaminated forest soils. J Appl Ecol 24:953–968

Vymazal J, Březinová T (2016) Accumulation of heavy metals in aboveground biomass of *Phragmites australis* in horizontal flow constructed wetlands for wastewater treatment: a review. Chem Eng J 290:232–242

Wamelink GW, Goedhart PW, Van Dobben HF, Berendse F (2005) Plant species as predictors of soil pH: replacing expert judgement with measurements. J Veg Sci 16(4):461–470

Welman JB (1948) Preliminary survey of the freshwater fisheries of Nigeria. Government Printer, Lagos, Nigeria

White E (1965) The first scientific report of the Kainji Biological Research Team. University of Liverpool, Liverpool, UK

WHO (1997) Guidelines for drinking water quality surveillance and control of community supplies, 2nd edn, vol 3. http://www.who.int/water_sanitation_health/dwq/gdwqvol32ed.pdf

Williams JB (2002) Phytoremediation in wetland ecosystems: progress, problems, and potential. Crit Rev Plant Sci 21:6–20

Wittig R (1993) General aspects of biomonitoring heavy metals by plants. In: Market B (ed) Plants as biomonitors. VCH, Weinheim, The Netherlands

Wood A, Pybus P (1992) Artificial wetland use for wastewater treatment—theory, practice and economic review. In: WRC Report No. 232/93, Water Research Commission, Pretoria. http://www.wrc.org.za/Lists/Knowledge%20Hub%20Items/Attachments/8147/232-1-93_CONTENTS.pdf

Xing W, Wu H, Hao B, Huang W, Liu G (2013) Bioaccumulation of heavy metals by submerged macrophytes: looking for hyperaccumulators in eutrophic lakes. Environ Sci Technol 47:4695–4703

Yabe K, Nakamura T (2010) Assessment of flora, plant communities, and hydrochemical conditions for adaptive management of a small artificial wetland made in a park of a cool-temperate city. Landsc Ecol Eng 6(2):201–210

Yarahmadi M, Hossieni M (2009) Application of *Moringa oleifera* seed extract and polyaluminium chloride in water treatment. World Appl Sci J 7(8):962–967

Yin CY (2010) Emerging usage of plant-based coagulants for water and wastewater treatment. Process Biochem 45(9):1437–1444

Yongabi KA (2009) The role of phytobiotechnology in public health: In: Biotechnology, Cited in: Encyclopedia of Life Support Systems (EOLSS). Developed under the auspices of the UNESCO, EOLSS Publishers, Oxford, UK. https://www.eolss.net/Sample-Chapters/C17/E6-58-10-18.pdf

Yongabi KA (2012) A sustainable low-cost phytodisinfectant-sand filter alternative for water purification. Ph.D. dissertation, The School of Chemical Engineering. Faculty of Engineering, Computer and Mathematical Sciences, University of Adelaide, Adelaide, Australia

Yongabi KA, Harris PL, Lewis DM (2009) Poultry faeces management with a low cost plastic digester. Afr J Biotechnol 8:1560–1566

Yongabi KA, Lewis DM (2011a) Application of phytodisinfectants in water purification in rural Cameroon. Afr J Microbiol Res 5(6):628–635

Yongabi KA, Lewis DM (2011b) Integrated phytodisinfectant-sand filter drum for household water treatment in subsaharan Africa. J Environ Sci Eng 5:947–954

Zayed A, Gowthaman S (1998) Phytoaccumulation of trace elements by wetland plants: duckweed. J Environ Qual 27:715–721

Zhang J, Zhang F, Luo Y, Yang H (2006) A preliminary study on cactus as coagulant in water treatment. Process Biochem 41(3):730–733

Chapter 10
Accumulation of Metals by Mangrove Species and Potential for Bioremediation

Kakoli Banerjee, Shankhadeep Chakraborty, Rakesh Paul, and Abhijit Mitra

Introduction

Coastal zones and estuaries in tropical and subtropical regions, which are often the habitat of mangrove forests, are frequently impacted by industrial effluents, domestic wastewater and aquaculture (Wu et al. 2008; Peng et al. 2009; Zhang et al. 2014), and spills of oil and fuel also find their way into these ecosystems (Santos et al. 2011). As a consequence, nutrients, metals, persistent organic compounds and hydrocarbons are transported into the intertidal mudflats of mangrove forests (Bayen 2012; Natesan et al. 2014; Sukhdhane et al. 2015). Studies in coastal wetlands have shown that water soluble and exchangeable metals are biologically available, while metal complexes with humic materials and metals adsorbed to hydrous oxides are less available for uptake into biota (Gambrell 1994; Williams et al. 1994; Bayen 2012). However, a significant portion of the metals in the sediments of coastal wetlands are available for accumulation in biota (Islam and Tanaka 2004). In India, about 1125 million liters of wastewater is discharged per day through the Hooghly estuary, which empties into the Bay of Bengal. Various metals are a major constituent of this wastewater (UNEP 1992). The Indian Sundarbans region located on the Bay of Bengal in eastern India is experiencing the impacts of this pollution, which

K. Banerjee • R. Paul
Department of Biodiversity and Conservation of Natural Resources,
Central University of Orissa, Koraput, Landiguda, Koraput, Odisha, India
e-mail: banerjee.kakoli@yahoo.com

S. Chakraborty
Department of Oceanography, Techno India University,
Salt Lake Campus, Kolkata, West Bengal, India

A. Mitra (✉)
Department of Marine Science, University of Calcutta, Kolkata, West Bengal, India
e-mail: abhijit_mitra@hotmail.com

is due to rapid industrialization and urbanization in the Gangetic delta region. The newly developed Haldia port-cum-industrial complex in the Bay of Bengal has contributed to this pollution. Ecological impacts have been previously reported for the coastal zone of the Indian Sundarbans, including effects on benthic molluscs (Mitra and Choudhury 1992; Mitra 1998).

Mangroves are woody plants that inhabit tropical and sub-tropical coastal areas around the world. They are major primary producers in the estuarine ecosystem and provide a variety of ecosystem services in terms of shoreline stabilization, pollution control and biodiversity, as well as providing food, fuel and fodder to human populations. Despite the fact that mangroves play a vital role in marine food webs, they are increasingly being affected by anthropogenic activities, including pollution (Saifullah 1997; Ma et al. 2011; Bayen 2012). The sediments act as a sink for metals due to their anaerobic, reducing properties, as well as their high organic loads that aid in metal accumulation (Harbison 1986; Tam and Wong 1993; Bayen 2012; Natesan et al. 2014). The mangrove community traps and stabilizes sediments, nutrients and persistent pollutants (Mitra 2013; Chakraborty et al. 2014a, b), and hence may help to improve water quality. The Indian Sundarbans, at the apex of Bay of Bengal is a unique location to study the impacts of pollution on mangrove ecosystems. Notably, this is the only mangrove-based territory of the Royal Bengal tiger (*Panthera tigris*). There are 34 true mangrove species and some 62 associated species of flora and fauna in the region. The dominant mangrove species are *Avicennia alba*, *Avicennia officinalis*, *Avicennia marina* and *Excoecaria agallocha*. There are data in the literature on the status of pollution in and around the Indian Sundarbans (Mitra and Choudhury 1992; Mitra et al. 1994, 2011; Mitra 1998; Banerjee et al. 2012; Mitra and Ghosh 2014; Chakraborty et al. 2014a, 2016; Mitra and Zaman 2015), but there is little information on the accumulation of metals in the vegetation of the mangroves from this region. Therefore, the objective of this study was to determine the accumulation of selected metals, specifically, zinc (Zn), copper (Cu) and lead (Pb) in the roots and vegetation of these mangrove species at various locations in the Indian Sundarbans. The potential for bioremediation of contaminated sites by mangroves is discussed in light of the data on bioaccumulation of the metals from sediments and water.

The Study Area

The Indian Sundarbans (between 21^0 30′ N to 22^0 30′ N latitude and 87^0 25′ E to 89^0 10′ E longitude) is located on the southern fringe of the state of West Bengal on the northeast coast of India. The Indian Sundarbans occupy an area of about 9630 km^2, of which the mangrove forest area is about 4200 km^2. The region is demarcated by the border with Bangladesh in the east, the Hooghly River in the west, the Dampier and Hodges line in the north, and the Bay of Bengal to the south.

The important geomorphologic features of the Sundarbans are beaches, mudflats, coastal dunes, sand flats, estuaries, creeks, inlets and mangrove swamps. According to West Bengal State Biodiversity Strategy and Action Plan (2002), the region is divided into three principal zones:

1. *Western Sundarbans*: The Digha—Junput coastal plain along the estuary and Bay of Bengal.
2. *Central Sundarbans*: Tidal sea water traversing along the Hooghly Rover, up to the Diamond Harbour Municipality in the south, and up to the Haldia port in the west.
3. *Eastern Sundarbans*: Starting from the mouth of Harinbhanga River, delineating the India-Bangladesh border to the mouth of Hooghly River, being essentially the Sundarbans delta area.

This deltaic complex sustains 102 islands, of which 48 are inhabited and 54 are uninhabited. The flow of the Hooghly River through the estuary in the western sector of the Indian Sundarbans contributes to a hydrological situation that is totally different from the central sector, where five major rivers have lost their connection with the Ganga-Bhagirathi hydrologic system due to heavy siltation. Twelve stations were selected for the monitoring program, namely Kakdwip (Station 1), Harinbari (Station 2), Chemaguri (Station 3), Sagar south (Station 4), Lothian island (Station 5), Jambu island (Station 6), Frasergunge (Station 7), Gosaba (Station 8), Chotomollakhali (Station 9), Bali island (Station 10), Sajnekhali (Station 11) and Bagmara (Station 12). The locations of these stations in the eastern zone (stations 1–7) and western zone (stations 8–12) of the Indian Sundarbans are shown in Fig. 10.1.

Fig. 10.1 Map showing the sampling stations as GPS points. Source: Google Earth

Experimental Design

At all 12 monitoring sites, samples of surface water, surficial sediments (1 cm depth) and the roots, stems and leaves of mangroves were collected in 2013 and 2014 during the pre-monsoon, monsoon and post-monsoon seasons. Shortly after collection, the water samples were filtered (0.4 μm) and aliquots of the filtrate were acidified with sub-boiling distilled nitric acid to a pH of about 2 and stored in cleaned low density polyethylene containers. Dissolved metals were preconcentrated using complexation with ammonium pyrrolidine dithiocarbamate and subsequent extraction into Freon TF, followed by back extraction into nitric acid. Sediment samples were dried in an oven at 105 °C for 5–6 h. After drying, visible shells or shell fragments were removed and the sediments were ground to a powder in a mortar and stored in acid washed polythene bags. Prior to analysis, a 1 g subsample was taken and digested with 0.5 N HCl, as described by Malo (1977).

Samples of the roots, stems and leaves of *A. alba*, *A. officinalis*, *A. marina* and *E. agallocha* were collected during low tide. Samples were collected from mangroves with a height of 3–5 m and a 25–40 cm diameter at breast height to reduce biases due to biomass differences that are a function of the age of the tree. The collected samples were washed in distilled water, oven dried at 60 °C for 24 h and homogenized, as described by MacFarlane (2002). Subsamples of homogenized samples (1 g) were prepared for metal analysis. Samples were digested with a mixture of concentrated nitric acid and hydrogen peroxide, as described by Krishnamurthy et al. (1976) and MacFarlane (2002), and made up to a 25 mL final volume.

Extracts were analyzed for zinc, copper and lead by flame Atomic Absorption Spectrophotometry (AAS) using a Perkin Elmer Model 3030 AAS instrument. Metal concentrations were calculated from absorbance values and expressed in μg/L (i.e. ppb) for water samples, and μ/g dry weight (i.e. ppm) for sediment and mangrove samples. Statistical analysis using correlation coefficients was done to determine whether there were relationships between the concentrations of the metals in water and sediments and the plant tissues. Analysis of variance was performed to assess whether heavy metal concentrations varied significantly between sites and mangrove species. All statistical analysis was performed with SPSS 21.0 for Windows.

Trends in Metal Concentrations

Metals in Water

The mean concentrations (±SD) of dissolved zinc, copper and lead in water samples collected in 2013 and 2014 at the various sampling stations during the pre-monsoon, monsoon and post-monsoon periods are illustrated in Fig. 10.2, where the monitoring stations are ordered from the seven western stations on the left to the five eastern

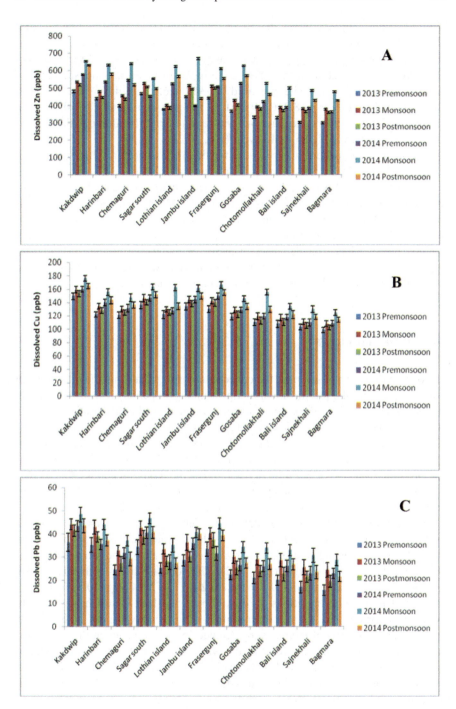

Fig. 10.2 Spatio-temporal variations in the mean (±SD) concentrations (μg/L) of dissolved zinc (**a**), copper (**b**) and lead (**c**) in water samples collected in 2013–2014 at the 12 monitoring stations. From left to right, the sampling sites are ordered from west to east in the study area

stations on the right. In general, dissolved Zn concentrations were slightly higher at the western monitoring stations. The mean concentrations also showed a trend of increasing slightly during the monsoon seasons. For instance, in 2013 at Kakdwip, the mean zinc concentration was 480.2 ± 7.3 µg/L during the pre-monsoon and the mean concentration at this site increased to 534.1 ± 6.1 µg/L during the monsoon, then dropped slightly to 518.2 ± 6.9 µg/L in the post-monsoon (Fig. 10.2a). This pattern in dissolved zinc concentrations was repeated in 2014, although the concentrations in water were higher in that year (Fig. 10.2a).

The same basic pattern was repeated for the mean concentrations of dissolved copper and lead. For instance, during the pre-monsoon in 2013, the mean dissolved copper concentration was 149.5 ± 4.9 µg/L at Kakdwip, and during the monsoon it increased slightly to 159.2 ± 5.2 µg/L (Fig. 10.2b). Mean dissolved lead concentrations in 2013 were generally highest during the monsoon; for instance, rising to 44.1 ± 2.4 µg/L at Kakdwip (Fig. 10.2c). As mentioned previously, this trend of slightly higher mean concentrations of dissolved metals during the monsoon was repeated in 2014 (Fig. 10.2). Concentrations in water were generally higher in the western part of the study area relative to the east (Appendix 1).

Metals in Sediments

In contrast to the water samples, the concentrations of metals in surficial sediments were high during the pre-monsoon and post-monsoon periods and low during the monsoon season (Fig. 10.3). For instance, the mean lead concentrations in sediments at Kakdwip during 2013 were 9.8 ± 2.0 µg/g dry wt. in the pre-monsoon, and dropped to 7.4 ± 1.0 µg/g dry wt. during the monsoon (Fig. 10.3c). Similarly, during 2014, mean sediment zinc values at Fraserguni were 117.3 ± 4.3 µg/g dry wt. during the pre-monsoon, 74.5 ± 3.9 µg/g dry wt. during the monsoon and 85.9 ± 3.3 µg/g dry wt. during the post-monsoon period (Fig. 10.3a). The magnitude of the concentrations in sediments were in the order of zinc > copper > lead.

This seasonal trend for the sediments may have been due to remobilization of metals from surficial sediments into the overlying water during the monsoon season, which would explain the slightly elevated concentrations of dissolved metals in the water during the monsoon (Fig. 10.2). The concentrations of metals in sediments were comparable during all seasons in 2013 and 2014 (Fig. 10.3). There was a trend of higher concentrations of zinc, copper and lead in sediments from the western stations relative to the eastern stations (Fig. 10.3; Appendix 1). In particular, the concentrations of metals were elevated in the sediments at the Frasergunj station (i.e. station 7 in Fig. 10.1) in the western part of the study area, with an especially obvious elevation of lead levels in the sediments (Fig. 10.3c). The metal contamination at this site may have been due to the presence of a fish landing station near the site. Fish from this region are known to have high levels of metals in their tissues (Mitra and Ghosh 2014), so metals could have been remobilized from decaying biological tissues as a result of disposal of fish offal into the water.

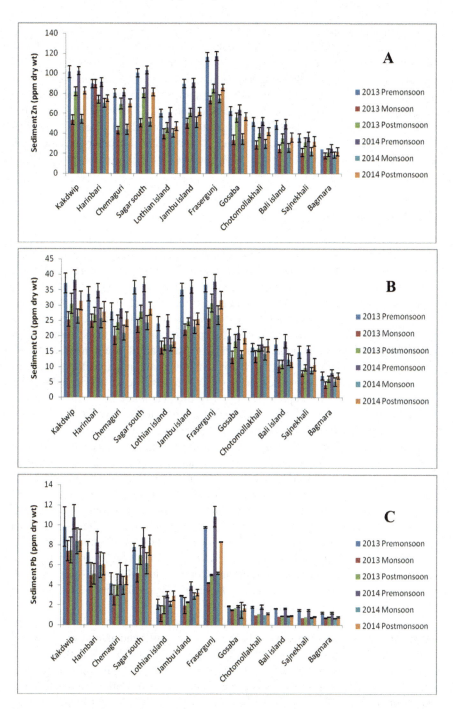

Fig. 10.3 Spatio-temporal variations in the mean (±SD) concentrations (µg/g dry weight) of zinc (**a**), copper (**b**) and lead (**c**) in samples of surficial sediments collected in 2013–2014 at the 12 monitoring stations. From left to right, the sampling sites are ordered from west to east in the study area

Data on the concentrations of metals in sediments and water were compared using two-way ANOVA for both the years 2013 and 2014 (Appendix 2). These analyses showed significant differences in metal concentrations between stations and seasons. For instance, the temporal variations in mean sediment concentrations of zinc computed for the three seasons showed a trend of: Pre-monsoon (72.2 µg/g) > Post-monsoon (57.2 µg/g) > Monsoon (43.2 µg/g). Sediment copper varied as Pre-monsoon (25.9 µg/g) > Post-monsoon (20.7 µg/g) > Monsoon (17.7 µg/g) and that for dissolved lead it varied in the order Pre-monsoon (4.6 µg/g) > Post-monsoon (3.5 µg/g) > Monsoon (3.0 µg/g) respectively.

Metals in Mangroves

The mean concentrations (± SD) of metals accumulated in the above ground structures (i.e. stem and leaves) and below ground structures (i.e. roots) of *A. officinalis* at the monitoring stations in 2013 and 2014 are illustrated in Figs. 10.4 and 10.5, respectively. The data on mean concentrations of these metals in the other mangrove species, *A. alba*, *A. marina* and *E. agallocha* are illustrated in the Appendix in Figs. 10.6, 10.7 and 10.8, respectively. The concentrations and the spatiotemporal patterns of metals in *A. officinalis* were very consistent across the samples collected in 2013 (Fig. 10.4) and in 2014 (Fig. 10.5). There was a trend of declining concentrations in mangrove structures across the sites from west to east, although this trend was least pronounced for copper. The order of the magnitude of concentrations of the metals in mangroves was zinc > copper > lead, which is consistent with the relative concentrations of these metals in sediments and water. The concentrations of all three metals in *A. officinalis* were highest in root samples, intermediate in stem samples and lowest in the leaf tissue. For instance, during 2013, the mean concentration of zinc in samples collected from mangroves at Kakdwip during the monsoon was 107.1 ± 6.3 µg/g dry wt. in root, 95.3 ± 6.4 µg/g dry wt. in stem, and 57.1 ± 3.01 µg/g dry wt. in the leaf. The bioaccumulation potential was highest in the root, followed by stem and leaves, irrespective of the species. A unique seasonal variation with respect to bioaccumulation of heavy metals was observed in the present study. The highest levels of metals in root, stem and leaves were observed in mangrove samples collected during the monsoon, followed by post-monsoon and pre-monsoon. This was observed in *A. officinalis* (Figs. 10.4 and 10.5) and in all other species (Figs. 10.6, 10.7 and 10.8).

Bioaccumulation factors (BAFs) between concentrations in sediments (µg/g) and mangrove vegetation (µg/g) and between water (µg/L) and mangrove vegetation (µg/g) were calculated for the four mangrove species (Table 10.1). The BAFs calculated for these species shows a decreasing trend from lead (Pb) followed by copper (Cu) and zinc (Zn), respectively, when calculated for accumulation from both water and sediments. The BAFs were highest in *A. officinalis* followed by *A. alba*, *A. marina* and *E. agallocha*, respectively (Table 10.1).

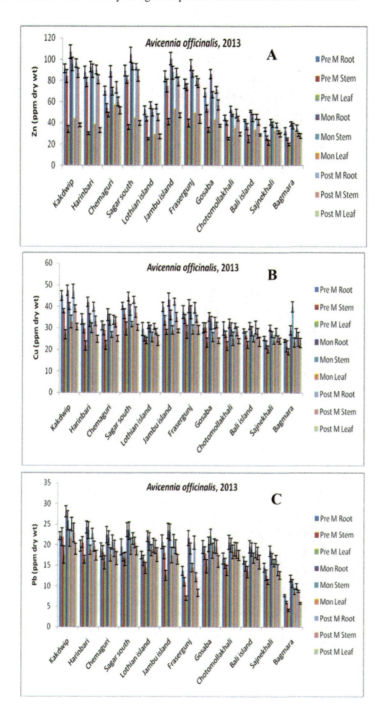

Fig. 10.4 Spatial-temporal variations in the mean (±SD) concentrations (μg/g dry weight) of zinc (**a**), copper (**b**) and lead (**c**) in root, stem and leaf samples of *A. officinalis* collected in 2013 at the 12 monitoring stations. From left to right, the sampling sites are ordered from west to east in the study area

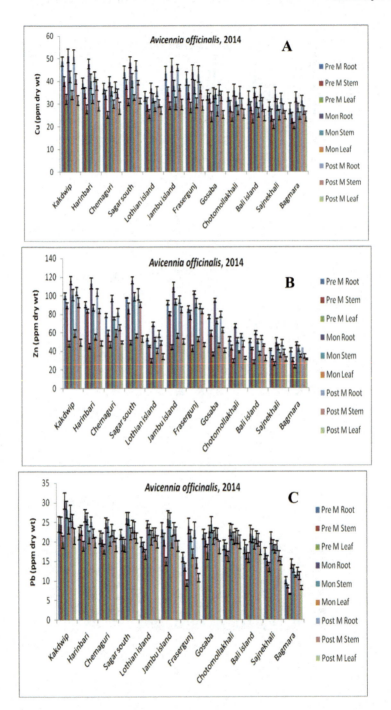

Fig. 10.5 Spatio-temporal variations in the mean (±SD) concentrations (μg/g dry weight) of zinc (**a**), copper (**b**) and lead (**c**) in root, stem and leaf samples of *A. officinalis* collected in 2014 at the 12 monitoring stations. From left to right, the sampling sites are ordered from west to east in the study area

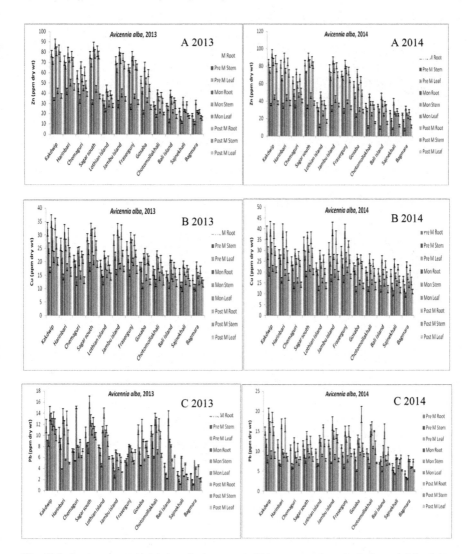

Fig. 10.6 Spatio-temporal variations in the mean (±SD) concentrations (µg/g dry weight) of zinc (**a**), copper (**b**) and lead (**c**) in root, stem and leaf samples of *A. alba* collected in 2013 (*left*) and 2014 (*right*) at the 12 monitoring stations

Correlation analysis of metal (Zn, Cu, and Pb) concentrations in sediment and water and the metal concentrations in the root, stem and leaf of *A. officinalis* in samples collected in the pre-monsoon, monsoon and post-monsoon seasons of 2013 and 2014 showed that there were generally significant correlations between the levels of the metals in the environmental media and the concentrations in the vegetative parts of this mangrove species (Table 10.2). However, some of the correlations between lead levels in ambient water and sediment and the vegetative parts of *A. officinalis* were not significant. The spatial-temporal data for the other three species

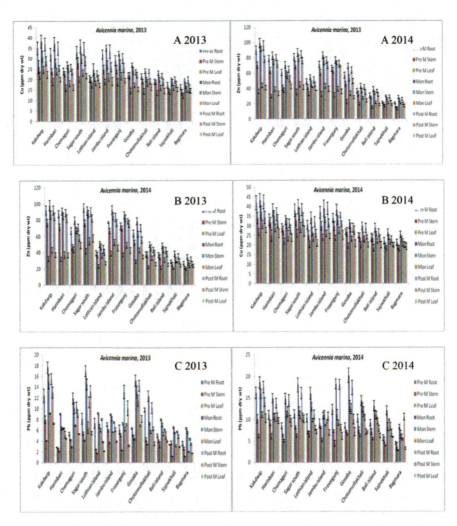

Fig. 10.7 Spatio-temporal variations in the mean (±SD) concentrations (μg/g dry weight) of zinc (**a**), copper (**b**) and lead (**c**) in root, stem and leaf samples of *A. marina* collected in 2013 (*left*) and 2014 (*right*) at the 12 monitoring stations

of mangroves were consistent with the trends observed for *A. officinalis* (Figs. 10.6, 10.7 and 10.8). The mean concentrations of zinc, copper and lead in mangroves were generally lower at sampling sites in the eastern part of the Indian Sundarbans relative to the western sites. The magnitude of the concentrations of the metals were in the order of zinc > copper > lead in the vegetative parts of the three species. Once again, the roots accumulated the highest concentrations of the metals, relative to intermediate concentrations in the stems and the lowest concentrations in the leaves.

However, the concentrations of metals were comparatively low in the vegetative parts of *E. agallocha* (Fig. 10.8) relative to the concentrations in the *Avicennia*

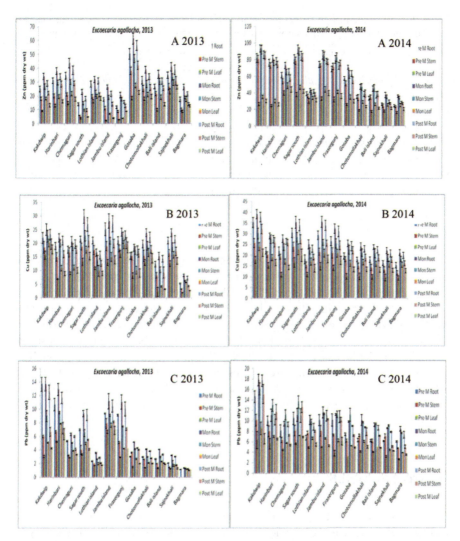

Fig. 10.8 Spatio-temporal variations in the mean (±SD) concentrations (μg/g dry weight) of zinc (**a**), copper (**b**) and lead (**c**) in root, stem and leaf samples of *E. aggalocha* collected in 2013 (*left*) and 2014 (*right*) at the 12 monitoring stations

mangroves species. This was especially the case for zinc concentrations in samples of *E. agallocha* collected in 2013, which were about two thirds of the levels in the other species. For instance, in *E. agallocha*, sampled during 2013, the mean concentration of zinc in the root during the monsoon at the Kakdwip site was 61.4 ± 3.9 μg/g dry wt., while the root samples collected from *A. marina* at the Kakdwip site during the monsoon had higher mean zinc concentrations of 99.2 ± 6.1 μg/g dry wt.

The reason for these interspecies differences is not clear, but could reflect the differences in the physiology of *Avicennia* and *Excoecaria* mangroves related to

Table 10.1 Bioaccumulation factors for metals into vegetation of mangrove species calculated for accumulation from water into mangroves and for accumulation from sediments into mangroves

		Dissolved metal				Sediment metal			
		A. officinalis	E. agallocha	A. alba	A. marina	A. officinalis	E. agallocha	A. alba	A. marina
Zn	Root	0.157	0.096	0.123	0.132	1.283	0.788	1.010	1.082
	Stem	0.134	0.085	0.106	0.115	1.099	0.697	0.866	0.941
	Leaf	0.083	0.048	0.060	0.071	0.680	0.396	0.494	0.583
Cu	Root	0.309	0.172	0.198	0.222	1.918	1.067	1.228	1.376
	Stem	0.244	0.142	0.161	0.191	1.516	0.881	0.999	1.187
	Leaf	0.199	0.101	0.115	0.155	1.238	0.628	0.713	0.962
Pb	Root	0.645	0.255	0.337	0.331	5.552	2.196	2.898	2.852
	Stem	0.602	0.213	0.275	0.257	5.176	1.829	2.370	2.211
	Leaf	0.512	0.133	0.214	0.190	4.407	1.146	1.842	1.634

regulation of essential elements, such as zinc. In any event, the results of correlation analysis of metal concentrations in water and sediment and the metal concentrations in the vegetation of *E. agallocha* (Table 10.3) showed that the concentrations in the environment were generally correlated with the concentrations in the vegetative parts of this mangrove species, except for some non-significant results for zinc levels in roots and stem samples collected in 2013. These observations are consistent with the correlation analysis for *A. officinalis* (Table 10.2). The correlation analysis for the other *Avicennia* mangrove species showed similar results (data not shown), although lead levels in water and sediment were not always correlated with the concentrations in the vegetative parts of the mangroves, as observed for *A. officinalis* (Table 10.2).

Sources of Metals

Metals contaminate the aquatic environment as a result of inputs from industrial and urban sources, and the Gangetic delta and the Sundarbans region in the Bay of Bengal, India are no exception. Rapid industrialization and urbanization in Kolkata, Howrah and the newly emerging port of Haldia complex in the maritime state of West Bengal have caused considerable ecological imbalance in the adjacent coastal zone (Mitra and Choudhury 1992; Mitra 1998). The Hooghly estuary, situated on the western sector of the Gangetic delta receives drainage from these adjacent cities, which all discharge sewage and industrial effluents into the estuarine system. The chain of factories and industries situated on the western bank of the Hooghly estuary is a major cause of the gradual transformation of this beautiful ecotone into a highly degraded environment (Mitra and Choudhury 1992).

In the lower part of the estuary, there are a variety of industries that produce paper, textiles, chemicals and pharmaceuticals, plastics, shellac, leather, jute, tires

Table 10.2 The results of analysis of the correlations between metal (Zn, Cu, and Pb) concentrations in sediment and water and metal concentrations in the vegetative parts (root, stem, leaf) of *A. officinalis* in samples collected in the pre-monsoon, monsoon and post-monsoon seasons of 2013 and 2014

Metals	Combination	2013 r-value			2013 p-value			2014 r-value			2014 p-value		
		PreM	Mon	PostM	PreM	Mon	PostM	PreM	Mon	PostM	PreM	Mon	PostM
Zn	Dissolved Zn × Root Zn	0.605	0.794	0.617	$p < 0.01$	$p < 0.01$	$p < 0.01$	0.970	0.957	0.926	$p < 0.01$	$p < 0.01$	$p < 0.01$
	Dissolved Zn × Stem Zn	0.573	0.729	0.575	$p < 0.01$	$p < 0.01$	$p < 0.01$	0.983	0.979	0.955	$p < 0.01$	$p < 0.01$	$p < 0.01$
	Dissolved Zn × Leaf Zn	0.637	0.793	0.274	$p < 0.01$	$p < 0.01$	NS	0.708	0.734	0.706	$p < 0.01$	$p < 0.01$	$p < 0.01$
	Sediment Zn × Root Zn	0.926	0.844	0.933	$p < 0.01$	$p < 0.01$	$p < 0.01$	0.924	0.739	0.941	$p < 0.01$	$p < 0.01$	$p < 0.01$
	Sediment Zn × Stem Zn	0.947	0.844	0.944	$p < 0.01$	$p < 0.01$	$p < 0.01$	0.947	0.819	0.944	$p < 0.01$	$p < 0.01$	$p < 0.01$
	Sediment Zn × Leaf Zn	0.918	0.793	0.921	$p < 0.01$	$p < 0.01$	$p < 0.01$	0.770	0.457	0.701	$p < 0.01$	$p < 0.05$	$p < 0.01$
Cu	Dissolved Cu × Root Cu	0.975	0.904	0.960	$p < 0.01$	$p < 0.01$	$p < 0.01$	0.968	0.939	0.942	$p < 0.01$	$p < 0.01$	$p < 0.01$
	Dissolved Cu × Stem Cu	0.963	0.814	0.939	$p < 0.01$	$p < 0.01$	$p < 0.01$	0.934	0.568	0.962	$p < 0.01$	$p < 0.01$	$p < 0.01$
	Dissolved Cu × Leaf Cu	0.972	0.864	0.942	$p < 0.01$	$p < 0.01$	$p < 0.01$	0.865	0.946	0.908	$p < 0.01$	$p < 0.01$	$p < 0.01$
	Sediment Cu × Root Cu	0.934	−0.115	0.891	$p < 0.01$	NS	$p < 0.01$	0.912	0.894	0.897	$p < 0.01$	$p < 0.01$	$p < 0.01$
	Sediment Cu × Stem Cu	0.939	0.940	0.946	$p < 0.01$	$p < 0.01$	$p < 0.01$	0.886	0.434	0.940	$p < 0.01$	$p < 0.05$	$p < 0.01$
	Sediment Cu × Leaf Cu	0.945	0.946	0.924	$p < 0.01$	$p < 0.01$	$p < 0.01$	0.835	0.912	0.818	$p < 0.01$	$p < 0.01$	$p < 0.01$

(continued)

Table 10.2 (continued)

Metals	Combination	2013						2014					
		r-value			p-value			r-value			p-value		
		PreM	Mon	PostM	PreM	Mon	PostM	PreM	Mon	PostM	PreM	Mon	PostM
Pb	Dissolved Pb × Root Pb	0.712	0.840	0.786	$p < 0.01$	$p < 0.01$	$p < 0.01$	0.626	0.841	0.800	$p < 0.01$	$p < 0.01$	$p < 0.01$
	Dissolved Pb × Stem Pb	0.682	0.768	0.503	$p < 0.01$	$p < 0.01$	$p < 0.01$	0.611	0.772	0.491	$p < 0.01$	$p < 0.05$	$p < 0.05$
	Dissolved Pb × Leaf Pb	0.579	0.585	0.375	$p < 0.01$	$p < 0.01$	NS	0.477	0.589	0.379	$p < 0.05$	$p < 0.01$	NS
	Sediment Pb × Root Pb	0.394	0.789	0.695	NS	$p < 0.01$	$p < 0.01$	0.364	0.760	0.716	NS	$p < 0.01$	$p < 0.01$
	Sediment Pb × Stem Pb	0.390	0.691	0.337	NS	$p < 0.01$	NS	0.364	0.667	0.459	NS	$p < 0.01$	$p < 0.05$
	Sediment Pb × Leaf Pb	0.270	0.522	0.221	NS	$p < 0.01$	NS	0.249	0.515	0.352	NS	$p < 0.01$	NS

Listed are the r-values for correlations, along with the level of significance ($p < 0.05$; $p < 0.1$) of the association; *NS* not significant

Table 10.3 The results of analysis of the correlations between metal (Zn, Cu, and Pb) concentrations in sediment and water and metal concentrations in the vegetative parts (root, stem, leaf) of E. agallocha in samples collected in the pre-monsoon, monsoon and post-monsoon seasons of 2013 and 2014

Metals	Combination	2013						2014					
		r-value			p-value			r-value			p-value		
		PreM	Mon	PostM	PreM	Mon	PostM	PreM	Mon	PostM	PreM	Mon	PostM
Zn	Dissolved Zn × Root Zn	−0.379	−0.368	−0.457	NS	NS	p < 0.05	0.616	0.733	0.603	p < 0.01	p < 0.01	p < 0.01
	Dissolved Zn × Stem Zn	−0.453	−0.384	−0.465	p < 0.05	NS	p < 0.05	0.568	0.691	0.563	p < 0.01	p < 0.01	p < 0.01
	Dissolved Zn × Leaf Zn	−0.628	−0.659	−0.698	p < 0.01	p < 0.01	p < 0.01	0.522	0.643	0.297	p < 0.01	p < 0.01	NS
	Sediment Zn × Root Zn	−0.379	−0.267	−0.226	NS	NS	NS	0.932	0.802	0.94	p < 0.01	p < 0.01	p < 0.01
	Sediment Zn × Stem Zn	−0.422	−0.244	−0.214	p < 0.05	NS	NS	0.952	0.825	0.94	p < 0.01	p < 0.01	p < 0.01
	Sediment Zn × Leaf Zn	−0.612	−0.501	−0.47	p < 0.01	p < 0.05	p < 0.05	0.737	0.555	0.734	p < 0.01	p < 0.01	p < 0.01
Cu	Dissolved Cu × Root Cu	0.677	0.664	0.675	p < 0.01	p < 0.01	p < 0.01	0.941	0.834	0.956	p < 0.01	p < 0.01	p < 0.01
	Dissolved Cu × Stem Cu	0.728	0.747	0.708	p < 0.01	p < 0.01	p < 0.01	0.97	0.844	0.921	p < 0.01	p < 0.01	p < 0.01
	Dissolved Cu × Leaf Cu	0.666	0.646	0.689	p < 0.01	p < 0.01	p < 0.01	0.925	0.835	0.929	p < 0.01	p < 0.01	p < 0.01
	Sediment Cu × Root Cu	0.722	0.691	0.73	p < 0.01	p < 0.01	p < 0.01	0.879	0.869	0.881	p < 0.01	p < 0.01	p < 0.01
	Sediment Cu × Stem Cu	0.752	0.803	0.785	p < 0.01	p < 0.01	p < 0.01	0.938	0.934	0.917	p < 0.01	p < 0.01	p < 0.01
	Sediment Cu × Leaf Cu	0.625	0.618	0.703	p < 0.01	p < 0.01	p < 0.01	0.901	0.84	0.845	p < 0.01	p < 0.01	p < 0.01
Pb	Dissolved Pb × Root Pb	0.906	0.93	0.921	p < 0.01	p < 0.01	p < 0.01	0.94	0.774	0.817	p < 0.01	p < 0.01	p < 0.01
	Dissolved Pb × Stem Pb	0.808	0.894	0.827	p < 0.01	p < 0.01	p < 0.01	0.819	0.917	0.944	p < 0.01	p < 0.01	p < 0.01
	Dissolved Pb × Leaf Pb	0.701	0.789	0.729	p < 0.01	p < 0.01	p < 0.01	0.894	0.892	0.867	p < 0.01	p < 0.01	p < 0.01
	Sediment Pb × Root Pb	0.858	0.869	0.838	p < 0.01	p < 0.01	p < 0.01	0.878	0.802	0.736	p < 0.01	p < 0.01	p < 0.01
	Sediment Pb × Stem Pb	0.618	0.801	0.704	p < 0.01	p < 0.01	p < 0.01	0.681	0.93	0.854	p < 0.01	p < 0.01	p < 0.01
	Sediment Pb × Leaf Pb	0.461	0.642	0.593	p < 0.05	p < 0.01	p < 0.01	0.673	0.855	0.897	p < 0.01	p < 0.01	p < 0.01

Listed are the r-values for correlations, along with the level of significance ($p < 0.05$; $p < 0.1$) of the association; NS not significant and cycle rims (UNEP 1992). These industries are point sources that contribute to the high levels of metal contamination in the sediments, water and mangroves at the Kakdwip (station 1), Harinbari (station 2) and Chemaguri (station 3) sampling sites. The main point sources of zinc in the present study are probably the galvanizing plants, paint manufacturing plants and pharmaceutical production facilities in the region. The main sources of copper in the coastal waters are probably antifouling paints (Goldberg 1975), as well as algaecides used in the aquaculture industry, pipe line corrosion inhibitors and oil sludges (Das et al. 2014). Lead finds its way into coastal waters through discharges of industrial wastewaters, such as from painting, dyeing and battery manufacturing plants. Antifouling paints used to prevent growth of marine organisms at the bottom of the boats also contain lead (Bellinger and Benhem 1978; Young et al. 1979).

Hydrologic processes may also influence the concentrations of metals in the region. The western part of the Indian Sundarbans is connected to the Himalayan glacier field through the Ganga-Bhagirathi River. Researchers report that the glaciers in the Himalayan range are melting rapidly (Hasnain 2002). This has resulted in a gradual freshening of the watershed, which has lowered the pH and therefore may be increasing the levels of dissolved metals in the system (Mitra et al. 2009; Chakraborty et al. 2013). The central sector of the Indian Sundarbans is deprived of a freshwater supply from the Ganga-Bhagirathi watershed because of siltation of the Bidyadhari River. The Matla River, in the central sector is now tide fed and is increasing in salinity (Mitra et al. 2009; Sengupta et al. 2013). The eastern sector of the Indian Sundarbans is mainly fed from fresh water originating from the creeks and canals of the Brahmaputra-Padma-Meghna basin. This sector is not as polluted by industrial and urban sources because it is more remote (Chakraborty et al. 2013).

To date, very little data have been available on the bioaccumulation of metals in the halophytes inhabiting the lower Gangetic delta, although several studies have been published on metal contamination of shellfish and finfish from the region (Mitra et al. 1992, 2011, 2012; Mitra and Choudhury 1993). Similar trends in the accumulation of metals in mangrove vegetation in India have also been documented in very recent literatures by Kannan et al. (2016) and Dudani et al. (2017). Little data are available on the bioaccumulation of heavy metals by mangroves in this coastal zone, although a report published by the Department of Environment, Government of West Bengal provides some relevant data (Mitra et al. 2004). This earlier report, however, does not provide information on the spatial and temporal distribution of metals in the mangroves from the region.

Bioaccumulation of Metals in Mangroves

There were significant positive correlations observed between the concentrations of the metals in water and sediment and the vegetative parts of all the selected mangrove species, except perhaps for zinc in *E. agallocha* collected in 2013. These correlations indicate that bioaccumulation of metals in these species reflects ambient concentrations. However, in the case of lead contamination, the levels in *Avicennia* spp. tissues did not show a positive correlation with ambient concentrations. However, there was a positive correlation between lead levels in the vegetative structures of *E. agallocha* and the ambient concentrations of lead. In all cases, the bioaccumulation potential of the mangrove root exceeded the bioaccumulation of the metals in the stem and leaf.

Mangroves must cope with high salinities and osmotic stress, so they have developed unique physiological mechanisms to regulate major cations, such as sodium. Therefore, several studies have investigated whether mangroves can also regulate levels of trace metals. McFarlane et al. (2007) reviewed the literature on the levels of copper, lead and zinc in mangroves and concluded that mangroves bioaccumulate these metals to concentrations equal or slightly lower than the adjacent sediment concentrations, and the concentrations in leaves are typically about half the concentrations in the roots. More recent studies have showed similar patterns of metal bioaccumulation in mangroves (Agoramoorthy et al. 2008; Bayen 2012). The results of the present study are consistent with these earlier studies, as the metal concentrations in the roots of the mangrove species were comparable to the concentrations in the sediments and BAFs were ~1. However, the BAFs for lead in the roots of *A. officinalis* were >1. Mangroves did show some evidence of temporal changes in metal concentrations across seasons, with levels being slightly higher during the monsoon season. The concentrations in mangrove leaves observed in the present study were approximately half of the concentrations observed in the roots; consistent with observations by McFarlane et al. (2007).

Overall, McFarlane et al. (2007) concluded that mangrove plants are "excluder" species for non-essential metals (e.g. lead) and regulators of essential metals (e.g. zinc, copper), and predicted that these plants are unlikely to bioaccumulate metals to levels that exceed concentrations in the ambient environment (i.e. sediments). However, in the present study, the BAFs in the vegetation of the *A. officinalis* mangrove species were >1 relative to sediment concentrations, so there is some evidence of bioaccumulation of this non-essential element above ambient levels. Because of the sheer mass of mangroves present in coastal ecosystems, there are considerable amounts of metals bound up in these vegetative structures. For instance, 1000 tonnes of roots from mangroves located at the Kakdwip site could potentially sequester approximately 25 kg of lead. Work is needed to determine whether these metals return to the marine environment when there is die-back of the vegetation. Remediation of metal contaminated sites may be an additional ecological function of mangrove forests in polluted environments.

In any event, monitoring of the leaves and other vegetative structures of mangroves is a useful bioindicator of metal contamination in impacted regions. (Chakraborty et al. 2014a). These studies on mangroves contribute to the effort to understand the distribution and fate of contaminants in the Indian Sundarbans region that has been ongoing over the past 30 years (Mitra et al. 1987, 1990, 1992, 1994, 1996, 1999, 2010; 2011; Mitra and Choudhury 1993; Trivedi et al. 1994; Niyogi et al. 1997; Mitra 1998, 2000; Saha and Mitra 2001; Bhattacharyya et al. 2001, 2013; Das et al. 2005; Mitra and Banerjee 2011; Chakraborty et al. 2014a; Ray

Chaudhuri et al. 2014). More work is needed to understand the dynamics governing the distribution of contaminants between water, sediment and biota in these mangrove ecosystems. The mechanism of accumulation and mobilization of metals in sediments is controlled by the characteristics of the substrate, as well as the physiochemical conditions in the substrate (Panigrahy et al. 1997; Bayen 2012).

Conclusions

The degree of metal accumulation in the vegetative structures of four species of mangroves in coastal areas of the Indian Sundarbans was in the order of root > stem > leaf at almost all monitoring stations, and the concentrations in the roots were comparable to the concentrations of the metals in the sediments. The degree of accumulation of the selected heavy metals in mangrove species was in the order: Zn > Cu > Pb; consistent with the relative concentrations of these metals in water and sediments. Even though mangroves cannot be typically classified as "hyper accumulators" of metals (Zayed and Gowthaman 1998), there is potential to use mangrove species for bioremediation, which adds a new dimension to the spectrum of ecosystem services offered by mangroves. Further research is necessary to confirm the magnitude of accumulation of metals in mangroves relative to the mass of these metals entering the Indian Sundarbans. There is a need for immediate protective measures to address the adverse impacts of metal contamination on coastal biotic communities.

Appendix 1

Spatial trends in mean concentrations of zinc, copper and lead in water (dissolved) and in sediments from all samples collected in 2013 and 2014

Dissolved zinc spatial trend
Station-1 (565.84ppm) >
Station-7 (520.30ppm) >
Station-2 (517.88ppm) >
Station-4 (500.82ppm) >
Station-3 (498.48ppm) >
Station-6 (493.92ppm) >
Station-8 (487.01ppm) >
Station-5 (479.72ppm) >
Station-9 (418.88ppm) >
Station-10 (400.70ppm) >
Station-11 (390.50ppm) >
Station-12 (384.20ppm)

Dissolved copper spatial trend:

Station-1 (160.46ppm) >
Station-4 (147.44ppm) >
Station-7 (147.10ppm) >
Station-6 (145.372ppm) >
Station-2 (137.14ppm) >
Station-5 (133.57ppm) >
Station-3 (131.77ppm) >
Station-8 (129.74ppm) >
Station-9 (124.46ppm) >
Station-10 (118.39ppm) >
Station-11 (113.08) >
Station-12 (109.74)

Dissolved lead spatial trend

Station-1 (42.86ppm) >
Station-4 (40.48ppm) >
Station-2 (38.88ppm) >
Station-7 (37.70ppm) >
Station-6 (35.22ppm) >
Station-3 (30.40ppm) >
Station-5 (29.53ppm) >
Station-8 (27.72ppm) >
Station-9 (26.86ppm) >
Station-10 (26.27ppm) >
Station-11 (23.59ppm) >
Station-12 (22.13ppm)

Sediment zinc spatial trend

Station-7 (92.06ppm) >
Station-2 (81.40ppm) >
Station-1 (79.19ppm) >
Station-4 (77.71ppm) >
Station-6 (67.30ppm) >
Station-3 (64.63ppm) >
Station-8 (50.99ppm) >
Station-5 (48.93ppm) >
Station-9 (40.55ppm) >
Station-10 (36.40ppm) >
Station-11 (29.60ppm) >
Station-12 (21.33ppm)

Sediment copper spatial trend

Station-7 (31.55ppm) >
Station-1 (31.46ppm) >
Station-4 (29.47ppm) >
Station-2 (28.95ppm) >
Station-6 (27.71ppm) >
Station-3 (24.58ppm) >
Station-5 (19.63ppm) >
Station-8 (17.58ppm) >
Station-9 (15.60ppm) >
Station-10 (13.42ppm) >
Station-11 (11.26ppm) >
Station-12 (6.19ppm)

Sediment lead spatial trend

Station-1 (8.69ppm) >
Station-7 (7.20ppm) >
Station-4 (7.11ppm) >
Station-2 (6.25ppm) >
Station-3 (4.17ppm) >
Station-6 (2.85ppm) >
Station-5 (2.19ppm) >
Station-8 (1.66ppm) >
Station-9 (1.27ppm) >
Station-10 (1.11ppm) >
Station-11 (0.99ppm) >
Station-12 (0.87ppm)

Appendix 2

Results of ANOVA analysis of comparisons between concentrations of metals in water (dissolved), sediments and mangrove tissues between stations and between seasons

Factor	Year	Variable	F_{cal}	F_{crit}
Dissolved Zn	2013	Between stations	117.5971	2.258518
		Between seasons	108.7961	3.443357
	2014	Between stations	13.97773	2.258518
		Between seasons	47.88598	3.443357
Dissolved Cu	2013	Between stations	633.1651	2.258518
		Between seasons	267.0763	3.443357
	2014	Between stations	45.3174	2.258518
		Between seasons	79.41899	3.443357
Dissolved Pb	2013	Between stations	348.9886	2.258518
		Between seasons	425.6385	3.443357
	2014	Between stations	53.78033	2.258518
		Between seasons	69.27794	3.443357
Sediment Zn	2013	Between stations	19.31241	2.258518
		Between seasons	28.35627	3.443357
	2014	Between stations	23.91596	2.258518
		Between seasons	53.57964	3.443357
Sediment Cu	2013	Between stations	56.0951	2.258518
		Between seasons	52.01096	3.443357
	2014	Between stations	59.34014	2.258518
		Between seasons	55.86655	3.443357
Sediment Pb	2013	Between stations	33.05731	2.258518
		Between seasons	15.1717	3.443357
	2014	Between stations	29.19174	2.258518
		Between seasons	11.33467	3.443357
Tissue Zn (*Avicennia alba*)	2013	Between stations	58.01375	2.258518
		Between seasons	38.88166	3.443357
	2014	Between stations	73.50654	2.258518
		Between seasons	66.13799	3.443357
Tissue Cu (*Avicennia alba*)	2013	Between stations	64.62678	2.258518
		Between seasons	79.25109	3.443357
	2014	Between stations	66.55449	2.258518
		Between seasons	139.0477	3.443357
Tissue Pb (*Avicennia alba*)	2013	Between stations	33.67818	2.258518
		Between seasons	23.12939	3.443357
	2014	Between stations	14.78311	2.258518
		Between seasons	29.37503	3.443357

Factor	Year	Variable	F_{cal}	F_{crit}
Tissue Zn (*Avicennia marina*)	2013	Between stations	44.37472	2.258518
		Between seasons	32.97058	3.443357
	2014	Between stations	40.86767	2.258518
		Between seasons	31.82557	3.443357
Tissue Cu (*Avicennia marina*)	2013	Between stations	55.92154	2.258518
		Between seasons	61.1669	3.443357
	2014	Between stations	69.01076	2.258518
		Between seasons	65.63433	3.443357
Tissue Pb (*Avicennia marina*)	2013	Between stations	17.46312	2.258518
		Between seasons	28.24426	3.443357
	2014	Between stations	11.94884	2.258518
		Between seasons	34.21577	3.443357
Tissue Zn (*Avicennia officinalis*)	2013	Between stations	32.16326	2.258518
		Between seasons	35.84972	3.443357
	2014	Between stations	53.29541	2.258518
		Between seasons	60.14736	3.443357
Tissue Cu (*Avicennia officinalis*)	2013	Between stations	34.2617	2.258518
		Between seasons	44.14824	3.443357
	2014	Between stations	11.29179	2.258518
		Between seasons	20.71317	3.443357
Tissue Pb (*Avicennia officinalis*)	2013	Between stations	38.02522	2.258518
		Between seasons	10.31111	3.443357
	2014	Between stations	85.51478	2.258518
		Between seasons	56.50448	3.443357
Tissue Zn (*Excoecaria agallocha*)	2013	Between stations	63.21846	2.258518
		Between seasons	66.35212	3.443357
	2014	Between stations	32.92302	2.258518
		Between seasons	33.55594	3.443357
Tissue Cu (*Excoecaria agallocha*)	2013	Between stations	16.649115	2.258518
		Between seasons	22.37503	3.443357
	2014	Between stations	61.61232	2.258518
		Between seasons	117.9371	3.443357
Tissue Pb (*Excoecaria agallocha*)	2013	Between stations	47.61722	2.258518
		Between seasons	19.31825	3.443357
	2014	Between stations	29.34293	2.258518
		Between seasons	64.41538	3.443357

References

Agoramoorthy G, Chen FA, Hsu MJ (2008) Threat of heavy metal pollution in halophytic and mangrove plants of Tamil Nadu, India. Environ Pollut 155:320–326

Bhattacharyya AK, Choudhury A, Mitra A (2001) Accumulation of heavy metals in commercially edible fishes of Gangetic West Bengal. Res J Chem Environ Sci 5(1):27–28

Banerjee K, Roy Chowdhury M, Sengupta K, Sett S, Mitra A (2012) Influence of anthropogenic and natural factors on the mangrove soil of Indian Sundarban wetland. Arch Environ Sci 6:80–91

Bayen S (2012) Occurrence, bioavailability and toxic effects of trace metals and organic contaminants in mangrove ecosystems: a review. Environ Int 48:84–101

Bellinger E, Benhem B (1978) The levels of metals in dockyard sediments with particular reference to the contributions from ship bottom paints. Environ Pollut 15(1):71–81

Bhattacharyya SB, Roychowdhury G, Zaman S, Raha AK, Chakraborty S, Bhattacharjee AK, Mitra A (2013) Bioaccumulation of heavy metals in Indian white shrimp (*Fenneropenaeus indicus*): a time series analysis. Int J Life Sci Biotechnol Pharma Res 2(2):97–113

Chakraborty S, Zaman S, Pramanick P, Raha AK, Mukhopadhyay N, Chakravartty D, Mitra A (2013) Acidifications of Sundarbans mangrove estuarine system. Discov Nat 6(14):14–20

Chakraborty S, Trivedi S, Fazli P, Zaman S, Mitra M (2014a) *Avicennia alba*: an indicator of heavy metal pollution in Indian Sundarban estuaries. J Environ Sci Comput Sci Eng Technol 3(4):1796–1807

Chakraborty S, Zaman S, Fazli P, Mitra A (2014b) Spatial variations of dissolved zinc, copper and lead as influenced by anthropogenic factors in estuaries of Indian Sundarbans. J Environ Sci Comput Sci Eng Technol 3(4):182–189

Chakraborty S, Rudra T, Guha A, Ray A, Pal N, Mitra A (2016) Spatial variation of heavy metals in *Tenualosa ilisha* muscle: a case study from the lower Gangetic delta and coastal West Bengal. J Environ Sci Comput Sci Eng Technol 3(4):1–14

Das S, Mitra A, Banerjee K, Mukherjee D, Bhattacharyya DP (2005) Heavy metal pollution in marine and estuarine environment: the case of North and South 24 Parganas district of coastal West Bengal. Indian J Environ Ecoplann 10(1):19–25

Das S, Mitra A, Zaman S, Pramanick P, Ray Chaudhuri T, Raha AK (2014) Zinc, copper, lead and cadmium levels in edible finfishes from lower Gangetic delta. Am J Bio-pharmacol Biochem Life Sci 3(1):8–19

Dudani SN, Lakhmapurkar J, Gavali D, Patel T (2017) Heavy metal accumulation in the mangrove ecosystem of south Gujarat coast, India. Turk J Fish Aquat Sci 17:755–766

Gambrell RP (1994) Trace and toxic metals in wet-lands: a review. J Environ Qual 23:883–891

Goldberg ED (1975) The mussel watch—a first step in global marine monitoring. Mar Pollut Bull 6:111

Harbison P (1986) Mangrove muds: a sink and a source for trace metals. Mar Pollut Bull 17:246–250

Hasnain SI (2002) Himalayan glaciers meltdown: impact on south Asian rivers. Int Assoc Hydrol Sci 7:274–286

Islam MD, Tanaka M (2004) Impact of pollution on coastal and marine ecosystems including coastal and marine fisheries and approach for management: a review and synthesis. Mar Pollut Bull 48:624–649

Kannan N, Thirunavukkarasu N, Suresh A, Rajagopal K (2016) Analysis of heavy metals accumulation in mangroves and associated mangroves species of Ennore mangrove ecosystem, east coast India. Indian J Sci Technol 19(46):1–12

Krishnamurthy KV, Shpirt E, Reddy M (1976) Trace metal extraction of soils and sediments by nitric acid and hydrogen peroxide. At Absorpt Newsl 15:68–71

Ma L, Cai LZ, Yuan DX (2011) Advances of studies on mangrove benthic fauna pollution ecology. J Oceanogr Taiwan Strait 22:113–119

MacFarlane GR (2002) Leaf biochemical parameters in *Avicennia marina (Forsk.)* as potential biomarkers of heavy metal stress in estuarine ecosystems. Mar Pollut Bull 44:244–256

Malo BA (1977) Partial extraction of metals from aquatic sediments. Environ Sci Technol 11:277–288

Mitra A, Ghosh PB, Choudhury A (1987) A marine bivalve *Crassostrea cucullata* can be used as an indicator species of marine pollution. In: Proceedings of national seminar on estuarine management, p 177–180

Mitra A, Mukhopadhyay B, Ghosh AK, Choudhury A (1990) Preliminary studies on metallic pollution along the lower stretch of the Hooghly estuary, West Bengal, India. Proceedings of the International Symposium on Marine Pollution, p 241–249

Mitra A, Choudhury A (1992) Trace metals in macrobenthic molluscs of the Hooghly estuary, India. Mar Pollut Bull 26(9):521–522

Mitra A, Choudhury A, Zamaddar YA (1992) Effects of heavy metals on benthic molluscan communities in Hooghly estuary. Proc Zool Soc 45:481–496

Mitra A, Choudhury A (1993) Heavy metal concentrations in oyster *Crassostrea cucullata* of Sagar Island, India. Indian J Environ Health 35(2):139–141

Mitra A, Trivedi S, Choudhury A (1994) Inter-relationship between trace metal pollution and physico-chemical variables in the frame work of Hooghly estuary. Indian Ports 2:27–35

Mitra A, Trivedi S, Chaudhuri A, Gupta A, Choudhury A (1996) Distribution of trace metals in the sediments of Hooghly estuary, Indian. Pollut Res 15(2):137–141

Mitra A (1998) Status of coastal pollution in West Bengal, with special reference to heavy metals. J Indian Ocean Stud 5(2):135–138

Mitra A, Bhattacharyya DP, Misra A (1999) The effect of environmental variables on Cr and Pb in brackish water wetland system. Indian. J Physiol 73(1):105–109

Mitra A (2000) The northeast coast of the Bay of Bengal and deltaic Sundarbans. In: Sheppard C (ed) Seas at the millennium—an environmental evaluation. University of Warwick, Coventry, UK, pp 143–157

Mitra A, Banerjee K, Bhattacharya DP (2004) The other face of mangroves. Department of Environment, Government of West Bengal, West Bengal, ndia, p 132

Mitra A, Banerjee K, Sengupta K, Gangopadhyay A (2009) Pulse of climate change in Indian Sundarbans: a myth or reality? Nat Acad Sci Lett 32(1/2):19–25

Mitra A, Banerjee K, Ghosh R, Ray SK (2010) Bioaccumulation pattern of heavy metals in the shrimps of the lower stretch of the River Ganga. Mesopotamian J Mar Sci 25(2):1–14

Mitra A, Chakraborty R, Sengupta K, Banerjee K (2011) Effects of various cooking processes on the concentrations of heavy metals in common finfish and shrimps of the River Ganga. Nat Acad Sci Lett 34(3 & 4):161–168

Mitra A, Banerjee K (2011) Trace elements in edible shellfish species from the lower Gangetic delta. Ecotoxicol Environ Saf 74:1512–1517

Mitra A, Chowdhury R, Banerjee K (2012) Concentration of some heavy metal in commercially important fin fish and shell fish of the river Ganga. Environ Monit Assess 184:219–230

Mitra A (2013) Blue carbon: a hidden treasure in the climate change science. J Mar Sci Res Dev 3(2):1–14

Mitra A, Ghosh R (2014) Bioaccumulation pattern of heavy metals in commercially important fishes in and around Indian Sundarbans. Glob J Anim Sci Res 2(1):33–44

Mitra A, Zaman S (2015) Blue carbon reservoir of the blue planet. Springer, New Delhi. https://doi.org/10.1007/978-81-322-2107-4

Natesan U, Kumar MN, Deepth K (2014) Mangrove sediments, a sink for heavy metals? an assessment of Muthupet mangroves of Tamil Nadu, southeast coast of India. Environ Earth Sci 72:1255–1270

Niyogi S, Mitra A, Aich A, Choudhury A (1997) *Sonneratia apetala*—an indicator of heavy metal pollution in the coastal zone of West Bengal (India). In: Iyer CSP (ed) Advances in environmental science. Educational Publishers and Distributors, New Delhi, pp 283–287

Panigrahy PK, Nayak BB, Acharya BC, Das SN, Basu SC, Sahoo RK (1997) Evaluation of heavy metal accumulation in coastal sediments of northern Bay of Bengal. In: Iyer CSP (ed) Advances in environmental science. Educational Publishers and Distributors, New Delhi, pp 139–146

Peng Y, Li X, Wu K, Peng Y, Chen G (2009) Effect of an integrated mangrove-aquaculture system on aquacultural health. Front Biol China 4(4):579–584

Ray Chaudhuri T, Fazli P, Zaman S, Pramanick P, Bose R, Mitra A (2014) Impact of acidification on heavy metals in Hooghly Estuary. J Harmon Res Appl Sci 2(2):91–97

Saha SB, Mitra A (2001) Concentrations of Zn, Cu, Mn, Fe and Pb in various sources and their bioaccumulation in shrimp (*Penaeus monodon*) during rearing in semi-intensive system. Aquaculture 2(2):159–164

Saifullah SM (1997) Management of the Indus Delta mangroves. In: Haq BU, Haq SM, Kullenberg G, Stel JH (eds) Coastal zone management imperative for maritime developing nations. Kluwer Academic Publishing, Amsterdam, The Netherlands, pp 333–347

Santos HF, Carmo FL, Paes JES, Rosado AS, Peixoto RS (2011) Bioremediation of mangroves impacted by petroleum. Water Air Soil Pollut 216:329–350

Sengupta K, Roy Chowdhury M, Bhattacharyya SB, Raha A, Zaman S, Mitra A (2013) Spatial variation of stored carbon in *Avicennia alba* of Indian Sundarbans. Discov Nat 3(8):19–24

Sukhdhane KS, Pandey PK, Vennila A, Purelshothaman CS, Ajima MNO (2015) Sources, distribution and risk assessment of polycyclic aromatic hydrocarbons in the mangrove sediments of Thane Creek, Maharashtra, India. Environ Monit Assess 187:274–289

Tam NFY, Wong YS (1993) Retention of nutrients and heavy metals in mangrove sediment receiving wastewater of different strengths. Environ Technol 14:719–729

Trivedi S, Mitra A, Gupta A, Chaudhuri A, Neogi S, Ghosh I, Choudhury A (1994) Inter-relationship between physico-chemical parameters and uptake of pollutants by estuarine plants *Ipomea pescaprae*. In: Proceedings of the seminar: on our environment: its challenges to development projects, American Society of Civil Engineers–India Section, p 1–6

UNEP (1992) Pollution and the marine environment in the Indian Ocean. UNEP Regional Seas Programme Activity Centre, Geneva, Switzerland

West Bengal State Biodiversity Strategy and Action Plan (2002) National Biodiversity Strategy and Action Plan, Department of Environment, Government of West Bengal and Ramkrishna Mission Narendrapur, West Bengal, India, executed by Ministry of Environment and Forest, Government of India, technical implementation by Kalpavrissh and administrative co-ordination by Biotech Consortium, India Ltd., funded by Global Environmental Facility through UNDP

Williams TP, Bubb JM, Lester JN (1994) Metal accumulation within salt marsh environments: a review. Mar Pollut Bull 38:277–290

Wu Y, Chung A, Tam NFY, Pi N, Wong MH (2008) Constructed mangrove wetland as secondary treatment system for municipal wastewater. Ecol Eng 34(2):137–146

Young DR, Alexander GV, McDermott-Ehrlich D (1979) Vessel related contamination of southern California harbors by copper and other metals. Mar Pollut Bull 10:50–56

Zayed A, Gowthaman S (1998) Phytoaccumulation of trace elements by wetland plants: duckweed. J Environ Qual 27:715–721

Zhang ZW, Xu X-R, Sun YX, Chen YS, Peng JX (2014) Heavy metal and organic contaminants in mangroves of China: a review. Environ Sci Pollut Res 21:11938–11950

Index

A
Adaptive management, OEW
    field-scale trial implementation
        Alum additions, 134
        fire, for removing organic matter, 135
        soil removal/de-mucking, 129, 130
    in long term, 127
    wintertime P spikes, 137
Aluminum sulfate (Alum) additions, 134
Amelia
    chlorination/dechlorination system, 42
    productivity, 43
Ammonium pyrolidine dithiocarbamate, 278
Antibiotic resistance genes (ARGs), 116
ArcMap software, 90
Arctic Summer, 86, 87, 92, 94, 95
Assimilation
    carbon sequestration, 36
    climate change, 36–37
    EBS, 21
    ecological monitoring, 28, 29
    EDCs, 26
    effluent disinfection, 25
    energy and economic savings, 37, 38
    enhanced productivity, 32, 33
    feasibility study, 21
    land ownership, 28, 29
    loading rate, 24, 25
    LPDES, 22
    metals, 26
    NPEOs, 27
    nutria management, 30
    nutrient reduction, 34–36
    PPCPs, 26, 27
    process, 19, 20
    regulatory and permitting, 21–23
    site selection, 20–21
    TMDLs, 28
    wastewater treatment plants, 22
    wetland restoration, 30–32
Assimilation wetlands
    fresh marshes, 71
    global change, 72
    pulsing, 71
Atomic Absorption Spectrophotometry (AAS), 257, 278
Australian Guidelines for Water Recycling, 181

B
Bagmara, 277
Beneficial management practices (BMPs), 116
Benthic invertebrate, 99
Bioaccumulation factors (BAFs), 257, 258, 282, 293
Biochemical oxygen demand ($_cBOD_5$), 92
Biodiversity Management Committees (BMC), 162
Bioindicators
    metal contamination, 255
    Mn pollution, 259
    phytoremediation, 255, 256
Biological oxygen demand (BOD), 245
Bio-shields planting
    coastal vegetation, 144
    development, 152
    development and implementation phase, 143
    EbA approaches, 143
    ecological benefits, 143

Bio-shields planting (cont.)
  invasive plants, 145
  mangroves, 155 (see also Mangroves)
  plant selection, 144, 145
  protecting plants, 145
  slope stability, 145, 152
  in tropical coastal ecosystems, 145–151

C
Calla lily, 244, 246, 247, 250
Canada
  constructed habitat wetlands, 222
  habitat and stormwater wetlands, 216
Canna flacida, 246
Carbonaceous biochemical oxygen demand ($CBOD_5$), 92, 94, 95, 97, 99, 100, 115
Carex aquatilis, 91
Cattails (*Typha* spp.), 130
Central Sundarbans, 277
Centre for Alternative Wastewater Treatment (CAWT), 86, 87, 92, 104, 109, 116
Centre for Water Resources Studies (CWRS), 86, 87, 116
Ceratophyllum demersum, 261
Chemaguri, 277, 292
Chemical oxygen demand (COD), 95, 245
Chloroflexi, 212
Chotomollakhali, 277
Cities of the Future
  Australian water industry, 175
  challenges facing cities, 194
  city spaces, 177
  community knowledge and engagement, 178
  embracing cities, 177
  IWA cities, 174
  New Urban Agenda, 173, 174
  principles, 174, 175
  resilience, 177
  SDGs, 173
  sustainability, 177
  sustainable urban development, 173
  urban water management, 174
  urban water transitions framework, 176
  water cycle city, 176
  water sensitive city, 176, 177
  water supply, sewered and drained cities, 176
City of Sydney's Decentralised Water Master Plan, 196
Climate change, 172, 177, 202
Coastal communities, 142, 155, 159, 163, 164
Coastal Community Development Program, 167
Coastal zones and estuaries

  coastal zones and estuaries, 275
Constructed wetlands
  benefits, 4
  continual accretion, P, 122
  Cup and Saucer Wetland (*see* Cup and Saucer Wetland)
  DOM-photo-demethylation, 215
  efficient operation of, 242
  functional microorganisms, 3
  hybrid systems, 4
  maintenance and operation, 4, 11
  microbial processes, 3
  natural physical, chemical and biological processes, 181
  phosphorus and mercury methylation, 222
  phosphorus removal, 121
  plant and microbial communities, 3
  quality of wastewater, 242
  stages of development, 243
  surface flow wetlands, 3
  treatment effectiveness, 122
  treatment of wastewater, 244
  vertical and horizontal flow, 249
  WSUD (*see* Water sensitive urban design (WSUD))
Cooks River Stormwater Management Plan, 184
Cooks River Urban Water Initiative, 183
Cup and Saucer Wetland
  annual pollutant loads, Cooks River and Botany Bay, 185
  before construction and wetland after, Heynes Reserve, 187, 188
  benefits
    community involvement, 190, 191
    economic analysis, 194
    habitat, biodiversity and local climate, 192
    liveability indicators, 192, 193
    organisational capability and appetite, 192
  cells, 186
  Cooks River and Botany Bay catchment, 183
  Cooks River Stormwater Management Plan, 184
  Cooks River Urban Water Initiative, 183
  effluent water quality, 188, 189
  environmental outcomes, 194
  environmental protection and improvement, 183
  features, 185
  Heynes Reserve, 184

MUSIC modelling, 184, 185
plants and species, 186, 187
primary aim, 183
urban catchment, 183
water cycle, 183, 184
water quality, 187
water sensitive city, 197
wetland influent water quality, 187, 188, 190

D

Deltaproteobacteria, 212
De-mucking, 129, 130, 134, 137
Digha—Junput coastal plain, 277
Disaster Risk Reduction (DRR)
 definition, 141
 exposure, 153
 hazard, 153
 plan and programme, 164
 situation analysis, Nbs approach, 162, 163
 training and capacity building, 164
 vulnerability, 154

E

East Joyce Wetlands (EJW), 60
Eastern Sundarbans, 277
Eco Development Committees (EDC), 162, 166
Ecological baseline study (EBS), 21
Ecologically Sustainable Development, 178
Ecosystem-based adaptation (EbA), 143
Endocrine disrupting compounds (EDCs), 26
Exposure, 153, 154

F

Fire, 135
Firmicutes, 212
Flood control, 210
Forest Conservation Committees (FCC), 162, 166
Frasergunge, 277
Frasergunj station, 280

G

Ganga-Bhagirathi hydrologic system, 277
Genito-urinary system disorders, 260
Giant bulrush (*Schoenoplectus californicus*), 130
Global positioning system (GPS), 90
Gosaba, 277

H

Haldia port-cum-industrial complex, 276
Hamlets, in Nunavut, 84
Hammond assimilation wetland
 baldcypress seedlings, 66, 67
 BR1 and BR2, 66
 denitrification, 69
 herbaceous vegetation, 64
 LPDES, 70
 Manchac/Maurepas ecosystem, 66, 69
 marsh deterioration and recovery, 65
 NH_4 and PO_4 concentrations, 68
 nutria herbivory, 67, 70
 RPM5, 66
 waterfowl herbivory, 67
Harinbari, 277, 292
Harinbhanga River, 277
Hazard, 153

I

Illinois pondweed (*Potamogeton illinoensis*), 130
Indian Sundarbans, 275–277, 286, 292–294
Indigenous community in Mexico
 academic staff and students, 246
 alkalinity, 250
 antibiotic-resistant bacteria, 242
 construct and conventional wastewater treatment plants, 242
 constructed wetland project, 243
 cultivation of Calla lily, 247
 employment and economic development, 242
 fish ponds, 246
 human health risks, 242
 individuals, 242
 Ixmiquilpan municipality, 245
 La Coralilla cooperative, 242, 250
 location of Mezquital Valley, 243
 Mezquital Valley, 241
 municipal and state governments, 246
 pisciculture, 244, 249
 project investment, 246
 stabilization ponds, 249
 untreated wastewater, 241
 wastewater irrigation system, 242
 wastewater treatment, 247, 249
 water quality, 244, 245
 water quality parameters, 248, 250
 water scarcity, 241
Inorganic mercury (IHg), 207, 211
Integrated Conservation and Development (ICD) approach, 166

Intergovernmental Panel on Climate
    Change (IPCC), 116
International Water Association's (IWA)
    Cities, 174
Invasive species, 145
Ixmiquilpan municipality, 245

J
Jambu island, 277
Joint Mangrove Management (JMM)
    definition, 166
    restoration and management, in Godavari
        and Krishna, 166
    village developmental activities, 166

K
Kakdwip, 277, 280, 287, 293

L
Leapfrogging, 201, 202
Long-term tree ring analysis, 48
Lothian island, 277
Louisiana
    Amelia, 42–44
    Breaux Bridge, 44, 46
    Broussard Cote Gelee, 49, 50
    Cypriere Perdue, 44
    Hammond, 59–64
    litterfall and woody productivity, 47
    Luling, 54–56
    Mandeville Bayou Chinchuba, 54, 57–60
    St. Martinville, 51, 52
    Tchefuncte Marsh, 54, 57–60
    Thibodaux (*see* Thibodaux)
    TN concentration, 47

M
Macrophyte, 257
Mandeville's wastewater treatment system, 54
Mangrove ecosystems, 9, 10
Mangrove species
    A. alba samples, 285
    A. marina samples, 286
    A. officinalis mean concentrations, 282
    A. officinalis samples, 283, 284
    AAS instrument, 278
    accumulation of, 276
    Avicennia, 286, 288
    BAFs, 282
    bioaccumulation, 288, 292–294
    correlation analysis, 285
    E. aggalocha samples, 287
    Excoecaria, 287
    Gangetic delta, 292
    humic materials and metals, 275
    hydrologic processes, 292
    "hyper accumulators" of metals, 294
    Indian Sundarbans, 292
    Indian Sundarbans region, 275
    at Kakdwip, 282, 292
    mean concentrations of zinc, copper and
        lead, 286
    metals
        in sediments, 279–282
        in water, 278–280
    in pre-monsoon, monsoon and post-
        monsoon seasons, *A. officinalis*,
        289–290
    pre-monsoon, monsoon and post-monsoon
        seasons, *E. agallocha*, 291
    research, 294
    samples of roots, stems and leaves, 278
    samples of surface water, 278
    sediment samples, 278
    sources of metals, 288
    study area, 276–278
    tropical and sub-tropical coastal areas, 276
Mangroves
    as bio-shields, 155
    community-based joint management (*see*
        Joint Mangrove Management
        (JMM))
    as effective coastal protection solutions, 156
    global distribution of areas, 157, 158
    global distributions, 155
    mangrove loss and rehabilitation, 157
    marshy area mangrove species, 156
    for pollution control, 156, 157
    restoration, conservation and management,
        166
    root system and dense growth, 155, 156
    socio-economic conditions, 159
    steps, restoration, 159
May Palmer Drought Severity Index (PDSI), 48
Mean concentrations
    zinc, copper and lead, 294
Mercury biogeochemistry
    anthropogenic and natural, 210
    Minamata Convention on Mercury, 211
    natural wetlands, 211
    particle-bound metals, 211
    phytoremediation, 211
    re-emission of, 210
    urban artificial wetlands, 211

Methylmercury (MeHg)
  adverse effects, 208
  artificial wetlands
    biogeochemical controls, 220–222
    geographic distribution, 216–220
    impacts of, 224–226
    management, 223, 224
    risks, 226–228
    temporal patterns, 222, 223
  biogeochemistry (*see* Methylmercury biogeochemistry)
  contamination, 228
  environmental detection and identification, 229
  fish/aquatic insect prey, 208
  the *hgcAB* gene pair, 229
  hydroelectric reservoirs, 208
  individuals, 207
  insectivorous birds, 208
  mercury biogeochemistry (*see* Mercury biogeochemistry)
  mercury methylation and demethylation, 229
  nitrate-related reductions, 228
  production and concentrations, 228
  stormwater and habitat wetlands, 208
  subclinical intellectual impairment, 208
  in surface-flow artificial wetlands, 207
Methylmercury biogeochemistry
  controls on mercury methylation, 212–214
  demethylation, 214, 215
  inorganic, 211
  mercury methylation, 212
  physicochemical and ecological parameters, 211
  production of, 211
  variability, 215, 216
Mezquital Valley (MV), 241, 243, 250
Mississippi Delta
  assimilation wetland (*see* Assimilation)
  flood control, 18
  sustainability, 19
  wetlands (*see* Natural wetlands)
Model for Urban Stormwater Improvement Conceptualisation (MUSIC) software, 185, 187, 188
Moringa oleifera, 262, 263
Muck removal, 129, 130

N

Natural disasters, 174
Natural hazards, 141
Natural wetlands
  assimilation systems, 16, 17
  freshwater resources, 17, 18
  long-term storage nutrient reservoirs, 16
  nutrients, 16
Nature-based Solutions (NbS)
  approaches, 143, 144
  bio-shields (*see* Bio-shields planting)
  definition, 143
  description, 7
  institutional and policy instruments, approaches
    actions plans, 164, 165
    community involvement, 161
    community-based project, in Kenya, 161
    EDC/VSS/BMC, formation of, 162
    micro-plan preparation, 164
    monitoring and evaluation (M&E), 165
    process documentation, 165
    programs, in Vietnam, 160
    situation analysis, to DRR, 162, 163
    step-by-step strategic approach, 160
    training and capacity building, in DRR, 164
  IPBES conceptual framework, 142
  mangroves (*see* Mangroves)
  in post-disaster planning, 7
  resilience
    assessing resilience, 153
    definition, 152
    Sendai Framework, 154, 155
    water security agenda benefits, 5
    wetland systems, 7
    Yangtze River basin, in China, 7
New Urban Agenda, 173, 174
Nonylphenol ethoxylates (NPEOs), 27
NOVA analysis
  concentrations of metals, 296
Nutria management, 30
Nymphaea lotus, 260

O

Oreochromis niloticus, 244
Orlando Easterly Wetlands (OEW)
  adaptive management, P removal performance, 128, 129
  area, described, 123
  longitudinal flow paths, 123
  methodical research program, 127
  operational and performance parameters, 124
  performance, 124
  spike-factor, 125
  time-dependent criteria, 124
  wetland treatment system, 122
  winter outflow TP concentrations, 125
  winter-time spikes, 125

P

Participatory Rural Appraisal (PRA), 162
Pharmaceuticals and personal care products (PPCPs), 26, 27, 116
Phosphorus
    adaptive management, OEW, 128, 129
    annual performance, OEW, 126
    as "legacy phosphorus load", 122
    removal of, in constructed wetlands, 121
    soluble reactive (SRP), 122, 134, 135
Phragmatis australis, 246
Phytomaterials
    biogeography, 255
    Moringa oleifera, 263
    natural disinfectants and coagulants, 263
    NbS approach, 266
Phytoremediation
    Africa, 256, 257
    in aquatic macrophytes, 255
    and bioindicators, 255, 256
    constructed wetlands, 254
    metal accumulation, 257–259
    metal contamination, 260–262
    national guidelines, 254
    NbS approach, 266–268
    NbS eco-model in Africa, 265–267
    NbS natural processes, 254
    phytomaterials, 255
    the Millennium Development Goals, 253
    water pollution, 254
    water resources, 253
    water treatment approach, 262–265
    wetland ecosystems, 254
Pisciculture
    wetland, 244
Pistia stratiotes, 261
Pollution abatement
    mangrove ecosystems, 9, 10
    NbS, 9
    parameters, 10
    pollution control, wetlands, 9
Pond Inlet wetland, 112
Population growth, 172
Pycreus lanceolatus, 261

R

Ramsar Convention, 1, 5
Rapid Rural Appraisal (RRA), 162, 163
REDD+ programs, 162
Rhodamine WT (RWT) fluorescent dye tracer, 91

S

Sajnekhali, 277
Sendai Framework, 154, 155
Sewage lagoons, 83, 99, 116
Social-ecological system (SES), 153
Southern naiad *(Najas quadalupensis)*, 130
Subsurface flow wetlands (SSFW), 244
SubWet model, 109
Sustainable development goal (SDG)
    17 SGD goals, 8
    objective, SGD 6.6, 9
    objectives, 8
    SDG 6, goals of, 8
    water security, 8
Sustainable Development Goals (SDGs), 173, 201–203
Sustainable urban development, 173

T

Tanks-in-Series (TIS) mathematical model, 109
Thibodaux
    baldcypress trees, 39
    decomposition and accretion, 40
    effluent, 40
    elevation/sediment accretion, 41
    ESLR, 41
    NPP, 40
    soils, 39
    surface water height, 39
    tree-ring analysis, 41
Total ammonia nitrogen (TAN), 92, 93
Total suspended solids (TSS), 94, 95, 97, 112–114, 116
Treatment wetlands
    chemical reactions, 122
    natural processes, 121
    "polishing" phase, 121
Tundra WTAs, in Canada
    BMPs, 116
    design and assessment, practices
        flow rates, 114
        hydraulic retention time, 115
        inlet structures, 114
        intersystem variability, 110
        loading rates, 114, 115
        monitoring, microbial indicators, 112
        proposed standardized assessment framework, 110, 111
        site-specific data collection program, 110
        timing and frequency, monitoring, 113, 114

Index   307

timing of discharge, 115
wastewater compliance parameter
   TSS, 112
emerging areas, 115
goals, 86
hydraulic and hydrological
   characterization, 91
in Kugaaruk and Grise Fiord, 96–98
IPCC, 116
physical characteristics, wetland
   boundaries, 90, 91
PPCPs, 116
site descriptions
   Arctic Summer, 87
   hydrology, 87
   interpolated mapping, 87
   wetland treatment areas, locations of, 87
SubWet 2.0 computer model, 109
territorial regulatory provisions, 85
TIS mathematical model, 109
treatment performance assessments, 91
treatment performance, factors
   discharge technique, influence
      of, 99, 101
   hydraulics and hydrology, 96–98
   pre-treatment, 94, 95
   seasonal influences, 92, 93
   spatial distribution, treatment, 106–108
   vegetation (*see* Vegetation)
Typha angustifolia, 246

U
UN Water for water security, 5, 6
UNESCO SWITCH benchmarking project, 202
Urban stormwater management, 179
Urban water management, 174
Urbanisation, 172, 173, 179

V
Vana Samrakshana Samithis (VSS), 162
Vegetation
   tundra WTAs, in Canada
      arctic-specific investigations, 103, 104
      Carex aquatilis, 104, 106
      climate change, 105
      constructed wetlands, 103
      Coral Harbour WTA, 102
      distinct classes, 101
      diversity of, species, 103
      dominant vegetation and land cover,
         distinct classes, 101, 102

extreme environment, 105
fertilization studies, 105
in natural wetlands, 103
organic detritus, 102
photosynthesis and
   evapotranspiration, 104
Salix richardsonii, 101
small scale fertilization studies, 105
soil depth, 101
wastewater, as hydrologic regime, 105
Vulnerability, 154, 164

W
Wastewater
   CAWT, 86, 109
   compliance parameter TSS, 112
   emerging areas, Canadian wastewater
      management, 115
   ice-free treatment season, 84
   in Northwest Territories, 84
   in Paulatuk and Taloyoak communities, 99
   parameters, 92
   passive treatment, in Northern
      Canada, 84
   in Paulatuk, treatment area, 107
   treatment system, in Grise Fiord, 98
   treatment system, in Kugaaruk, 97
   treatment, in Arctic regions, 83
   treatment, in Canada (*see* Tundra WTAs, in
      Canada)
Wastewater stabilization ponds (WSPs), 83, 84
Wastewater treatment plant (WWTP), 137
Wastewater treatment plants, 17, 19, 22, 27,
   51, 70
Water cycle city, 196, 197, 202
Water Quality Standards (WQS), 23
Water security, 5, 6, 8
Water security agenda
   framework, on water security, 5
   "Nature-based Solutions" (NbS), 5
   wetland solutions, 4
Water sensitive cities
   application, transition framework, 201
   benchmarking, city's water management
      state, 202
   constructed wetland (*see* Constructed
      wetland)
   creation of, 197
   practice, water management, 177
   sustainability, 177
   urban water managers, in Australia, 175
   wetland outcomes, 203

Water sensitive urban design (WSUD)
  in Australia, 179, 180
  ecologically sustainable development, 178
  natural water cycle, 179
  principles of, 181
  stormwater treatment technologies, 181, 182
  typical water quality parameters, 180
  urban stormwater management, 179, 180
  water cycle management strategies, 179
  water quality guidelines, 181
Waterways city, 176, 195, 196, 202
Water-Wise City vision, 174, 175
Western Sundarbans, 277
Wetland treatment areas (WTAs)
  Arviat WTA, 95
  performance model, 107
  sizing tools, 108
  tundra (*see* Tundra WTAs, in Canada)
  variable–order and compartmental models, 108
  Whale Cove, 95
Wetlands
  constructed wetlands, 3, 4
  continuum, 2
  ecological services, 2
  international efforts, 2
Wintertime P spikes, 137, 138

Z
Zantedeschia aethiopica, 244

CPSIA information can be obtained
at www.ICGtesting.com
Printed in the USA
LVOW05*1613191217
560259LV00001B/43/P